OUTSIDE IN

OUTSIDE IN

Minorities and the Transformation
of American Education

PAULA S. FASS

New York   Oxford
OXFORD UNIVERSITY PRESS
1989

Oxford University Press

Oxford   New York   Toronto
Delhi   Bombay   Calcutta   Madras   Karachi
Petaling Jaya   Singapore   Hong Kong   Tokyo
Nairobi   Dar es Salaam   Cape Town
Melbourne   Auckland

and associated companies in
Berlin   Ibadan

Published by Oxford University Press, Inc.,
200 Madison Avenue, New York, New York 10016

Oxford is a registered trademark of Oxford University Press

Library of Congress Cataloging-in-Publication Data

Fass, Paula S.

Outside in : minorities and the transformation of American education / Paula S. Fass

p.   cm.
Includes index.
ISBN 0-19-503790-1
1. Education—United States—History—20th century.
2. Minorities—Education—United States—History—20th century.
3. Americanization—History—20th century.   4. High schools—United
States—History—20th century.   I. Title.
LA217.F38   1989   88-31953
370'.973—dc19

Grateful thanks is given to the University of Chicago
Press for permission to use the author's article
"The IQ: A Cultural and Historical Framework," published
in *American Journal of Education*, 88 (Aug. 1980): 431–458,
as a basis for part of Chapter 2 of this book.

2 4 6 8 9 7 5 3 1

Printed in the United States of America
on acid-free paper

For Jack
mon amour et mon confrère

Acknowledgements

In researching and writing this book, I have had the good fortune to receive various kinds of assistance. A Fellowship in the Humanities from the Rockefeller Foundation allowed me to begin in 1976–77 what was destined to become a long period of study and investigation of twentieth-century American educational institutions and experience. Subsequently, fellowships and grants from the National Institute of Education (project number NIE-G-81-0068), the Humanities Research Committee of the University of California at Berkeley, and the National Endowment for the Humanities (1984–85) permitted me to take the teaching leaves without which the work on this book could not have proceeded. I have also benefitted from summer funds and research money which greatly facilitated my progress—a summer stipend from the National Endowment for the Humanities; money for computer services and research assistance from the Institute for Human Development at Berkeley, through the kind offices of Guy Swanson; repeated grants for research and typing assistance from the Committee on Research at Berkeley; and a special subvention during the summer of 1986 from the Spencer Foundation, administered through the auspices of Dean Bernard Gifford of the Graduate School of Education at Berkeley. I am extremely grateful for this assistance and to those who made it possible.

The kindness and hospitality of individuals and several institutions made my work a pleasure as they provided necessary and invaluable access to special resources. I would like to thank the staffs of the principals' offices and of the libraries of the following New York City high schools: Evander Childs High School, Theodore Roosevelt High School, Seward Park High

School, George Washington High School, Louis Brandeis High School (formerly the High School of Commerce), New Utrecht High School, and Bay Ridge High School. I am also grateful for the courtesy extended to me by the staff at the archives of Teachers College, Columbia University, especially Robert Morris; the librarians at the Education-Psychology Library at the University of California, Berkeley; the staff at the U.S. Army Center of Military History in Washington, D.C., especially Morris MacGregor who gave me a good, quick introduction to the nature of military organization; the staffs of the Military Reference Branch and of the Judicial, Fiscal, and Social Branch of the National Archives in Washington, D.C.; the librarians at the Catholic University of America in Washington, D.C.; and the faculty and staff of the Center for Studies in Higher Education at Berkeley, especially Janet Ruyle. I am also indebted to the members of the History Department of Teachers College, Columbia University, especially Hazel Hertzberg and Frederick Kerschner for sharing their offices and their lives with me during an important period in the development of my project and my ideas. The staff of the History Department of the University of California at Berkeley often made my life livable, and I thank them for their efficiency and friendliness. I am especially grateful to Janet Weitz who did a splendid job typing the manuscript and to Ramona Levi who is simply splendid.

My friends and colleagues have been generous with their time in reading this manuscript at various points in its development and in encouraging my efforts. Sheldon Rothblatt, Reginald Zelnik, Richard Abrams, Lynn Hunt, Winthrop Jordan, and Randolph Starn gave me good advice after reading all or portions of this book during its evolution. Several conversations with Carl Degler helped to sharpen my ideas. Richard Salvucci allowed me to pick his well-stocked mind when I was trying to figure out what my numbers meant. David Kirp, with whom I was fortunate to participate in an excellent Berkeley-Stanford seminar on the finance and governance of education, gave me the courage to change the form of my original project and to move in new directions. The late Allan Sharlin shrewdly suggested that I exchange the pencil for a computer in analyzing my data in Chapter Three and introduced me to the resources of that technology. Michael Katz read the manuscript with care and concern despite the fact that we sometimes disagreed about interpretations. Geraldine Clifford read Chapter Five with her usual insight and extraordinary knowledge of the course of American educational development. Steven Schlossman read and commented on the book with great acumen at a significant time, and I have benefitted from his advice. Harold Wechsler, whom I delight in calling an old friend, has given of himself so often in this enterprise that my thanks to him must be couched in the broadest terms. My stay in New York during a critical time in my research and in my life was facilitated by another old friend, Howard Leon. Lizabeth Cohen, as a student, a friend, and a colleague, has been fundamental in my development as a historian. Her generosity in reading and commenting on my work and her stimulating suggestions and unwavering support have been invaluable. I have also been extremely fortunate in my choice of research assistants. Their help has taken many forms, from a rigorous demonstration of research skills and impeccable work habits to encouragement in periods of sagging morale, and it is a real treat to be able to thank them all: Julia Liss, Barbara Loomis, Robyn Lipner, Bruce Nelson, Theodore Feldman, Eyal Naveh, Steven Leikin, Kelly Schrum, and Jesse Berrett.

Sheldon Meyer of Oxford University Press encouraged me from the inception of my research. He has seen the project change form and direction, and he has continued to support me with patience, kindness, and friendship. I appreciate this deeply and hope his trust has not been misplaced. Scott Lenz has been an excellent editor. Careful, judicious, and courteous, he also has the skills and humor which make an author's task easier and the book better.

Chapters 2 and 3 contain materials which I have published elsewhere and in different form. "The IQ: A Cultural and Historical Framework," was published in the *American Journal of Education,* 88(August 1980): 437–458; "Without Design: Education Policy in the New Deal," appeared in the *American Journal of Education,* 91(November 1982): 36–64; "Before Legalism: The New Deal and American Education," was included in *School Days, Rule Days: The Legalization and Regulation of Education,* edited by David L. Kirp and Donald N. Jensen (The Falmer Press, 1986).

This book was composed in the eight years defined by my pregnancies and by the births and infancies of my two children, Bluma (Bibi) and Charles. Though it is the usual fashion to absolve others of responsibility for an author's shortcomings, I would like to reverse that tradition. My children have played a critical role in all aspects of the following book, good and bad. They have reshaped my life while they have gotten into my hair and into my heart. This book is inseparable from that experience. My lovely Bibi, especially, has shown forbearance well beyond her years and has given me one more reason to love and admire her. Unfortunately, my father did not live to see this book completed. Formally uneducated, but knowledgeable and wise in many things, he came to this country in middle age, committed to providing his children with the schooling he lacked. He probably could not have read this book, but he would have understood and supported its pleas for equal education and for social justice.

Jack is so much part of my life and of the creation of this book that any description of his contribution would underestimate it by defining it. I give him this book as a small token of my appreciation for everything he has done, of our life together, and of our future.

Berkeley, California   P. S. F.
October 1988

Contents

Others take finish, but the Republic is ever constructive and ever keeps
    vista,
Others adorn the past, but you O days of the present, I adorn you,
O days of the future I believe in you—I isolate myself for your sake,
O America because you build for mankind I build for you.
O well-beloved stone-cutters, I lead them who plan with decision and
    science,
Lead the present with friendly hand toward the future.
(Bravas to all impulses sending sane children to the next age,
But damn that which spends itself with no thought of the stain, pains,
    dismay, feebleness, it is bequeathing.)

WALT WHITMAN, "By Blue Ontario's Shore"

OUTSIDE IN

Introduction

By the early twentieth century, the appeal of schooling was already irresistible in the United States. Philosophers proclaimed its unparalleled importance, sociologists integrated it into the basic institutions of their analysis, reformers called upon its aspirations for improvement, and politicians were becoming adept at invoking its name as part of the rhetoric of democratic promise. In fact, schooling was then only marginally a factor in social life, more an incanted ideal than a strategic part of national experience. Since then, schooling has become a familiar part of the American dialogue about the meaning and fate of democratic society. Indeed, our attention to schooling, our discussions, and our dissatisfactions have helped to define us as a people in the twentieth century. Today, schooling does matter, and education as a massive enterprise has become an integral component of the economy, culture, and state. While Americans in high places still incant schooling as part of the liturgy of American ideology, the experience of almost all Americans has made the school part of their own intimate past and vital to their children's future. The domestication of schooling, from the realm of inspiration to the ghettos of daily life, has been part of the social transformation of the twentieth century.

This transformation is the subject of my book. It is my contention that the shape of American education in the twentieth century, its forms, functions, and contradictions, is crucially related to the problems associated with American diversity, and that to understand the role that the schools have come to play in social life as well as in social thought, we must examine in some detail the history of the education of outsiders in the society. In this century, outsiders, first immigrants and then various racial, gender, and religious groups, have challenged the simple and direct beliefs about citizenship that were embodied in the nineteenth-century common-school ideal and the sufficiency of the limited and fragmentary education captured in its aims. The massive educational enterprise with which we have become familiar in the late twentieth century, with its elaborate interconnections from primary to graduate school, intricate hierarchies of grades and certifications, and complicated governmental regulations, has developed with the American economy and growing state power. But the specifics of this development are not simply a function of accelerated modernization. On the contrary, its present forms are expressions of the particular history of American society and culture in the twentieth century as that history has been affected by problems associated with pluralism.

To grapple with that story is a formidable task, and this book is neither a full social and cultural history of the schools nor a complete examination of outsiders in the schools. Rather than a general survey, I have chosen to examine in depth several groups at particularly significant historical moments in their education and at points of special social consequence. The experience of these groups—European immigrants, blacks, women, and Catholics—and their impact on school development have been especially potent in this century. But my analysis might have been extended to other groups, and Asian-Americans, Chicanos, and Latinos as the most recent outsiders demand similar attention. By focusing on particular groups at especially problematic periods of their schooling, I have been able to look at various levels and kinds of schooling and to stretch my analysis along a long portion of the twentieth century. Since I believe that our understanding of the nature of schooling in American society requires that the enterprise be examined in the widest possible context and at all levels—in private schools as well as public, in institutions of higher education and secondary schools, by listening to shapers of policy as well as to students— a careful selection of cases guided by these concerns provided me with an effective entry to the larger problem. Finally, these diverse cases allowed me to capture the historical tension that has characterized the American experience of schooling throughout the twentieth century—the tension between the commitment to universal education in the context of uncertainty about its means—while they forced me to reckon with the very many forms in which this tension has been expressed within the restraints of the moment.

By examining specific problems in the evolution of modern education, I have been able to focus on the very large differences in the historical experience of groups as well as the broader cultural issues within which these experiences were manifested. How pluralism is defined and how we have organized the different groups within the whole has affected both their experience and their treatment in the schools. Indeed, pluralism in the United States has been fraught with ambiguity. This is in part because pluralism is at once a fact and an idea, a fundamental social reality and a question of social policy. Throughout the twentieth century, the schools have participated in these two different but related issues, and often, as critical social institutions, they have mediated the experience of pluralism for the culture. In fact, while school policies have always organized social perceptions, the different groups whose schooling was at issue did not always act in expected ways. And the schools' attempts to delimit how pluralism was expressed were often met by an equally powerful tendency for different groups to use schooling for their own ends. This dialectic between perception and behavior—the schools as cultural instruments and the schools acted upon by their populations—integrates the various episodes of this study.

I begin with an examination of progressive social reformers who looked at the problem of pluralism through the blinkers of their many concerns, especially the issue of work in an industrial society. The themes of progressive education and the problems to which it responded are still both urgent and perplexing, and the progressive solutions and their definitions continue to shape our understanding. In Chapter 1, I place the social reformers of the period in their own intellectual and social context to argue that their ideas about education must be read as part of the reconsideration of the nature of socialization in a changing society. As progressive social reformers began to examine the cultures of the many immigrant groups then peopling the nation, they increasingly invested the schools with renewed purpose and a larger role. In appraising the difficulties involved in schooling the wide range of immigrant groups, they redefined the meaning and content of schooling as a process of socialization, because this was the only adequate correlative to the complexity of their perceptions about the problems that immigration posed for the culture. Progressive education has been the subject of wide and searching historical attention, and my arguments have been anticipated by others, especially Marvin Lazerson. What this chapter attempts to do is to resynthesize our understanding of progressive education as a response to the issue of social pluralism and to highlight the manner in which the schools were enlisted in the travails of a culture besieged by outsiders.

In Chapter 2, I continue this discussion, looking at the schools themselves rather than at social reformers who turned to schooling to solve larger problems. By the 1920s, the schools were forced to confront the reality of pluralism about which reformers had theorized. The solutions that the schools offered—testing, tracking, and socialization—were specifically framed in the light of the complex issues of student heterogeneity and strongly informed by contemporary perceptions about immigrants. The rigid order imposed by the schools as they dealt with their pluralistic populations provided both a pedagogical structure and a social framework that defined American educational efforts during the period of greatest expansion. I argue that IQ testing and the social perceptions this codified were fundamental to the specific forms of this expansion, especially in the reformulation of secondary education in the twentieth century. While some of these matters have been discussed elsewhere, this chapter provides a more pointed discussion of the way in which the ambiguities of pluralism in the twentieth century were handled by the schools and of the consequences of their actions—both for school development and for the society.

In Chapter 3, I move from perceptions and policies to student experiences. This chapter is directed to a period (the 1930s and '40s) that has been largely ignored by historians, and I attempt to address head on the question of how ethnic students in high schools used the schools for their own social ends. In the thirties and forties, the second generation came into its own in the high schools; nowhere was this truer than in New York City. By then the children of immigrants had overwhelmed the schools of this premier immigrant city. In examining the extracurricular participation of students in the high schools of New York, this chapter attempts to penetrate the web of choices and associations made by descendants of turn-of-the-century immigrants and to assess the consequences, particularly for assimilation. While schooling in New York cannot be presumed to be the same as schooling elsewhere, the importance of New York as an ethnic center and the significant variation among the city's schools enable me to argue more broadly about the relationship between schooling, ethnicity, and social change in the twentieth century. In the thirties and forties, the children of immigrants began to make choices and to adopt values that have left a lasting imprint on the schools and the culture they helped to shape.

Together, the first three chapters (Part I) emphasize the interplay between social definitions, school policies, and student experience to describe how the schools were re-created in the context of their pluralistic populations. But it was not schools alone that were changed. The changes in the schools and the effect they had on students altered the culture in general. In this sense, the transformation of the schools was not only a product of social change but also affected the nature of twentieth-century social life.

Beginning with Chapter 4, I turn from European immigrants, who had been the prevailing problem in school and society until World War II, to other groups and issues. Since the 1950s, the place of blacks in American society has been among the most urgent problems that the schools have been asked to solve. This issue came to a climax in the 1960s and '70s, and it came together with constitutional questions about the role of federal regulations for school policies. But the origin of the problem lies much further back. In Chapter 4, I look at the origin of the federal government's concern with black education during the New Deal and the crisis of World War II. It is my contention that in the 1930s and '40s, black education was moved from the periphery to the center of discussions about education in the context of the new and aggressive role that the federal government began to play in matters of schooling. The specific experiences of blacks with federal educational programs during the depression and the war helped to confirm for blacks that change would come from direct federal involvement. Indeed, it was only in the context of federal policies that blacks became part of the definition of American pluralism and a problem for the schools. Before the depression and the war, the effective segregation of blacks had so marginalized their concerns that this most significant component of American diversity hardly existed as a school problem. The New Deal and the war made segregation untenable and set the stage for blacks to become the central dilemma for school and social policies in a pluralistic society.

Before World War II, the education of women, like the education of blacks, was rarely considered problematic, but for vastly different reasons. While blacks were so far from the mainstream that as a matter of social policy their schooling was not considered urgent (whatever the issues of justice involved), women had been so much part of the mainstream of educational development since the mid-nineteenth century that the growing disparity between women's educational achievement and its utilization had been relegated to the outer fringes of educational discussion. Since at least the late nineteenth century, women had acquired a significant portion of the educational goods available in America up to and including higher education. That right had required a struggle initially, but by the mid-twentieth century it was largely taken for granted. In the 1950s, however, the question of why American women were being educated in colleges and universities in a society where their skills were vastly underutilized became an educational issue and a social dilemma. Chapter 5 discusses the anomalous position of women in a culture where educational access was available without commensurate social opportunity and the educational problem this posed. It was this disparity and the conflicts it involved for educated women that set the stage for the women's revolution of the 1960s and '70s.

Although colleges and private schools are rarely examined together with public schools and are usually considered apart from the common-school tradition, they have been in this century an integral part of the whole educational enterprise. Colleges and Catholic schools have, moreover, par-

ticipated vigorously in the debates and problems associated with mass instruction, and they have been affected by problems associated with their
diverse populations. It is for this reason that I have chosen to examine, in
two instances, the problem of outsiders in these contexts. But, while Chapter 5 looks at women as outsiders inside mainstream institutions of higher
education, Chapter 6 looks at schools that have been organized by outsiders apart from the mainstream culture. Catholic education, like higher
education, when it is studied at all, is usually set apart as a subspecialty.
Yet, American education in the twentieth century—and specifically the education of outsiders—is simply incomprehensible, or in the very least incompletely rendered, without an understanding of how Catholic schools
have operated within American culture alongside the public schools. In
Chapter 6, I examine in broad strokes the Catholic alternative to the public schools. Although Catholic schools have paralleled public schools in
many ways, they have endured as separate and distinctly different agencies
that reflect the aspirations of a large minority not thereby obviously disadvantaged. In this chapter, I examine how they have been organized and
in what ways they have succeeded both in protecting minority integrity
and in providing Catholics with an entrée to American success.

In the largest sense, I use education in the following pages as a way of
looking at American culture and society in the twentieth century. Certainly, the manner in which social institutions deal with outsiders is a fundamental expression of the culture. This is because American self-identity
has ideologically and practically been formed around issues related to the
multiple sources of its population. Since schools are always a strategic instrument of culture formation, how they have dealt with outsiders and
have been shaped through this response is a crucial historical issue. As the
following chapters will show, and as many others have demonstrated before, schools reflect, enunciate, and often symbolize the commitments and
perplexities of American life. They can and do perpetuate the status quo,
the class structure, gender definitions, and the social status of minorities
and racial groups. But schools, and education more broadly, also create
culture. I argue that this was the case among second-generation high-school
students (Chapter 3), as well as for blacks during the New Deal and the
war (Chapter 4), and I also explore the unexpected way in which higher
education provided some women with the means to challenge gender definitions (Chapter 5).

The way a culture deals with the stranger is also a measure of its strength
and of its quality. It is because of this that American institutions—the
schools especially—have often been judged by the degree to which they
have provided outsiders with the resources for social success and personal
opportunity. Americans have made much of this dimension of schooling

and have long congratulated themselves on this basis. As enormously powerful and largely trusted social agencies, schools have the potential to refashion social relationships, redefine social objectives, and provide new ideological commitments. While the love affair with American schools—the unqualified satisfaction with their democratic success—was probably never more than a figment of educators' imagination, certainly the American faith in education as a force for ameliorative change has been strong. It has usually been this political dimension of schooling that has held out the most promise to social reformers and caused the most historical bitterness. In the early twentieth century when John Dewey proposed to reform the schools, he saw this as the beginning of social regeneration. Since then, especially in the 1960s and more modestly in the 1930s, others too have sought to use the lever of the school for social reconstruction. Thus when social reformers and historians who followed in their wake discovered the extent to which the schools have been instrumental in social solidification and cultural maintenance, they have often registered their anger and even a sense of betrayal. And in the past generation as historians documented the degree to which the schools have not fulfilled their promise, the manner in which schools have prevented mobility, perpetuated inequality, and confirmed racial and gender definitions, they have sometimes ignored the changes that schooling did bring about. In fact, the record of American schools, like that of the society, has been mixed. American schools have demonstrated both the strengths and weaknesses of the American commitment to pluralism—or better, the way pluralism has in fact functioned. In expanding the definition of who shall be educated, the schools have tried to incorporate and control at the same time as they have been altered and expanded. They have also generated conflicts that often led to significant change. The record of schooling in these matters has been neither black nor white, but complex, and it remains incomplete.

I have worried over what term to apply to the outsiders who are the central concern of this study. Non-normative is descriptively correct, but ponderous and ungainly. Minorities captures certain dimensions of the issues and is especially appropriate for groups like blacks or Catholics, but not entirely suitable for women or immigrants. I have used it in the title because it provides readers with a sense of my subject. But in the end, the concept of outsiders most closely approximates the problem all the essays contained here investigate. Immigrants, blacks, women, and Catholics were outside the mainstream culture as it was defined in the early twentieth century. They were outside the power networks that organized school systems and ran school organizations. They did not devise the ideology which connected schooling to success in modern America. But they were all very much part of the problem to which schooling has been addressed in this century, and in the end they have, often through schooling, redrawn the boundaries of the culture which had initially defined them apart. In the case of blacks, the pressure for inclusion has redrawn the channels of control over the schools. In many ways this is the story I have to tell here. My title is meant to describe not only the process by which outsiders were drawn in, but also that redefinition.

I
From Other Shores:
European Immigrants
and American Education

I

The Progressive, the Immigrant, and the School

These present-day problems of the child—the cities, the coming of immigrants, the collapse of homelife, the yardless tenement—are all due to one underlying cause. There has been an entire revolution in industry during the last century, and nearly all the problems of child life have grown up as a result of this revolution.

ROBERT HUNTER, *Poverty* (1904)

While still imperfect, our understanding of the emergence of the common school in the nineteenth century has become increasingly clear. Through the vigorous endeavors of a generation of gifted and energetic historians who have examined the successful common school campaigns, the early organization of the schools, as well as the intense conflicts that success often masked, we can now speak with confidence and considerable grace, as Carl Kaestle recently has, about the critical ways in which schooling in the nineteenth century was entwined with the developing capitalist economy, the ethos of republicanism, and the values of an ebullient Protestant middle class. As Kaestle has demonstrated, schooling for early nineteenth-century Americans was intended to be a broadly socializing experience. The hoped for outcomes included high-minded character, a religiously derived morality, and industriousness oriented to social progress—virtues understood to be essential to effective republican citizenship. "The chief end is to make GOOD CITIZENS," an Illinois superintendent maintained at mid-century. "Not to make precocious scholars . . . not to impart the secret of acquiring wealth . . . not to qualify directly for professional success . . . but simply to make good citizens." Toward this end, reformers labored to make at least a limited exposure to publicly supported schools the common possession of most Americans. Even the exceptions—blacks, Asians, Indians, whose relation to the republic was highly problematic—emphasized the inclusive goals of the ideology for those who were potential constituents of the civic society and contributed to its polity.[1]

By the late nineteenth century, the fate of the now triumphant and much extended republic was no longer in question. The issues of morality, character, and virtue, which had galvanized efforts in the 1830s and '40s, were partly eclipsed by other problems, which, while not new, seemed vastly more urgent—economic success, loyalty to the state, and the incorporation of a heterogeneous population that made the pluralistic society of the early nineteenth century seem simple by comparison. Even scholarship, once largely insignificant, had become much more important to a society where science and learning contributed to state progress and social welfare. The social problems of the late nineteenth century were often, in fact, old problems in more insistent form. And just as problems associated with the newly democratized republic of the 1830s often found their keenest expression as well as hoped for solutions in proposals for common schools, the problems of an industrial, urban, and immigrant swollen America with imperial aspirations were expressed in a new wave of reform efforts that focused once again on the schools. But while our understanding of the concerns that issued in the common school of the early nineteenth century seems to have come to a state of enviable historiographical fruition, the same is not true for what has normally been identified as "progressive education." That problem has provoked a rich and contentious scholarship for more than twenty-five years. In recent years, a spate of books, many of them quite compelling in themselves, has left us with a picture of progressive educational reform that is not so much unfocused as overexposed. A multitude of social groups, in addition to the old cast of professional pedagogues and middle-class bureaucrats, appear to have aspired to changing the schools in self-interested ways that they defined as progressive. One is tempted to reach for a scorecard as labor unions, socialists, city managers, mayors, teachers' organizations, middle-class reformer types, journalists, women's suffrage advocates, women's clubs, social workers, school principals, and parents voice their special interests through the label progressive. In the light of this clamoring chorus, progressive education becomes no more than a lance in the local political battles of the period. Progressive education has lost the coherence once assigned to it through the educational philosophy of its most determined spokesman, John Dewey, and luminously portrayed in the landmark study by Lawrence Cremin, *The Transformation of the School*. Nevertheless, as we have come increasingly to appreciate the many actors who contributed to reform efforts and their different agendas, the relationship between aspirations for a reformed school and the underlying economic and social problems of turn-of-the-century America has taken on renewed force. While nowhere as clear or cogent, our new understanding has once more linked the reform of the schools directly to the reform of the society as John Dewey had intended.[2]

It is not my intention to resolve the many issues or bring order to the perspectives that surround the problem of progressive education. To do so would mean to untangle not only the historical knot within which it emerged but also the many contemporary educational issues that are still bound up in its elongated skein. I would like instead to look once again at a narrow segment of its proponents, professional reformers, and to suggest how the fundamentals of its meaning can be recaptured in part, not by attending to motives or to the many political conflicts through which different groups attached themselves to the progressive label, but by comprehending the cultural investments with which reformers endowed schooling. The solutions that reformers hoped would issue from school reform connected their efforts both to the social problems to which they unanimously responded and to older concerns about republican citizenship that defined the original common school impulse. In recreating that double commitment, I am trying to recapture certain essential ambiguities, even contradictions, latent in educational values at the turn of the century that can help us to understand why progressive education has been both compelling and problematic in the twentieth century. In that context, perhaps, many of the contradictory resonances which the idea of school reform had for various groups and interests may be better understood.[3]

The reformers were important not only because they articulated the social concerns of the time in particular form, an understanding that linked disparate groups through the common template of culture, but also because they helped to set the agenda for the century to come. At the root of the problem they addressed was a basic paradox of American culture: American existence, expansion, and prosperity depend on the continuous infusion of outsiders, but outsiders threaten to dissolve the culture and its links to the past by their presence. That paradox led reformers to their most vital question—how could a culture that had newly stabilized itself in the nineteenth century maintain its historic roots while responding democratically to its diverse population? When they turned to education, and to the schools specifically, to answer that question, reformers brought to their solutions two unquestioned assumptions. The first was that education was transparently a social good; the second, that everyone not only should be educated but was also capable of being educated. Those assumptions were fundamental to progressive educational endeavors, and they reflected the degree to which schooling had become by the early twentieth century not only an instrument of social policy but also a basic cultural commitment.

I

The society to which reformers directed their attention at the turn of the century had experienced the repeated shocks of the industrial change we have come to associate with the gilded age—disorderly cities, corrupt pol-

itics, enormous disparities in wealth, sobering depression, industrial regi-
mentation and disturbance, and civil strife. But the mess of American life
in the late nineteenth century was also a function of perceptions, of a lost
sense of deep moral order and social meaning. And it was to an attempted
re-creation of meaning as well as to the general urban cleanup that social
reform was most critically directed. By the twentieth century, much of this
cleaning up, both physical and psychological, was centered on the immigrant, who was both a source of disorder and its magnifying lens.

Decade after decade in the late nineteenth century, vast numbers of immigrants arrived at American ports, three-fourths of them at New York.
Between 1860 and 1890, thirteen and one-half million new immigrants
arrived. In the first three decades of the twentieth century, nineteen million
more made their way to the United States. Between 1890 and 1920, immigrants and their children formed between one-half and three-fourths of
the population of cities like Cleveland, Milwaukee, Boston, San Francisco,
and St. Louis, as well as the great metropolises of New York and Chicago.
In many industries one-half to two-thirds of the workers were immigrants
and their children. To the problem of numbers was added the complica-
tion of origins.[4] In the early twentieth century, analysts and journalists
emphasized that immigration had become newly problematic because the
sources of immigration had changed as Italy, Poland, Russia, and the Bal-
kans replaced the British Isles, Germany, and the Scandinavian countries
as the home of the majority of newcomers. The groups composing the so-
called "new immigration" were portrayed as more alien, more transient,
from more autocratic societies, less family-oriented, less skilled, less literate. In addition, they were often visibly darker and rarely worshipped any
recognizably Protestant God. The perceptions of difference and the astonishing size of the immigrant population led to considerable alarm.[5]

In the context of industrial convulsion, this unprecedented immigration
threatened all aspects of American political, cultural, and institutional life,
from the family to the political process and aesthetic experience. In the
late nineteenth and early twentieth centuries, social reformers addressed
the many problems associated with this experience.[6] Although their visions
and motives were diverse and complex, they were responding individually
and as a group to what might best be described as the changed context of
civic socialization—that is, the manner in which individuals were trained
to a responsible American adulthood. In the early nineteenth century,
Americans had envisaged the republic, epitomized in the appeal of the
common school, as the expression of personal virtues supported by institutions which encouraged morality and order. But citizenship itself was
inseparable from autonomy. Since the early 1820s and '30s, American ideology had made a critical connection between work and politics. Work

made a man independent, and suffrage reflected and protected that independence. Work was the effective basis of self-identity and self-worth, and as significantly, given the dominant belief in opportunity, the foundation for status, wealth, and advancement.[7] Early nineteenth-century reformers, most of them Whig, had never entirely trusted free-market forces and had invested schools, churches, and families with important auxilliary functions. Nevertheless, work was the dominant agency of social responsibility. Independent work, especially on the land, remained a potent ideal long after the reality had been altered. "The dangerous effects of city life on immigrants and the children of immigrants," John R. Commons, the progressive labor historian and reformer, declared, "cannot be too strongly emphasized. This country can absorb millions of all races from Europe and can raise them and their descendents to relatively high standards of American citizenship in so far as it can find places for them on the farms. 'The land has been our great solvent.' But the cities of this country not only do not raise the immigrants to the same degree of independence, but are themselves dragged down by the parasitic and dependent conditions which they foster among the immigrant element." Commons's statement reflected more than a pastoral romanticism.[8] It pointed to the tenacity of the relationship between independence, work, and citizenship which was often hidden in discussions of immigration, industrialization, and city life.

The cities, the factory regime, and industrial capitalism raised profound questions about the relationship between work and citizenship that were never effectively resolved. Instead, by the early twentieth century, the issues were rerouted into questions about the private lives of the men who tended the machines and the composition of city populations. As social reformers in the late nineteenth century began to attend to poverty and urban squalor, they gave increasing attention to the quality and influence of leisure, family life, health, housing—in short, to the private lives of workers rather than to the socializing effect of work. Certainly, they understood that poverty was the result of economic deprivation. "Of course," Mary Simkovitch declared, "many measures for the improvement of living conditions are a virtual admission of injustice."[9] But work itself figured increasingly as the place in which wages were generated, or at best as a physical environment, rather than as the moral force for character formation. Thus, in Margaret Byington's *Homestead* family nurture not work became the crucial determinant of social habits despite the fact that what dominated the lives of workers was the mill. In *Homestead,* work remains the ghost at the table, while family budgets, family relations, recreations, and community events are elaborately described.[10]

The recognition that work was no longer reliable for social cohesion was forcefully expressed in Mary Simkovitch's perceptive, sensitive, and unsentimentalized study, *The City Worker's World*. After describing in detail the family life of urban laborers, she observes, "The education of school and home is deeply supplemented by the education of one's work. . . . In general, work keys one up. A pace is set, a new kind of standard is inaugurated. . . . It is WORK that enlarges. At this point two things happen. Other groups, their way of living, their outlook, begin to exist for the worker who previously had never dreamed of them and second, the worker realizes that he is an integral part not only of a Family but also of Industry." Thus work socializes to larger concerns—to a broader vision of the social body and to an identification with social life in general. Then Simkovitch, as if suddenly realizing that it is the industrial worker's work she is describing and not some ideal and fanciful version of what WORK ought to be, retreats and continues in a different vein. "Education ought never to stop. With the privileged, it goes on from one form to another. . . . What is it that blights the education of the industrial family? The fact that work absorbs the vitality and time of the worker. This would not be so objectionable, if the work itself were educational. But, in highly subdivided industries, this is, generally, not the case." Rarely was the progressives' conscious and unconscious more pointedly revealed. Work—the capitalized ideal—ought to educate and to serve as a force for personal growth and social identification, but industrial labor, as it is, cannot.[11]

Another illustration of this dilemma came in Felix Adler's definition of and solution for the social and moral crises of the time. Adler, best remembered as the founder of the Ethical Culture Society and proponent of ethical religion as a modern-day alternative to God-centered faiths, was searching for nonpunitive foundations for community ethics and responsibility. Adler, who was associated with a wide variety of reform causes and activities, found the solution in what he called "calling." As a form of intense identification with work, calling was an "intermediate form of devotion. . . . and through the medium of one's calling to society as a whole." "I do not hesitate to say," Adler continued, "that the choice of a vocation or calling has a more decisive and far-reaching effect on character than any other act." But Adler also knew that the character forming quality of work and his hoped for reconstruction of community identification through application to calling were impossible in view of the type of work that dominated the lives of most industrial workers. "In developing a social ideal on vocational lines the gap that separates factory labor from the requirements of such an ideal must not be covered over."[12] Significantly, Adler started and ran for a short time an "industrial school" in New York City.

For most reformers, the recognition that industrial labor failed to fit an older framework of socialization and did not serve as a force for social cohesion paved the way for investigations that tended to divest work of meaning and to redefine socialization in other terms. These investigations, the social surveys, provided ballast for social reform efforts. With the evidence of their investigations, reformers in the late nineteenth and early twentieth centuries called upon the government to intervene, to remedy, to control, or to compensate for conditions detailed in the surveys—family disorganization, unsanitary conditions, child neglect, impoverished recreational facilities, squalid tenements. It has been demonstrated often enough by historians how the thrust of such reforms was toward an interventionist state that would exercise authority in unprecedented ways over the private lives of citizens.[13] It has less often been made clear how this resulted, not from some sudden paternalism, but from the subtle reorganization in perceptions that progressive investigations enforced. As progressives turned from the workplace and the adult citizen to newer problems and remedies, they made the child as future citizen of the state the fundamental concern. By addressing the conditions of poverty, reformers observed how children were socialized and how habits were formed in impoverished and alien homes where work was no longer integral to child development. Thus, Jacob Riis's opening statement in *The Children of the Poor* is a clarion call to action: "The problem of the children is the problem of the state. As we would mould the children of the toiling masses in our cities, so we shape the destiny of the state which they will rule in their turn, taking the reins from our hands."[14] This explicitly paternalistic attention to the children of the poor helped to complete the shift from the workplace to other social agencies and instruments whose objective would be early and prophylactic socialization.

Americans had invested the family with great purpose and importance throughout the nineteenth century, as they had the school and the church. In the nineteenth century, however, these institutions had a moral mission, to refine habits, instill virtues, provide ennobling ideals. They provided a balance to the entrepreneurial energies and individualism developed through work. They gave children a foundation which allowed them to work successfully, efficiently, skillfully. The family provided Christian sensibilities; work, republican character. As we have seen, the school had already been enlisted to help in character formation, and the family certainly had a crucial role to play. It is significant, however, that by the late nineteenth century, as reformers grew increasingly uneasy about work in general and in their nostalgia for an earlier and simpler time, they portrayed the historical family not as the bosom of Christian morality but as above all a workplace which trained children to social responsibility. "In past generations," Frank T. Carlton noted in the journal *Education*, "the home produced and prepared nearly all the food consumed by its members; much

of the work which is now carried on by the factory was then performed
in the home. The home was the scene of diversified industry as well as the
center of the child's social life; the school was merely the place where the
three Rs were expounded to the youth; the playground was broad and
spacious, often consisting of an entire farm."[15] In this revealing misperception, reformers betrayed their sense of the declining efficacy of work
throughout society.

This view of how the family had changed would soon form the basis of
an entire school of family and urban sociology.[16] For reformers, however,
it was less a theoretical problem than the grounds for remedial action. In
the reformers' fanciful past the school was strictly limited to literacy training while other social agencies provided socialization. The loss of that past
was the basis for a reevaluation of the potential of schools for much broader
social instruction and for a much more effective empowering of the school
as an agency of social life. In the penultimate paragraph of *The Delinquent
Child and the Home,* Sophonisba Breckinridge and Edith Abbott made "A
strong plea . . . for the adaptation of the school curriculum to the actual
demands of industrial and commercial life, the multiplication of uses of
the school buildings, the prolongation of the school year by means of vacation schools, the establishment of continuation schools, the further development of industrial and trade training, and the perfection of the machinery for apprehending all truant children and securing their regular
presence at school. . . . " It was a statement which in brief form outlined
a good deal of the progressive program for the schools and a plea with
which most reformers could have concurred.[17]

The family divested of work and the workplace of meaning had contributed to the anxieties of the late nineteenth century. The optimism of
progressives resulted from their ability to transmute concerns about inadequacy into a new faith in social reconstruction that would issue from
alternative forms of socialization. The school, newly reformed and expanded, would become a primary instrument of that faith. David Rothman
has described progressives as optimistic positivists whose renewal of confidence came as the social surveys allowed reformers to turn their attention
from broad questions to specific and solvable problems.[18] In place of work,
reformers proposed to relocate the basis of social life in the new areas they
had begun to investigate—in leisure and play, in family relationships, and
in neighborhoods—areas that schools especially, but also social settlements, playgrounds, and specific legislation could reform or reinvigorate.
One could argue that reformers sought to reform schools and neighborhoods precisely because they could not or would not reform work. But
this is too simple. Increasingly, in the early twentieth century, while still
troubled by the inability to integrate industrial work into an older frame-

work of values, progressives replaced the connection between work and civic life with a new commitment to formal education of all kinds as the strategic basis for adult preparation and community survival. Indeed, with their renewed confidence and positivistic solutions, the school became for some reformers like John Dewey not merely a substitute or an auxilliary but a primary sphere, a protean force for democratic citizenship.

II

Most institutions had been shaken by the crises associated with the industrial transformation of the late nineteenth century, but the public schools (despite their contentious beginnings) had weathered the shocks remarkably well. David Tyack noted some time ago that the value of common schools in the late nineteenth century was largely unquestioned even when other institutions were objects of corrosive criticism or neglect. That support was, moreover, very broad-based as working-class parents, labor unions, and many in opposition parties—like the Socialists and the Populists—looked to the schools as a source of relief and opportunity. American support for public education was already a fact of the culture well before reformers attached themselves to the schools as democracy's hope, and the schools seem to have enshrined an American (and modern) faith in education for personal improvement and social progress. Enrollment figures, at least for young children between seven and thirteen, were high, even in the absence of effectively enforced compulsory education laws.[19] Despite the portrayal of reformers like Carlton, the schools had always had broader goals than instruction in the three Rs. Training in morality, citizenship and industriousness, obedience and orderliness accompanied even the most rudimentary literacy training. And literacy itself was a far more significant aim than simple alphabeticism in a Protestant culture with a wide-based suffrage. Still, the fulfillment of these aims was limited throughout the nineteenth century by brief, sporadic, and attenuated attendance. Thus, while the objective of universal schooling had been largely fulfilled, the effectiveness of the schools had remained limited. Elementary instruction in basic subjects often in very abbreviated form was all that most children, and certainly most working-class children, could expect. The effective expansion of the schools, longitudinally toward longer, more regular attendance and horizontally in the accretion of social services, activities, and subjects, was very much an expression of the school-reform efforts of the late nineteenth and early twentieth centuries.

The public secondary school still competed with the private academy throughout the post-Civil War period, and was disproportionately the preserve of the middle class, of the children of proprietors, professionals, managers, and of some privileged shopkeepers, artisans, and clerks. Most high schools were small, uncomplicated institutions which directed their efforts to preparing boys for the university or clerical professions and girls for teaching. In 1870, only 2.7 percent of all students eligible to attend four-year high schools were in public schools. That proportion rose to only 4.2 percent by 1890. Except in isolated instances like the Boston Latin School, American publicly supported high schools had never offered an exclusively classical curriculum with ancient languages at its base, but had proffered more functional English or science-based curricula. But they had been overwhelmingly academic. Rather than the strict adherence to a sequence of Latin texts, the public high schools provided a wide range of courses in history, mathematics, English, science, foreign languages, and philosophy. Some also offered commercial subjects like bookkeeping.[20] But only with the reforms of the early twentieth century and in the context of the much larger student body to which they aspired was the academic high school replaced by the comprehensive high school of the twentieth century.

These were the reforms, enforced truancy laws, school attendance beyond the primary grades, more relevant courses, and the public school as a social center—what might best be described as the intensification of education—that progressives hoped to bring to the schools. The reasons were related both to the reevaluation of work in industrial America and to the shock of recognition that reformers experienced when they set out to discover the sources and consequences of the poverty of industrial workers.

All about them in the slums, the mill towns, and the mines, reformers discovered not only industrial poverty but also the specific poverty of the immigrant masses. "In the poorest quarters of many great American cities, and industrial communities," Robert Hunter concluded, "one is struck by a most peculiar fact—the poor are almost entirely foreign." Jacob Riis observed the same thing. "Here in New York to seek the children of the poor one must go among those who, if they did not themselves come over the sea, can rarely count back another generation born on American soil." And Mary Simkovitch came to the same conclusion. "Our wage-earners are mainly foreigners and the life of the industrial family is the family of the immigrant or of his children. Their traits, customs, habits, desires," she added pointedly, "must therefore be of interest and concern to us all."[21]

This double discovery, that most workers' lives were impoverished and that most of the poor were foreign, forced reformers to deal not only with change but also with diversity and to confront its consequences for American social life. As a result, their demands on the schools became even more significant and onerous. The schools were to provide not only a substitute form of socialization but also a means for remedial socialization.

The schools had to attempt what work could no longer achieve. But the schools also had to overcome what alien families and foreign communities had already begun.

Common school advocates in the nineteenth century had also dealt with heterogeneity, but their commitment to clear-cut citizenship training, their confident moralism, and their pedagogical faith in rote learning made this problem far less complicated than that which faced social reformers by the early twentieth century. By then, the regnant evangelical morality of early nineteenth-century Protestantism had been called into question by science and secularization, and the simple didacticism had been largely undercut by a reformed pedagogy and new views of childhood learning.[22] In addition, the challenges mounted against attempts at cultural hegemony through schooling by Catholics, foreign language groups, and some radical critics in labor and socialist organizations had taken their toll on the unambiguous assimilation implicit in the original common school.[23]

But the reformers too had contributed to the problem. By the time of the First World War, the industrial crisis of the late nineteenth century had also become a cultural crisis, a crisis that by then involved not republican character but the meaning of American identity and state loyalty. Reformers contributed to this not simply because they were suspicious of immigrants, as many were, but also because in helping to redefine how an individual was socialized in cultural rather than political terms, they drew attention to the vast differences among ethnic groups and began to see how complex the process of learning was. Their attention to the many differences that underlay immigrant life came, moreover, at just that point in the early twentieth century when discussions about immigration took on a markedly racist slant.[24] While racist thinkers and social reformers were by no means coincident, both emphasized the tenacious characteristics of immigrant groups. And even though reformers hoped to change the home conditions of immigrants and to provide institutions for their social integration, their efforts highlighted just how different immigrants were while deflating the common experience of work and politics as sufficient for citizenship.[25] So too, their writings were riddled with confusions about the role of inheritance on character, on poverty, and on talent. They used misleading terms like "race traits" and traded in invidious group descriptions.[26] To the modern sensibility, Riis's portrayal of various immigrant groups, or the essays by settlement workers at Boston's Southend House, or even Mary Simkovitch's usually sensitive descriptions of workers read like racial slurs, full of stereotypes about alcoholic but affable Irish, uncivilized blacks, thrifty and hardworking Germans, stingy Jews, and sturdy but dull Poles. Mary Simkovitch summed up the dilemma of her reform generation: "The anthropologists will no longer allow us to use the expres-

sion 'race traits' and any characterization of a race group is bound to be
partial and limited. Due as these qualities doubtless largely are to the environment, we can yet not fail to observe the differentiation that marks off
one group from another."[27]

In fact, while their language and concepts were often confused, reformers were groping toward a way of identifying different groups by what we
would today call culture—habits that form from within the dynamics of
group participation rather than from individual heredity. The most sensitive reformers, like Jane Addams or Grace Abbott, certainly understood
the historical circumstances within which group characteristics evolved and
how they functioned for group survival and personal strength. Others were
less careful or caring, but reformers were almost without exception before
World War I concerned with social conditions, not genetic endowment.
Their lapsing into personal bigotry or the confusions of contemporary language ought not to disguise their dominant concern.

In shifting attention from the workplace to the lives and habits of immigrant workers—what they ate, how their children played, how the family functioned—reformers, although they did not name it, began to observe
the complex realities of immigrant culture. "In the largest cities of America
there are many things which separate the rich and the poor," Robert Hunter
declared, "language, institutions, customs, or even religion separate the
native and the foreign. It is this separation which makes the problem of
poverty in America more difficult of solution than that of any other nation." The discovery of immigrant culture—neither genetic nor strictly environmental (because it could not be quickly altered)—posed special problems. "Americans have perhaps too readily assumed that all immigrants
can be assimilated with equal ease," Frances Kellor declared. "We now
realize . . . that some races, unfamiliar with our language, form of government, industrial organizations, financial institutions, and standards of
living require much more aggressive efforts toward assimilation." Not until the 1920s would Americans' insistence on the tenacity of ethnic differences inspire immigrant exclusion, 100 percent Americanism, and group
pride. But before reform enthusiasm gave way to war and the 1920s solutions that predefined America in terms of a racial and ethnic status quo,
reformers managed to invest one institution with their hopes for social
change. The school was the only institution that could hope to alter immigrant culture where it was environmentally most permeable—the care
and instruction of children. "Down in the worst little ruffian's soul, there
is, after all, a tender spot not yet preempted by the slums," Jacob Riis
observed. And in that still unblemished potential, the reformers placed their
hopes. "Most of the questions affecting the immigrant in America," one
reformer noted, "his relationship to the national life and its relationship

to him, come back at the last to the problem of education. Before anything else must come education."[28]

III

In the light of their implicit discoveries about culture, education had become a much subtler matter for reformers than the simple exposure to common schooling had once been. The experience of reformers with the intricate cultural networks that they found in visiting the neighborhoods of the foreign-born helped to inform the view held, almost universally, that education was nothing less than the whole process of socialization, the process by which the child became a part of society. Their experience made John Dewey's approach to education (also formed in the complex ethnic environment of late nineteenth-century Chicago) as the product of and the preparation for life as a whole urgent and compelling. Thus, when reformers spoke of education, they usually did so in capacious terms. "Nothing less than education is powerful enough to save the child," Robert Hunter announced. "And, 'to prepare for complete living' is the function education has to discharge."[29] Hunter's use of the term "education" is deceptive here, for despite their broad language, most reformers, Hunter included, were satisfied to identify their hopes for education with the school. The school required for this task had to be reconstructed and expanded so as to provide an effective correlative to the culture from which poor, foreign children came. Thus, when Hunter and others spoke of education "for complete living" they did not have in mind the totality of childhood socialization, which after all was the source of the problem for immigrant children, but a reformed and expanded schooling. Only the schools could save the child from the life he was destined otherwise to lead. That schooling now had to replace the "complete living" which had never before required a simple solution.

Once, Hunter explained, "the children received their entire education either in the home or in the adjoining fields. Certainly in those days the child received his best education under the supervision of his own parents."[30] That informal education was not only no longer sufficient but also no longer trustworthy. For the children of immigrant workers to be educated entirely under their parents' aegis was to leave them in unsanitary slums exposed to the lures of the street and to perpetuate the very cultural differences and often undemocratic practices reformers hoped to overcome. Though not without ambiguity and some regret, reformers aspired to replacing one form of "life" with another for immigrant children.[31] The reformers' vision of education was paradoxical. They wished the schools to be like life and to prepare for life, while they hoped the schools would replace objectionable forms of living and prepare the young to live better (and more American) lives than they would have lived without schooling. They hoped the schools would substitute American forms for the cacophony of alien cultural modes, but they also hoped the schools could become an integral part of the communities from which the children came. The paradox resulted, in part, from the fact that progressives intended at once to reform the schools and to make the schools an instrument of social reform. They wanted to make the schools more expansive, more protean, more socially instrumental. While the schools needed to become more like communities, the communities they hoped the schools would prepare for were to become more like the schools of their fantasy.

John Dewey captured the nature of that school and the mechanisms of this reciprocal reform in *School and Society* (1915). Like other progressives, Dewey hoped to create in the schools the modern-day equivalent of democracy in nineteenth-century America, and in so doing to confirm democracy by re-creating it in revised form. "When the school introduces and trains each child of society into membership within such a little community," Dewey observed, "saturating him with the spirit of service, and providing him with the instruments of effective self-direction, we shall have the deepest and best guarantee of a larger society which is worthy, lovely and harmonious."[32] For Dewey, this vision of the school was itself a type of reform, for it constituted a consciously created community that was a vast improvement over the atomistic society that Dewey, like other progressives, saw in the industrial and urban world around them. And it would lead to reform in that larger society by providing children with the skills and the tools of democratic participation and social responsibility. Dewey's vision was both a recipe for and an epitome of what progressives hoped reformed schools would provide.

A good part of the progressives' vision of what the schools were to become came from the work-centered republican values they inherited from the past. "We are apparently entering upon a period of cheap standardized production upon an enormous scale, which will multiply commodities and increase leisure," Charles Horton Cooley, the progressive sociologist, noted, "but will make little demand upon the intelligence of the majority of producers and offer no scope for mental discipline. Work is becoming less than ever competent to educate the worker, and if we are to escape the torpor, frivolity, and social irresponsibility engendered by this condition, we must offset it by a social and moral culture acquired in the schools and in the community as a whole." Cooley was not alone in looking back to the old ideal behind the common school. His view of the school was very much part of a tradition in which schools provided a common and moral experience. But Cooley, like Dewey and other progressives, was not content to make the school a didactic instrument of morality and citizenship. Instead, schools had to be democratic and participatory, the equivalents of once richly functioning work-centered communities which industrialization had impoverished and replaced with the fragmentation of the slums, the tenements, and the city streets. "The children in public schools," Mary Simkovitch observed, "lead more nearly a life of real democracy than any other group. They are instilled with a common point of view. They lead a united life of work, play, and hope."[33] Not only did children in this view lead a more "real" life in school than out, but they lived it more democratically. Again the schools were paradoxical, democratic but not diverse, like life only better. Theirs was a democracy of a "common point of view," the democracy of a fancied American small-town past now so obviously missing from the immigrant-swollen cities. It was, of course, not America's real life but its fancied past that Simkovitch hoped to create in the schools, just as it was the forms of work reformers believed had once been the common underpinnings of democratic citizenship that Cooley now missed in the workplace and sought in the schools.

At just the time progressives were turning to new methods like the schools, their beliefs remained informed by traditional values which those methods were to refashion and replace. The schools were deeply part of those confusing, even contradictory, cultural aspirations, and many of the confusions that marked progressive education must be understood in that light. The progressives hoped to invest the traditional common school with new functions and a new significance as part of a larger social reconstruction, but the visions they brought to that reconstruction were limited both by the past order for which the schools were to provide a substitute and by the reformers' particular diagnosis of the disorder they saw about them. The attachment of the progressives to manual and vocational instruction illustrates many of their dilemmas. Even though the school was the modern-day replacement for work, most progressives still hoped to preserve some of the character-building agency of work in the school.[34] While the fervor for manual education had passed with the old century, vocational and industrial education, which took its place, was still highly craft oriented. Work, for reformers who rarely labored in field or shop themselves, still usually meant manual work, and the reformers' attachment to manual labor at just the point when they found such labor empty of social and civic value suggests not the insincerity of their beliefs but incomplete transformations in their thinking. Like their faith that schools would be like life only better, democratic but productive of a common culture, so too reformers hoped to preserve an old-fashioned experience of physical labor in an environment intended as both a substitute for work and an escalator out of the working class.

Part of their dilemma was real, the result of their heightened sensitivity to the problems of the alien poor. Reformers were not naive about the dangers of industrial education. Most understood how easily it could degenerate into class education that fitted the masses to menial roles in life. As Mary Simkovitch noted, "The education of an aristocracy where some are to be leaders and others are to be workers would lead to a classification in education to correspond to social class, and in fact, that is the danger of the industrial education movement in this country. If this movement becomes a tool in the hands of employers merely to produce more effective workers, public funds are being diverted from the proper function of creating citizens, to that of creating an industrial army." Nevertheless, Simkovitch insisted on the desirability of industrial education. "The new education, seen in the development of industrial training, if properly understood will not lead to class education but rather will be a new mode of educating all." Training in work and an understanding of the work process, Simkovitch believed, had to become part of the total "socializing process" and part of everyone's education.[35]

Certainly John Commons did not wish to see industrial education become a captive of employer interests. He hoped that manual instruction, as part of a much broader curriculum, would develop the worker's intelligence and generate creativity and innovation—producing not drones or an industrial army but citizens. "The foundations of intelligence for the modern workingman is his understanding of mechanics. Not until he learns through manual and technical training to handle the forces of nature can the workingman rise to positions of responsibility and independence. . . . intelligence in mechanics leads to intelligence in economics and politics."[36] Commons, like many progressives, could not divest himself of the nineteenth-century ideal: the mind informed by meaningful work was politically trustworthy.

For some reformers, vocational education also promised economic and social mobility to the children of the immigrant poor. "Vocational direction," Lillian Wald noted, "has been in the minds of many social workers. It is inevitable that men and women, as they become acquainted with the fortunes of the children in an immigrant neighborhood, take a look ahead and rebel at repeating in the lives of the next generation the misfortune of the proceeding." For Henry Moskowitz, a New York settlement worker, labor activist, socialist, and an immigrant himself, this repetition of misfortune was exactly what old-fashioned education meant for the children of the immigrant. Moskowitz had nothing but contempt for the shortsightedness of a purely academic curriculum: "The immigrant child who must leave school at fourteen is fit for nothing. The curriculum was not designed to meet his needs. . . . He must face the work of industry without having received the slightest equipment imparting that manner of capacity required for a beginner. Our democracy has been lavish in its expenditure of wealth on this undemocratic education."[37]

Moskowitz's anger came from his perceptions about the nature of education as it then existed. All around them in the schools, reformers saw a highly academic curriculum, geared to subject progressions culminating in the university. That schooling had allowed "the former cultural educational system . . . to decline from its old high estate of creating cultivated persons to a plan for educating clerks." In its place, many reformers hoped to institute an education whose aims would be the education of the laborer rather than the elite. Edward Divine, a leader of charity reform, presented one of the most provocative statements of the hopes of reformers for vocational education. Aware of the class dangers of an insincere incorporation of vocational training into a school system geared to college entrance, Divine proposed to reunify education by standing the classical curriculum on its head, making the education of the workingman, not the professional, the organizing principle and fulcrum of the school curriculum. "The great body of workers—industrial, agricultural, and commercial—should furnish the unifying element rather than the vocational needs of the few who are to enjoy a higher education. . . . What I urge . . . is that in the very interest of higher education itself the elements which are needed for industrial and agricultural workers are valuable . . . that lawyers will be better lawyers, teachers better teachers, preachers better preachers, and businessmen certainly infinitely better businessmen, if they can bring it about that secondary and higher and professional education will take adolescents who have already been deliberately grounded in the things which workers should know in order to be good workers and in order to lead a good life, and from such material will develop the national leaders, democratically-minded leaders in tune with the life of the nations' workers."[38] This radical proposal rejected not only a too early differentiation but also the usual antidote to separate education—a uniformly academic curriculum. Divine was emphasizing the progressives' expectation that vocational and manual education would make work and the worker the schools' central commitment and that this vision would unite all Americans through the bonds of reformed schooling.

Reintegrating Americans through schools, with or without workers' education, posed a difficult agenda, however. Progressives hoped to make the schools a vital community force, best epitomized in the social-center ideal where schools would serve a wide range of neighborhood purposes.

But the relationship between the immigrant child and his own family and community presented special dilemmas for reformers. In the first place, immigrant children had special academic difficulties. Progressive reformers like Grace Abbott understood that immigrant children were often educationally misplaced because of their difficulties with language. "Newly arrived immigrant children," Abbott observed, "are put in classes organized for backward or subnormal children, or subnormal or backward children are put in the 'steamer' classes [for recently arrived immigrants]; and grave injustice is thus done to both groups." As seriously, the handicaps the immigrants generally labored under affected the schooling of their children. The children attended irregularly because they were needed to care for younger children at home or because they were put to work underage when parents could not afford to keep them at school. And parents resented the schools' undermining of their authority which "widen[ed] the gap between the parent and the child." Some reformers, like Jacob Riis, saw these problems largely as the result of parental greed; but others, like Abbott and Simkovitch, understood that need, not greed, kept children from school, and thus impeded their academic progress. Despite compulsory school requirements (even when these were effectively enforced), immigrant children rarely went to school before the age of seven and were usually at work by the time they were fourteen. "No adequate training for life can be obtained in this pitifully brief period," Simkovitch concluded.[39]

The very concept of "training for life" was problematic in light of the alien population whose children's lives were at stake. When this was combined with the ideal of school and community integration, the mixture could be explosive. Progressives hoped to enrich the schools and adapt them in whatever way necessary to provide the most complete education possible for children who attended irregularly and left early so that they could be prepared for living and to make a living. Unlike school administrators whose concern was with an efficient school order, an integrated curriculum, effective evaluation procedures, the progressives were often deeply concerned with the relationship between the immigrant child and his parents and community. "We must not detach the child from [his parents]," Lillian Wald insisted, "or from the traditions which are his heritage." Henry Moskowitz made the same point. "In an immigrant neighborhood . . . the welfare of the child is determined by two important factors, the inner environment represented in the social world and religious traditions, the historical background brought over to the new country by his parents; and the outer environment expressed in the physical, social and economic conditions, over which the parent has no control."[40] But how could the schools both respect the child's alien culture—and not detach him from it—yet replace it with more American experiences which would serve to unify child life through school?

The more progressives appreciated immigrant diversity and the more they called on the schools to pay heed to the special backgrounds of the pupils, the less acceptable a simple uniform educational system became. "Every system of education that attempts to train the 'whole man' must," Simkovitch insisted, "plan to relate the child to his family, his neighborhood, his state . . . as well as to relate him to his prospective work." Grace Abbott forcefully rejected the "steamroller approach" to schooling, an approach that tried to mould old immigrants "into true Americans as fast as possible. . . . this cannot be accepted as an educational end either for children or for adults. The 'moulding' process is contrary to sound educational standards. It means ironing out individual, as well as group, differences. It means that the native Americans set themselves up as the true American type to which the immigrants must conform. This would, of course, be reckless in its disregard of the talents and capacity of other people."[41]

But this very appreciation of the diversity of immigrant cultures and the specific talents and contributions of the groups from which children came could mean that certain kinds of education, like vocational and industrial training, would be preferred for some children, especially the poor and the foreign for whom it seemed so much better suited.[42] Abbott's fine expression of understanding for the cultures of immigrants and broadly sympathetic view of their gifts and talents could, with a slightly different emphasis, become a plea for a differentiated education fitted to the special talents, traits, inborn characteristics, or what have you, of different groups. Thus Robert Hunter noted, "One familiar with the homes abroad from which the immigrants have come, and familiar, as well with the parents and the environment of the child here, knows that the child of one of these races may be almost as unlike the child of another as Caliban is unlike Prospero." Hunter continued, "each race and class has, in more or less degree, a certain peculiar essence or flavor of mind . . . which, if given its proper bent and lovingly cultivated, would yield to the world untold values in specially powerful aptitudes. This is an aspect of education which is peculiarly important to America with its mixture of races."[43]

Despite Hunter's best efforts to extricate himself from the obvious difficulties of this perspective and his plea that "the individual child of whatever class or race should be given every opportunity to yield the best in any walk of life which his natural gifts make possible,"[44] neither he nor others could avoid the dilemma posed by the recognition of cultural differences and their implications for an education committed to socialization

broadly defined. The problem was vastly complicated by the frequent confusion among reformers between culture and race. That confusion was
exacerbated by the invention of mental testing that took place at just
that time, but whose impact would come with explosive force only in the
1920s.

For some the dilemma found an easy resolution in a hierarchical view
of education. Robert Woods believed that an enlarged democratic vision
aimed at the welfare of all the people and an efficiently ordered society
would define vocationalism so as to fit the masses for specific and limited
future roles and occupations. Woods hoped the schools would determine
what each child could be expected to do for the rest of his or her life and
give that child only the education he or she needed to prepare for that end.
"There is reality in the charge that the public schools are educating children beyond their station, though that is a very mean way to put it."
Woods declared that the "system of universal education must become a
system of universal vocational education. Somehow or other every student
who passes through the public-school system must have some measure of
such applied exercise of his wits and his skill as will enable him to enter
at once into productive industry. The shop does not give that training any
more. The home, even in the country, gives it but very little. The school
therefore must undertake the work."[45] This was education for life in its
most static form, and Woods took the easy path to a resolution of the
problems implicit in the combined belief that the schools had become the
most important instrument of instruction and that different individuals both
had different backgrounds and different tasks in life. Woods hoped to use
the schools directly in the interests of efficient social order. His solution
would be a continuing component of twentieth-century attempts to deal
with the issues that progressives began to explore early in the century.

But Woods's conclusion was neither the only possible conclusion to the
problem nor the one adopted by all progressives. Struggling with a wide
range of social issues whose essential difficulties still confound us today,
reformers' attitudes toward the education of immigrants were complex and
often contradictory. On the one hand, they were inching their way to a
radical appreciation of the fact of social diversity. Weaned on values derived from cohesive, organically integrated communities, the progressives
discovered a cacophony of different cultural modes, family economies, attitudes, and traditions. All the progressives found this disturbing, but some
also found vitality and hope in this diversity, and their curiosity was titil-
lated about how a new and better society could grow from the energy of
new forces and new contributions.[46] On the other hand, reformers looked
increasingly toward the unifying power of the schools to create a comprehensive social environment, not only to provide literacy but also to retrain

habits, encourage personal growth, provide economic mobility, and serve as a basis for a safe democratic citizenship.

To appreciate the culture of the immigrants, as many reformers did, could not mean to leave them alone, for the experience of schooling was, and progressives intended it to be, a process by which something positive was done to the immigrant child, something that would change him and possibly introduce deep conflicts in loyalty and socialization. To leave immigrant children alone was to freeze the status quo, not only by riveting the young into the social status inherited from their unskilled working-class fathers and mothers and forced upon them by an early entry into the labor market, but also by separating them into an unacceptable caste defined by language, habits, and beliefs. But to transform the children of immigrants into carbon copies of native Americans, as Grace Abbott feared, was not only to destroy the strengths provided them by their culture but also to assume that what an American was at any one time was and should be what an American always would be—a choice, it is well to remember, that was adopted by those who wrote the immigrant quotas of the twenties but which most progressives, following the lead of John Dewey, would not accept. This denied the very possibility of change and growth, of progress and evolution, which reformers, whatever their specific fears about disorder and reservations about diversity, hoped the schools would encourage.

Similarly, if schools remained strictly academic and didactic, the children of immigrants would be denied schooling that made sense to their lives. Their all-too-brief exposure would provide nothing more than a rudimentary literacy. That kind of schooling, brief and irrelevant, could hardly compete with the educational life of the neighborhood. But, if the schools were to attend to the specific needs and probable goals of their students, how could these be judged without predefining their futures by an over-scrupulous adjustment to the present? Schooling could open possibilities, or close them; it could be alert to students' needs, or ignore them in the interests of uniformity or equal opportunity. One thing was clear. In being redefined and repositioned in the matrix of socialization, schools could no longer ignore the populations they served. They had become more public than common as they had been invested with both new power and new hope—and also with a multitude of new problems.

IV

The concepts and values that progressive reformers brought to the problem of schooling framed not only its functions but also its definition. Schools were to be at once instruments of remedial socialization and primary agents of culture; they were to connect the democratic potential of an enormously diverse population to the unities of an ancient citizenship; they were to educate for future success but be attentive to present needs. These fissures were not contradictions but the essence of progressive school ideology, and they connected that ideology to the paradox of American society in the twentieth century. As reformers transformed the idea of common school education into a commitment to socialization, they exposed the schools to the centrifugal forces of that concept. The commitment to socialization as a substitute for work linked the values of progressive reformers to the democratic impulse of common school citizenship but increasingly eroded the uniformity of the common school idea. As a democratic reform, the idea of socialization captured the reformers' sense that all children could be educated. Since socialization is a process that all people experience through growth, then the schools as socializers could be extended to all and extended for longer and longer periods of development. If schools merely educated in fixed subjects, then it might be argued that not all children could be expected to comprehend or needed to comprehend those subjects, especially at their more advanced levels in the upper grades. But socialization necessarily democratized schooling at higher levels. At the same time, however, socialization required that schools become sharply attentive to the differences among students, since it connected schools to families and neighborhoods at one end and to social outcomes at the other. Children are not like academic subjects with clear definitions and limits, and their infinite variety required that schools become especially attuned to their differences. This became the predominant concern of schools in the 1920s as they adjusted to the populations that progressive reforms, especially child-labor and school-attendance laws, left as a legacy.

But the progressive solutions to the twentieth-century paradox have been more far reaching than the specific historical circumstances they were meant to address. This is because progressive visions of education have become part of the cultural construction of the problem of mass education. It is also because the visions that underlay progressive educational reform were complex as well as limited. While their views were limited by the particular circumstances and often still traditionally derived values of the time, the progressives' understanding of the problem of citizenship, of the nature of socialization, of the complications of culture and its meaning for individual development was much more sophisticated than those of their common school predecessors. Not only have the issues addressed by progressives been recurrent, but their concepts, language, and their mode of action has become an epitome of twentieth-century educational reform. The progressives saw the schools as instruments of social change. Despite variations in their views, all progressive social reformers expected that the school would help to reconstruct the society. In that sense, progressives were the forerunners of all subsequent twentieth-century reformers for whom schooling is a lever of social change. That fact reaches beyond the specific and contradictory quality of their social objectives, whether these were conservative or radical, liberating or paternalistic.

Progressive education was more than a touchstone of subsequent reform activities, however. It set the terms of educational debate throughout the twentieth century. That debate would last long after the initial conditions to which progressives responded had been forgotten. And it would define the issues for school administrators when the creative energy for reform and the urgency for social reconstruction had passed.

2

Education, Democracy,
and the Science
of Individual Differences

The question of mental capacity from the standpoint of race has become of particular interest for America on account of the immigration situation and the presence of the Negro. In this connection it becomes important to determine how a race rises from one level of culture to another, whether by internal stimulation and native ability, or by accepting and imitating the culture of the higher level of society; and more particularly, which races are fit to progress and which are not, and why.

W. I. THOMAS (1912)[1]

It has been recognized that equality of opportunity is not provided when all children must take precisely the same work, that what may be a significant opportunity for one child is a relatively valueless opportunity for another. Rather, the term has come to mean that every child should have equal opportunity to develop his particular abilities and aptitudes for successful and happy living in a democratic order. It is under the pressure of this demand that the curriculum has been broadened, special classes formed, varied types of materials introduced and flexibility of school organization increased.

HOLLIS CASWELL AND DOAK S. CAMPBELL (1935)[2]

When W. I. Thomas, the University of Chicago sociologist, posed his questions about race, he did so in order to encourage research that would dispel popular ideas about the innate inferiority of some races. In 1912, at the height of the progressive era, Thomas used a variety of sources, including Franz Boas and W. E. B. Du Bois, to argue for the primacy of culture and social environment in the formation of group character and to insist on the capacity of all races to learn. Thomas was not specifically addressing the problem of the schools, but his article contained a significant quote which underscored the meaning of the American faith in education as it had hitherto been understood and expressed by social progressives. In Poland, in the mid-nineteenth century, according to one of Thomas's sources, "the feeling of the nobles with regard to education of the peasant was expressed in the opinion . . . that culture not only did not become the peasant, but for the most part he was incapable of it."[3] The belief that certain segments of the population were incapable of being educated underscored, by contrast, the firm commitment initiated by the common school ideal and confirmed by social progressives that all Americans could be educated.

In the 1920s, views like those of Thomas, which explained racial characteristics in primarily cultural terms, grew in certain academic circles, especially in anthropology and sociology. But these views competed with and were frequently eclipsed by another and increasingly powerful explanation that provided very different answers to Thomas's concerns. In answer to Thomas's inquiries about mental capacity in the context of immigration, this view brought a sharp and unadorned emphasis on differences that were inborn, unlearned, and impermeable. And it was this set of beliefs, deeply informed by a psychology of measurement and expressed in IQ testing, that exerted the most powerful influence on the schools. The emphasis on inborn differences cast a pall on traditional American assumptions about the educability of all, and throughout the 1920s it existed in deep tension with the continuing pressure for more and longer education. The tension between fuller education for all Americans and the implied limitation on the educability of many outside the American mainstream characterized school development during its crucial period of expansion and is fundamental to an understanding of the particular meanings that schools gave to their adoption of progressive concepts and language. For while progressives had challenged the schools to expand and deepen their commitments to democratic education, they left an unclear legacy, incomplete definitions, and ambiguous challenges to an institution already deeply troubled by the practical realities of immigration.

The reasons for the increased reliance of the schools on forms of thinking that emphasized the inborn rather than the learned were complex as I hope the following discussion will reveal. But certain underlying influences can be more simply suggested. First, by emphasizing schooling as socialization, progressive social reform had impressed upon the schools the necessity to define the child whose schooling was their object, and this made the schools turn to those tools, above all the IQ, which could provide an efficient and cheap definition. Secondly, progressive school reform, which exerted an increasing influence as school systems grew in size and complexity, was part of another face of progressivism—managerial reform, from which it drew significant inspiration. And while social reformers expressed progressivism's most generous impulses and humane directions, managerial reformers, who were enamored of efficiency, expertise, and a systematic approach to institutional development, cast a long shadow over the entire period. Thus while social reformers' expectation of the new role that schooling would play in society added gravity to the schools' burdens and renewed seriousness to their purpose, it was the emphasis on system and order that defined the direction of school development in the 1920s. When the pedagogical emphasis on socialization and the institutional pressure for systematic expansion were combined with the reality of millions of school-age children from foreign homes and cultures, the product was both the fulfillment and the devaluation of the democratic faith in schooling.

I

The expansion of American education in the early twentieth century was often described in language drawn from the theater or the circus: "dramatic," "spectacular," "amazing," "extraordinary." Though overdrawn, the adjectives capture certain dimensions of the phenomenon. How else describe a system that within the half century from 1890 to 1940 saw the proportion of all children five to seventeen years of age attending school soar from 44 to 74 percent? How else portray an industry in which the annual expenditure per child climbed from 17 to 105 dollars in the same period and for which the average number of attendance days rose from 60 to 130?[4] Ordinary words fail to convey the scope of an enterprise which from a rudimentary dedication to teaching reading, writing, and citizenship prided itself by the 1920s and '30s on its medical services, vocational guidance programs, mental hygiene clinics, social dances, orchestras, gymnasia, free lunches, and community centers. Small wonder that the words of description were often also words of praise, suggesting that like the triumphant economy of which it was part, the development of education in the United States was simply larger than life.

Larger than life too were the problems with which the schools were forced to deal. Progressive reformers had articulated as an imperative and in theoretical form the burdens the schools were already beginning to carry— the burdens of a heterogeneous population and a rapidly changing environment. While John Dewey and those he influenced and represented challenged the schools to define democracy's future, school systems throughout the nation were, for better or worse, struggling with what had become an importunate democratic presence. Above all, the schools were confronted by the problem of heterogeneity, as a dramatically expanded population, which in cities meant overwhelmingly the children of immigrants, carried their different backgrounds, aptitudes, and behaviors into the heart of the schools and forced educators to seek pedagogical solutions to what were often social problems.

To the problem of heterogeneity, educators increasingly brought what is best described as the organizational solution—a remedy whose essential component was efficient instruction. While it was part of what Michael

Katz described as the bureaucratic tradition already implicit in nineteenth-century school development, the organizational solution drew heavily upon contemporary social developments, especially a faith in science and a new cultural orientation to hierarchical thinking that had a specially pungent influence on matters concerning race and immigration. The concept of IQ and testing for mental capacity which developed in this context became for the schools the most efficient organizational solution to the pedagogical problems posed by heterogeneity.

Throughout the nineteenth century, the schools had been moving toward more systematic forms of organization as they responded to the growing professional self-conciousness of school administrators, especially in urban settings. At the same time, public schools were from the mid- to the late nineteenth century witness to repeated and often explosive conflicts (wars, Diane Ravitch has called them) as different groups hoped to bend the schools to their needs in order to achieve or retain power over the schools' programs and directions. Politicians, businessmen, church leaders, labor unions, teachers, pedagogical reformers, parents, all projected their own perceptions and demands onto the schools. By the late nineteenth century, those demands often appeared under the progressive umbrella. In the end, the conflicts encouraged growth, not only in size, but in significance as the schools became the arena in which various aspirations, for power, for status, for training, for order, were necessarily to be realized.[5]

By the early twentieth century, however, as social reformers added their demands and visions to the schools' burdens, public schools had already taken on basic characteristics that would structure their subsequent development. Most significantly, as David Tyack has demonstrated, American public schools were by the early twentieth century already integrated institutions, administratively centralized, professionally self-conscious, and geared toward systematic expansion. Indeed, most of the school reforms of the period tended to accelerate this process. The reorganization and centralization of the massive New York City school system in 1896 illustrated the phenomenon, and this date may be taken as symbolic of the changes taking place nationwide, especially in the context of accelerating urbanization. By the second decade of the twentieth century, schools had developed sufficiently as complex institutions so that investigators were exploring issues of social efficiency and designing elaborate school surveys modeled on the better-known social surveys of the progressives.[6] The surveys assumed the desirability of professional control and administrative integration and held the central school officers responsible and accountable for the functioning of the schools in their districts. The school survey was meant to be "a study by an impartial outside expert thus freeing the schools from lay domination of professional matters," in order that "an educational pro-

gram might be designed to meet the future needs of the community. . . .
More than a method or a technique of inquiry, the survey idea is a part of
our education system which had developed in a fundamental way with our
whole educational organization."[7] The key words in this description, as
they were for managerial reformers generally, are "professional," "expert," "system," "organization." In the schools, this form of progressive
reform, in which experts and techniques freed the schools from public in-
terference, often overshadowed the more problematic and visionary demands of social reformers.

The systematic rationalization of the schools that began with administration in the nineteenth century turned toward issues of program development in the twentieth. By then the word "science" had been added to
"organization" as the guiding spirit of development. As one investigator
noted, "When the science of education shall have become fully formulated
we shall have ready at hand complete and verifiable conclusions relating
to three important aspects of the educational process: the child, the significant characteristics which mark stages in its growth; the demands of the
social group into which the child is born and in which he must live; and
the teaching method, whereby economy of time and of effort in teaching
and learning is secure." In short, a modern science of education needed to
coordinate the psychology of child development, the social context, and
the curriculum. The need to rationalize the learning process through mi-
nute attention to age, interest, ability, and socially useful learning underlay
most discussions of schooling at all levels throughout the early twentieth
century.[8] This rationalization was in part the result of the organizational
requirements of schools as they grew in size and complexity and a further
expression of the bureaucratic systematization which had begun in the
nineteenth century. But the attention to a science of education was not
simply the result of developments within educational thought or school
administration. Rather the schools' choices and concerns reflected their
intimate connection with the society which they hoped to serve. Two features of that society were especially crucial to the evolution of the schools,
and these developments had also significantly affected progressive social
reformers. The first was the reorganization of the work process; the second, the immigrant presence, especially in American cities.

Although the nineteenth century saw the beginnings of an American educational enterprise, both the scope and the significance of schooling were
fairly limited for the majority of the American population. Except for the
extremes of those who received no schooling and the very few for whom
schooling was a fundamental part of professional training or an expression
of elite status, the large majority of Americans received little more than
the fundamentals of literacy and the rudiments of what was believed nec-

essary to the exercise of responsible citizenship. Certainly, some students went to school longer than was required for the basics, and some did so because they expected to reap various social and economic rewards, but, in general, minimal attendance—from three to five years—was all that was either required or considered desirable throughout most of the nineteenth century. Indeed, before 1880 no state even had an effective compulsory education law,[9] and education was ambiguously and marginally related to an individual's future opportunites. Some Americans certainly did achieve a higher status and position than that occupied by their parents, but this was not normally the function of schooling. Instead, status and success were related to a host of factors that grew from the network of community, kin, politics, and church which dominated nineteenth-century social relationships during the early stages of industrialization. Mobility usually depended on connections, marriage, capital, skills acquired in a variety of ways, or the demonstration of work habits, industry, sobriety, and ingenuity. Similarly, inheritance, speculation, luck, and grit often defined the much more dominant agrarian economy of the time.[10]

In many ways, the late nineteenth century saw an erosion of this social world. And just as reformers responded to the loosening of the relationship between work and citizenship by placing their hopes in a more socially responsive education, educators, businessmen, and the public began to turn to the schools for their placement services and for the training in skills required by the increasingly impersonally organized modern world. The skills provided by the schools, then or now, ought not to be exaggerated. Nevertheless, in a society where business management, communications, and industrial integration on a large scale were beginning to subordinate entrepreneurship or mechanical know-how, a more refined literacy which the schools, and especially secondary education, could provide became increasingly desirable. Since the schools were also expected to transform unruly youngsters into citizens with regular and dependable habits, they became logical loci for concentrating the informal and promiscuous training which in the nineteenth century had been a function of general community participation.

To respond to these changes, the schools needed to refocus education by replacing the concentration on rudimentary literacy on the one hand and scholastic mastery on the other. They needed to divest themselves of older programs whose logic of development was internal to the disciplines and whose unstated but generally recognized progression was toward greater knowledge in those areas which eventuated in university admissions. The discussions centered on high-school education most acutely reflected these changes, but the rejection of the older perspectives affected all schooling and was fundamental to the reevaluation of educational endeavors at the turn of the century.[11] This reevaluation required that the schools, above all, shift their attention from the task to the child.

The child was, of course, the central concern of educational progressivism. The progressives expressed the new orientation in education most consistently and called upon a long pedagogical tradition to legitimate their views. Progressive educators hoped to reform education by first of all rejecting what they considered the dull traditional routine whose guiding spirit was subject matter. Instead, they proposed to substitute the needs of the child broadly defined, needs that included physical, mental, and emotional growth, interests and aptitudes, present and future relationships, and the "realities" of the world into which he or she would fit in later life. John Dewey had most carefully described this complex of social, political, work, and personal relationships toward which the schools had an obligation to educate their charges. But almost all school reformers, and indeed by the early twentieth century most educators, expressed their commitment to what they called schooling for "life." In many ways, educational progressivism provided articulate expression, often couched in highly serious pedagogical terms, to the new perspective that the schools had to adopt in order to respond to the realities of the world they hoped to serve. This is not to say that battles did not have to be fought against more conservative forces but only that progressivism tended to confirm and accelerate, not to challenge, the important new role the schools began to assume and the new concentration on child development as the core science.

In addition to the task of designing a modern curriculum to educate all children more attentively, the schools confronted a second and related problem that also emerged from the social changes of the period—the composition of the American population. Wherever schoolmen and women looked in the late nineteenth and early twentieth centuries, they saw the children of immigrants and the specific educational problems they posed. According to the 1911 Dillingham Commission investigation of selected American cities, 57.8 percent of all the pupils in thirty-seven cities investigated in 1908–9, were children of immigrants. In that year, 71.5 percent of New York's 500,000 public schoolchildren had foreign-born fathers. Of Duluth's 11,000 schoolchildren, 74.1 percent were of foreign parentage; and in Cleveland, Cincinnati, Detroit, Minneapolis, Buffalo, Boston, San Francisco, and most other large cities, two-fifths, and often much more, of all children in schools were of foreign parentage. The vast majority of these children were in the primary grades, and a very large proportion of them (40.4 percent) were what contemporaries called "retarded"—two or more years older than they should have been for the grade level they had attained in school.[12] The Dillingham Commission, in line with many investigations during the first two decades of the twentieth century (notably those of Leonard Ayres), showed that school children were overage, not performing, not progressing, not learning. While the problem was not peculiar to the children of immigrants, it was most prominent among these children and most worrisome.

The obvious solution was to keep children in school longer, and that solution was embodied in the two archetypical progressive drives—the campaign against child labor and the concurrent efforts for more effective and stringent school attendance laws. By 1920 the two related battles had resulted in marked success. By that date, only 8.5 percent of all children in the age group ten to fifteen were gainfully employed, while 90.6 percent of all children seven to thirteen and 79.0 percent of those fourteen to fifteen were in school. And the laws were most effective in increasing the attendance of the children of immigrants. By 1930, 90.0 percent of all fourteen- and fifteen-year-old children of native-white parents were in school, while 91.3 percent of the children of foreign or mixed parentage and 92.6 percent of all those who were themselves immigrants were at school.[13] But the child-labor laws brought into the schools precisely those children (of the laboring class and the immigrants) whose frequent academic failure, apparent lack of academic interest, and future economic status were most troublesome and with whom traditional school programs were least able to cope. The issue of retardation had emerged precisely from this context, and it posed a paradoxical problem. As the schools succeeded in incorporating more children, they seemed least successful in educating them.

Retardation underscored several early twentieth-century concerns and was fundamentally related to "scientific" curriculum development. Above all, retardation emphasized age-grade standards, the carefully calibrated grade system, and the accurate coordination of grades into a hierarchy of schools—elementary, junior high, and high school. And it reflected the infatuation with a kind of scientism of numbers which was also part of the school survey movement and the basis of mental measurement. This led to an obsession, not only with efficiency, but with the specific efficiency that seemed to inhere in age-appropriate education. Retardation also led to two very significant conclusions: something was wrong with the schools, and something was also wrong with some of the children in the schools.

While these were school issues and educators addressed them through traditional and revised pedagogical means, the most significant solution came from outside the world of education. The solution, like the problems, came from the peculiar confluence of immigration, science, and reform that affected most aspects of society in early twentieth-century America.

II

Progressive social reformers hoped to use education to revitalize democracy through the reconstruction of the elements of individual political responsibility. To this faith in democratic renewal through education, progressives added a new faith—in science, technical expertise, and the symbolic power of numbers. Amidst the disorders of the late nineteenth-century and early twentieth-century world, science and numbers seemed to promise just that fundamental knowledge and access to order that would result in control. Science was more than a form of data-collecting; it would also provide a method for solving social problems and making social policy.[14] In the early twentieth century, education and science went hand-in-hand with the euphoria of reform.

The science that had the most profound effect on educational practice was psychology, a hybrid calling which was part biology, part philosophy, and in good part linked with the evolving profession of education. Most significant for education was the fact that American psychology had by the first two decades of the twentieth century become deeply involved in mental measurement, especially in devising tests to measure mental capacity. "Mental testing," as the movement was universally called throughout its early years, was rooted in various currents of nineteenth-century European science. Its foundation lay in the laboratory techniques of German Wundtian psychology, powerfully augmented by the development of statistical methods in England by Karl Pearson and Charles Spearman. The specific form and content of mental testing finally bore the indelible signature of the French psychologist and educator Alfred Binet who, in attempting to solve some of France's own educational problems, had designed a technique for identifying feebleminded children in the growing French school system. Binet transferred the center of gravity of the science from the laboratory and statistical graphs to the school and the child.[15]

In the United States, mental testing first surfaced in a relatively minor way in the 1890s in a report by the pioneer psychologist James Cattell. But the early interest among psychologists and a few educators was nothing compared with what was soon to follow. For it was only as the French concern with personality and abnormality was joined to the English preoccupation with the measurement of individual and group differences in the form of aggregates and norms that mental testing as an American science was born. Despite its European ancestry, testing became a major preoccupation only in American psychology, where, in the words of an early historian of testing, the "movement . . . swept everything else before it" during the second and third decades of the twentieth century. As Harlan Hines observed in his book popularizing intelligence testing, "It is doubtful whether so much public interest has been aroused since Darwin propounded his theories about the descent of man."[16] Mental testing is important in the United States not because it dominated American psychology for a generation, although this is interesting, but because it crystallized the preoccupations of a social generation. It provided Americans with a powerful way of organizing perceptions and a means for solving pressing institutional problems.

The Americanization of mental testing began in 1911 when Lewis Terman, the Stanford psychologist, adopted Alfred Binet's age-mentality scale to test, sort, and classify American schoolchildren. Terman was not the first American to use Binet's tests, but he was the first to use them with normal children.[17] With Terman's adoption of the Binet scale and his subsequent modification and introduction of the concept of intelligence quotient (IQ) in 1916, the history of IQ testing in America properly began. Terman's work with normal children clearly anticipated the use of IQ tests in the schools, and he was a vigorous proponent of their large-scale adoption. His enthusiasm for the method was virtually unbounded, since it was, in his words, "from the practical point of view . . . the most important in all the history of psychology." By 1911, Terman had already claimed that "by its use it is possible for the psychologist to submit, after a forty-minute diagnostication, a more reliable and more enlightened estimate of the child's intelligence than most teachers can after a year of daily contact in the schoolroom." Educators like Elwood Cubberly, probably the most influential educational spokesman of the time, also foresaw the educational implications. In introducing Terman's book in 1916, he observed that "The present volume appeals to the editor [Cubberly] of this series as one of the most significant books, viewed from the standpoint of the future of our educational theory and practice, that has been issued in years."[18]

At the same time, Terman's tests and the IQ scale were not quite fitted to the needs of the American school system. First, they required costly individual administration on a one-to-one basis. Second, Terman and other early testers jealously guarded their own expertise in this scientific procedure and insisted on expert administration.[19] American entrance into the First World War changed that. As Americans prepared to fight a war to save democracy, they also found themselves with the kind of heterogeneous fighting force that resulted from democratic immigration policies. That army needed to be organized for maximum efficiency and effectiveness. For this task, a group of psychologists volunteered their services. This committee of the American Psychological Association at once hoped to assist the war effort and to raise psychology's still shaky status as a scientific profession in the eyes of the public.[20] Mental testing seemed to provide a way to fulfill both the needs of the army and of the profession. As

a science, testing could call on the enthusiasm for the exactitude of numbers for practical results. Mental testing seemed like the perfect instrument
in a progressive crisis: it was socially useful, numerical, and had reform
implications.[21]

Although the army tests were based on and still linked to the individual
Binet scale, the army test was a "group test," organized and routinized
around pen and paper exercises which required only one administrator to
provide general instructions to a large group. Over 1,700,000 men were
eventually tested in this way. The army tests demonstrated the feasibility
of mass testing, and as one textbook on testing noted, "The possibility of
measuring an individual's intelligence by a short and simple test has captured the imagination of school people and of the general public." In the
long run, the army tests were a significant administrative breakthrough,
but the headline-grabber and contemporary interest involved the results.
The early discussions of the army tests revolved around a pseudo-issue—
the apparent fact that almost half the American draftees had scarcely the
mentality of thirteen-year-olds. Doomsayers and gloomsayers quickly latched
on to this fact to add fire to the many forecasts about the decline of American civilization to which the period was prone. Walter Lippmann, in a
now famous essay, handily disposed of this issue by noting that in their
haste the testers had failed to translate into adult terms an age-grade scale
that had originated with the testing of schoolchildren.[22]

Neither Lippmann nor other critics were able so easily to dispose of
another feature of the results which would have long-lasting social consequences. For, as they proceeded to refine the test results, psychologists
went beyond comparing and ranking individuals within a large population. They also correlated differences in scores with social categories like
region, education, race, and country of origin of the draftees.[23] While there
were marked differences within all categories, the attention of testers and
the public was riveted by the markedly lower scores registered by blacks
and the recent immigrants, like Italians and Slavs, who did worse than
native-white Americans on both tests administered to those literate in English (Alpha) and those not literate or non-English speaking (Beta). The
army tests made the crucial link between mental ability and race, not for
the first time, but on a mass scale and in the full glare of public attention.

What had begun as a way of eliminating the feebleminded, proceeded
to a ranking of individuals according to talent, and finally became a means
for ordering a hierarchy of groups. Mental testing as a measuring device
was a defining and sorting instrument, a way of distinguishing and differentiating which, given the cultural concerns of the time and in the context
of statistical techniques like normal distributions, correlations, and factor
analysis, led predictably to racial comparisons. In so doing, of course, in-

telligence testing sharpened and confirmed the cultural awareness of individual and group differences.[24] Since Americans were, in the first thirty years of the century, inundated by dozens of immigrant groups who were so radically different first from "Americans" and then from each other, the emphasis on differences, now measurable by scientifically validated tests, was translated from the realm of the senses to that of statistics. At a time when democracy seemed threatened by heterogeneity, counting, sifting, and ranking provided a form of order and containment.

All of this was true regardless of whether the mentality measured by the tests was innate or learned. In fact, as far as the army was concerned, it mattered not at all whether the tests measured ability or achievement, whether they reflected what an individual could learn or had learned. What the army needed was an objective means to locate talent and to make assignments. Its concern was suitable to an organization drawn from a democratic mass which was to be organized for maximum efficiency.

It did, of course, matter very much that mental tests purported to measure innate ability rather than learning, but it mattered less for what the army functionally needed than for what it reveals about American preoccupations. For what the designers and promulgators of intelligence tests proposed was that they could arrive at an absolute measure of intelligence and not merely a relative way of discriminating among individual abilities. By the time of the army tests, American testers, especially Lewis Terman, were eager to demonstrate the infallibility of the tests as absolute yardsticks. In so doing, they emphasized the constancy of intelligence, the fact that once measured, an individual's intelligence would thereafter remain measurably the same. In part, this reflected the specific needs of education which, once the tests were used with sample "normal" children, obviously became the primary field for the future application of tests and was so recognized even before 1917.

But the stress on the IQ as an absolute measure of an unchanging quality was more than a response to the needs of American schools. It was especially attractive to Americans because it necessarily brought the whole issue of inherited endowment into sharp relief. It provided numerical confirmation to ideas that Thomas had hoped to dispel but with which fluid concepts of culture could hardly compete. Alfred Binet had never been overly concerned with defining tests as a measure of native capacity or of distinguishing between test performance which resulted from innate as opposed to environmental factors.[25] But in the United States, Binet's tests were greeted with an enthusiasm that they failed to arouse anywhere else, including his native France, precisely because Binet's tests, as distinct from Binet's views, offered the possibility of measuring innate mentality and unchanging potential. Most testers were careful to describe their results in such a way that they could always claim that they had not excluded environment as part of actual performance on the tests. But the clear direction of American interpretation and the construction of experiments with tests were toward the conclusion that intelligence tests measured something that was pure and inborn. If still imperfect, the hope was that the tests could eventually measure the innate, untutored capacity of the individual.[26]

The IQ, as an expression of what intelligence was and as its measure, epitomized this predilection. As developed by Terman, the intelligence quotient was more than a measure of performance on tests. It was a personal ratio (mental age/physical age) that exaggerated the static, individual, and physical quality of what was being measured. This distinguished it from more traditional evaluations of performance which were always implicitly comparative and time-linked.[27] It also emphasized age-grade standards in a manner increasingly important to the schools. In binding qualitative and comparative evaluations of mentality to the more concrete concept of calendar age, IQ seemed to stand apart from culture and to measure something that existed prior to learning. As a result the IQ appeared to capture some essential dimension of individual development as environmentally impermeable as eye color and as easily described as age. Terman's results were related to school developments, not only because they answered the schools' needs for evaluative procedures, but because both IQ and the concepts underlying school development at the turn of the century sought to make age an objective correlative of development.

IQ captured with great precision two dominant contemporary concerns, race and education, which were from the beginning most crucially intertwined in its evolution. It could effectively be argued that, had Americans not been racially oriented and educationally obsessed, neither the enthusiasm for mental testing nor the specific form of its evaluation would have arisen. I have already noted how the social concern with differences made testing to measure differences attractive. To this it need only be added that an evolving racial consciousness (in the early twentieth century attached to white immigrants as well as black natives) tended to look for and emphasize inborn qualities to explain those differences. The search for a biological means to explain differences in character and habits was not new in the twentieth century—Americans had measured skulls in the nineteenth century. But the presumption that what distinguished groups as well as individuals was not only some innate potential but also an unchanging quality invulnerable to environmental modification was not entirely developed in the United States by the time American psychologists began to work on mental tests in the early twentieth century.[28] Progressives were often confused about the relationship between race and culture, and many social scientists, like Thomas, were concerned to discuss that relationship.

The incipient search for a racialist explanation for differences informed American perceptions of the usefulness of testing, and perhaps more significantly helped to determine what the tests would be interpreted to mean. For if Alfred Binet could not have cared less whether intelligence tests measured inheritance or environment, Americans contemplating closing the door to open immigration could not have cared more.[29]

Once these needs and perceptions framed the form in which test results were expressed and explained, the two gave a powerful boost to a racism hitherto lacking scientific precision. Where nineteenth-century evolutionary science had provided Americans with the presumption that some races were superior to others, the IQ provided an absolute calibration and a ranked hierarchy as demonstrated by performance differences on tests. Unlike Binet who had generally been unconcerned about attaching some invariant number to general intelligence, Americans eagerly sought a number as precise and abstract as possible. As Kimball Young appropriately explained, "The grip which apparent exactitude in numbers has upon us ought not to need mention to psychologists. One of the strong appeals to educational and psychological workers using tests . . . is undoubtedly the fact that an array of averages and correlations gives them a sense of definiteness and finality."[30] In the context of the American search for a basis for evaluating racial differences, the newly developed science of statistics provided psychology with the gift of numerology.

This had very specific social consequences. In the early twenties, nearly every study of group differences in IQs was introduced or concluded by a scholarly *obiter dictum* on immigration policy.[31] Americans, and psychologists among them, were profoundly concerned with the effect that uncontrolled immigration was having on democratic institutions, values, and the moral fiber of the nation. With the introduction of intelligence testing, Americans also became obsessed with mentality. This had been foreshadowed by the panicked headlines surrounding the results of the army tests. While the racial conclusions that accompanied intelligence testing were not the cause of immigration restriction, testing provided measurable evidence that greatly enriched the alarmed context in which the immigration exclusion legislation of the 1920s took place. It would be a mistake to conclude as a result of this instrumental function of IQs that they were merely a way to support a racially informed social policy. Race influenced the development of mental testing in far more complex ways. Racial exclusion was a by-product of a whole manner of thinking, perceiving, and ordering that the IQ organized into a science.

One of the truly remarkable aspects of the early history of IQ testing is the rapidity of its adoption in American schools nationwide. In the twenties, the tests were taken from the laboratory, where they were still being tested, retested, and modified, and were fast on the way to becoming an entrenched part of educational administration. As early as 1919–20, New York City's *Bulletin of High Points* (a teacher's journal) contained numerous articles in IQ testing, including a symposium on the subject and an account of the successful application of testing at Evander Childs High School. In Mt. Vernon, New York, all students from elementary through the first year of high school had already been tested by 1920. One Detroit educator boldly announced in the *Yearbook of the National Society for the Study of Education* in 1922 that "the adoption of the group method [of testing] by hundreds of school systems is now an old story."[32]

Immediately after the war, the same group of psychologists who had constructed the army Alphas developed the National Intelligence Test. Over 575,000 copies were sold within a year of its issuance—800,000 copies the following year, 1922–23. By 1922, it was competing with numerous other tests of a similar kind. In 1922–23, over 2,500,000 intelligence tests were sold by just one firm which specialized in their development and distribution. Forty different group intelligence tests were by then available on the national market. Many more were rapidly being produced as psychologists suddenly found themselves in a lucrative business, possessing skills very much in demand. Indeed, tests were rushed into press before they were adequately evaluated. Some were a hodgepodge of different forms, completely unintegrated and uncoordinated. Edward Thorndike, éminence grise of psychological measurement and objective testing, was moved to declamation: "In the elementary schools we now have many inadequate tests and even fantastic procedures parading behind the banner of educational science. Alleged measurements are reported and used which measure the fact in question about as well as the noise of the thunder measures the voltage of lightning. To nobody are such more detestable than to the scientific worker with educational measurements."[33]

While the rapidity with which IQ tests were adopted was remarkable, it was not really surprising. From the beginning, Terman and others had targeted the schools as the principle beneficiaries of the tests. And school administrators, eager to adopt scientific tools to cope with problems of curriculum development, a heterogeneous population, and the progressive challenge to tailor school programs to individual needs, responded enthusiastically to an instrument that seemed to slice through their problems efficiently and democratically. Retardation had rung an alarm in schools committed to expansion, and the IQ allowed educators to shift the blame for inadequate schooling to the inadequacy of the pupils. More significantly, IQ could allow the schools to expand and incorporate ever larger and more diverse groups for longer and longer periods, while adjusting to progressive principles.

Education as it was being redefined and reformed in the early twentieth century was turning away from defining learning as the inculcation of information toward issues of socialization that emphasized the acquisition of knowledge in the total process of individual development and growth. John Dewey had most idealistically represented this new direction by passionately rejecting the view that education was the acquisition of the accumulated wisdom of the past and by urging instead that the unfolding of the child's (and society's) potential should be the mission of the schools. By shifting the charge to the schools from the traditional one of instilling an agreed upon body of knowledge to an active development of understanding pegged to individual talent and instrumental in a changing society, Dewey and progressive pedagogy sharpened the challenges facing the schools. Given the schools' design, their dependence on structured grades, central administration, their emphasis on order, and cost efficiency, progressive educational theories propelled the schools to seek ways to define children, not individually, but according to the range of their educability. As Josephine Chase observed in her study of New York schools, "It is to the task of individualizing education, of making the school program elastic enough to fit the needs of each child that the progressive school leaders of New York are bending every effort."[34] To this effort, the IQ was immediately recognized as a powerful ally.

For the schools, the IQ was a concept that seemed ideally suited to their new goals and problems. It seemed to establish a stable educational center by assuming an unvarying constant within each child—his inborn capacity—and was based on a simple testing method designed to discover that potential. On this basis, educators could design programs and curricula that would make education more individually usable and more socially relevant. The IQ thus provided multiple blessings as it was brought to bear on a host of institutional matters. As Elwood Cubberly correctly predicted in his introduction to Terman's *The Measurement of Intelligence,* "The educational significance of the results to be obtained from careful measurements of the intelligence of children can hardly be overestimated. Questions relating to the choice of studies, vocational guidance, schoolroom procedure, the grading of pupils, promotional schemes, the study of retardation of children in schools, juvenile delinquency, and the proper handling of subnormals on the one hand and gifted children on the other—all alike acquire new meaning and significance when viewed in the light of the measurement of intelligence as outlined in this volume."[35]

The IQ categorized students and made it possible for the schools to deal with them in group, class, and hierarchical terms. That the IQ ultimately also predefined children so that students would thenceforth learn only so much as they were at the outset judged able to learn and ironically limited the function of the school by establishing the primacy of innate ability over environmental stimulation was also a blessing. As one judicious authority on tests, Frank Freeman, observed, "The usefulness of measures of intellectual ability depends in part upon their stability and the possibility of predicting the individual's future intelligence from the measure of intelligence made at a given time. If the purpose of the test is, for example, to classify pupils so that the demands made upon them may be adjusted to their abilities, it is necessary that their abilities shall remain more or less constant."[36] In a final irony, the very needs of the schools thus helped to confirm the meaning of IQ.

The growth of American education and specifically the requirements of educational administration, like the American sensitivity to racial differences, meant that whatever instrument was used to place and locate the individual in the increasingly complex school structure had to provide a constant measure of some inherent and unvarying potential. And just as the immigrant presence was crucial to the cultural network that created IQ as a form of organizing perceptions, so the presence of vast numbers of immigrant children in the schools was basic to the educational situation which seemed to make IQ testing an instrumental necessity. The problems facing the schools—the need for better organization, selection, and a curriculum at once more fully tailored to the individual and more socially alert—were not entirely the result of the presence of new immigrants. But the pressures on education to expand beyond the three Rs, the stricter school attendance laws, and the concurrent child-labor legislation which vastly expanded school populations cannot be separated from problems associated with immigration during this period. Immigration made the schools grow exponentially in scope, size, and complexity. At the same time, it made the schools' problems vastly more complicated.

Immigration helped to frame the context for the schools' search for procedures to define children and to order the curriculum, but the tests were not aimed at immigrants alone. They were a convenient necessity to the schools because they made possible a learning process tailored to individual potential, organizable by classes, while still allowing for progression of instruction by age. The best way to describe this network is a "track." Just as age subdivided schools horizontally and permitted instruction to be developed according to progressive levels of advancement, the track subdivided the system vertically and channeled instruction more precisely as determined by individual abilities. The organizational grid that resulted from the intersection of age and ability permitted instruction to be technically more individualized. If IQ is a ratio that compares age and ability, the tracked class system was the map projected from that concept.

One of the expressed concerns of educators as they adopted progressive concepts was how the schools could best serve each child's needs and talents, and the early proponents of the IQ used this argument constantly. If the school system was not to swamp the individual in a monolithic curriculum which intimidated slow learners, held back fast learners, and failed to take into account future jobs and needed skills, factors like personal potential, ability, and interest should be taken into account. The vertical track could do just that. It was the peculiar virtue of the IQ that it could at once provide the organizing mechanism around which a complex school system could be built, while purportedly protecting the individual child against the impersonality of that system. As the progressive educator, Carleton W. Washburne, explained in 1925, tracking or ability grouping "is but a step toward individualization—a step which makes individual instruction easier both to initiate and to incorporate . . . ability grouping is one of the best first steps toward individual instruction."[37]

In many ways, IQ tracking was an administratively workable but also apparently enlightened and progressively informed way of answering Dewey's challenge to education. The track was a much more efficient way to personalize instruction than the individual-specific education that Dewey adopted in his own model school. The track brought together a group of individuals of similar IQ and tailored instruction to the group. But the IQ-determined ability group limited the definition of what was significant in that instruction. Whereas Dewey was concerned with the education of the whole child, the track brought together children within a similar IQ range. Since the IQ tested only selected forms of ability or accomplishment as expressions of general intelligence (and this was admitted by even the most vociferous proponents of testing) the track not only switched the orientation from Dewey's individual to a group but also defined that group by the specific qualities tested in the IQ.[38] In the process, it transformed Dewey's ever malleable child into one with a constant and determinate IQ capacity and translated the open-ended and constantly changing environment into a narrow-channeled curriculum. The different curricula had different goals and could either propel students toward richer academic prospects or severely limit their horizons.

It would be unfair and mean spirited to argue that educators used IQs simply to exclude the newer immigrants from the lines of advancement that the schools now promised. It was never so simple. The schools continued to be theoretically and actually a force that facilitated access to society's rewards both as an agency incorporating immigrant children into the mainlines of the culture and as a lever for social and occupational advancement for individuals. At the same time, educators were certainly not surprised when immigrant children scored significantly and consistently lower on the examinations. They had, after all, been well prepared for these results by the army tests and by a whole battery of tests often administered at the very wharfs of Ellis Island. They certainly did little or nothing to guard against the conclusions the tests offered. "If the stream of immigration from Southern and Eastern Europe continues to inundate us," Kimball Young declared, "the schools must take into account the mental abilities of the children who come from these racial groups. . . . The present situation is already causing a revamping of the curriculum and the general educational policy in many school systems." Initially, at least, immigrant children were often channeled into programs that either stopped before the high-school door or, if they entered, ended in technical and vocational programs which provided little basis for effective social mobility through education. From their earliest application, IQs were used to guide students to programs suitable to their "needs" and "talents," and as early as 1923, one popularizer of tests warned that "Many administrators . . . have expanded this idea of vocational guidance through the use of the test results until, if it were generally accepted, we would have nothing short of a type of Prussian control." Immigrants were especially vulnerable to this kind of channeling.[39]

The IQ permitted the systematic rationalization of the modern urban school, and its tenacious hold on the schools resulted from the critical function it served. At the same time, its introduction into the schools was the result of specific historical circumstances—the fervent belief in progress through exact science; the increasing responsibility of schools for training special skills and aptitudes in the twentieth century; the challenge posed to traditional education by progressive educational theories; and the critical role of immigration in compounding the problems of a democratic schooling. IQ was not only a measure which seemed to solve problems that immigration introduced into school systems, it was a concept in whose terms educators could comfortably think and behave. This was because it spoke in terms—"mentality," "development," "age," "grade," "educability"—which defined the educational agenda of the 1920s. It was also because its vision was colored by racial filters. As school administrators turned to the classroom and problems of learning, they saw before them children of many complexions and tongues who needed to be understood in universal, "scientific" terms. They had to be fitted into classrooms so that they could learn, and they needed to be classified so that they could be understood and controlled. The IQ was a form of understanding which made school efficiency thinkable and possible.

III

Why this insistence in educational discussions of the 1920s on adapting the curriculum and school programs to "individual differences" which IQ was to measure? In the context of the sheer growth in school population, overcrowded classrooms, and inadequate facilities, the striving to redefine education according to the needs of students seems more than a little strange.[40]

Certainly educators of the early twentieth century were not the first to discover differences among students. School teachers in any rural or urban setting during the nineteenth century and well into the twentieth must have been aware of and responsive to the extraordinary differences in talents, interests, and goals of their students. And yet the educators of the early twentieth century spoke as if they had discovered the idea, and they elaborated it into an ideology. Part of the concern with differences was stimulated by Dewey and the progressive education movement and was based on pedagogical principles which redirected attention from the acquisition of knowledge to learning as a process of individual growth. In part, educators alarmed by the maladjustment implied in retardation statistics turned their attention to individual differences as an explanation of learning disabilities. So too, the attention given to measurement itself accentuated the perception of and meaning given to differences thus recorded. As the editor of the National Society for the Study of Education yearbook devoted to individual differences put it: "The individual differences existing among children and the failure of the traditional methods of instruction to make adequate provision for them have been recognized to a greater or less degree for many years. With the spread of the measuring movement, however, the differences have become more obvious and the consequent attempts to adjust the schools more numerous." In large part too, the reason for the attention to what were called individual differences resulted from perceptions about group differences and the whole network of social perceptions defined by massive immigration in the context of an increasingly rationalized school regime. The ideology of individual differences was an administrative concept fashioned as an institutional substitute for teacher sensitivity. In 1931, a writer for the *New York Tribune* described the situation that made educational differentiation necessary: "The children of diverse races cannot be treated alike in the schools and are not in New York. Courses of study are shaded and varied to the peculiar needs and limitations of white people, yellow, black, brown and mixed within the walls of New York's schools, as those needs and limitations are seen by the individual teachers and principals dealing with swollen classes wherein the vari-colored and vari-national pupils are indiscriminately channeled." [41] The very problem to which the expansion of schooling had seemed so clearly the answer to progressive social reformers—the need for schools to become the loci of common socialization in an urban, industrial, immigrant culture—thus defined the search for measures of differentiation within the schools.

In line with their vision of the role of the school in industrial society, the progressives had urged the schools to become social centers, to attach a variety of social services, and to expand their function from the traditional realm of instruction to remedial socialization. Nathan Peiser of the New York Educational Alliance saw it in the following way: "The school building will become the meeting place, the public forum, the recreation house, the civic center, where contacts are made, where newcomers are welcomed, where troubles are told, and where organized action is taken for neighborhood improvement." In short, the school was to become the heart of the community. As William Reese has argued, this vision of the school as social center captured the widest range of progressive sentiment and was perhaps its noblest expression. But, like other features of the progressive vision, its operation in the context of the realities of school administration exposed how even humane policies were re-created when applied to complex situations. [42]

The schools never became the vital centers of community life envisaged by Peiser and others, but under the stimulus of a variety of reform groups and the tutelage of social settlements, they did expand to become something larger and more ambitious than the simple classroom-centered schools of the nineteenth century. After-school centers, visiting teachers, school nurses, free lunches, child guidance clinics, vocational counseling, speech therapy, medical inspection, summer recreation programs, and numerous other small and large services were incorporated into the school plant to become part of educational administration. As in so many phases of early twentieth-century development, the high school represented these developments in their most completed form, but the incorporation of a variety of services to meet the needs of a complex population took place on all levels of public education. Schools and school districts throughout the teens and twenties did this largely on an ad hoc basis. [43] But by the late thirties, the notion that expanded school services were essential to the fulfillment of an ideal of equal educational opportunity was promulgated in the highest national councils and defined as a national obligation (see Chapter 4).

The belief that individual differences required a differentiated curriculum and an expanded educational regime was rooted in perceptions about the nature of the population with which the schools had now to deal. Thus, for example, Jordan High School in Los Angeles, situated in a working-class suburb with a population composed of "Negroes, Orientals,

Mexicans, and members of various European races, as well as native-born whites," was a "striking example of the application of a consistent policy of offering education in relation to local need." According to Francis T. Spaulding, O. I. Frederick, and Leonard V. Koos, the school offered a "chef's course and a girls' course in beauty culture . . . provided to meet the special needs of pupils in this particular school." In addition, "Jordan has had to meet special needs in the community outside the school." One of the features of the school was that "it has sought to become a center for various phases of adult life in the community. By providing meetings of interest to adults, school exhibitions, athletic contests, 'shows' of various types, it has done much to make itself a civic and recreational center." The school offered as well "special evening courses for adults—particularly 'upgrading' courses in various skilled and unskilled occupations for both men and women."[44]

In some ways, including the respect and close contact between Jordan's principal and the surrounding community, this high school in Los Angeles was effecting what Leonard Covello was fighting for at Benjamin Franklin High School in Manhattan—closer cooperation between school and community. Covello, an Italian immigrant best known for his classic study of the cultural maladjustment experienced by Italian children in American schools, sought to enrich the high school and the community through closer contact between the institution and its host community. Covello used his own experiences at Benjamin Franklin as an example of how the problems of immigrant youths, including delinquency, learning disabilities, and generational tension could be relieved through a more humane and empathetic school regime. His efforts were praised by contemporaries and have remained a model of cultural contact aimed at the alleviation of the conflicts between the school and the local community within which it was often an alien presence.[45]

Covello had translated what he had learned as a youth at social settlements, his knowledge of Italian culture, and his long experience as a teacher into his plans for a community school to produce impressive results on behalf of lower-class boys in Harlem. Among his successes was the introduction of Italian as a language option in high schools. Another Italian schoolmaster, Angelo Patri, also welcomed an activist school: "The school must open its doors. It must reach out to spread itself, and come into direct contact with all its people. Each day the power of the school must be felt in some corner of the school district. It must work so that everybody sees its work and daily appraises that work."[46] This idea of a community school required more than an adaptation to the perceived needs of its immigrant constituents. It required that the school respect the culture of the students and act in such a way as to produce pride in self, family, and past. Nevertheless, this open-door school policy was ambiguous. Like the social settlements upon which it was modeled, it was at once an expression of democratic aspiration and an intrusion upon the community it was meant to serve.

The visiting teacher movement is a good example of how the expansion of the school into the community produced an extension of school controls. The visiting teacher, "who has for her chief function the removal and prevention, as far as possible, of those handicaps of school children which are the result of their social environment," gained a certain popularity in the late teens and twenties in school systems throughout the country. Often stimulated by the efforts of women's reform groups and especially by the progressive New York Public Education Association, visiting teachers were employed in Boston, Philadelphia, New York, Chicago, Los Angeles, Minneapolis, Kansas City, and a host of medium-sized cities. Their aim was to fuse social work with pedagogy by remedying those home conditions that obstructed learning. According to one of its early historians, the visiting teacher movement was committed to "helping the school make such adjustments as will meet the needs of individual children and groups of children. . . . Since compulsory education laws are getting children into the schools, it becomes the business of the schools to provide for them according to their varying needs. In order to provide greater equality of opportunity, various means have been adopted. Classification according to abilities has made a beginning. . . . The visiting teacher extends the service of the schools by making adjustments for children who are handicapped in their social environment." Social environment, he continued, "includes more than unfavorable homes, school or neighborhood influence . . . it includes the field of unfavorable social attitudes, ideals, and habits."[47]

This description richly illustrates both the logical and problematic implications of extending school services to meet the needs of the community as well as the ideology of individual differences within which these services functioned. On the one hand, visiting teachers were an extension of the schools' commitment to teaching and to devising means for educating all children. On the other, they were used by the schools to overcome individual learning disabilities whose roots seemed to lie in the home, in the community, and in those foreign attitudes that conflicted with the schools' influence. Appropriately, the description links the idea of individual differences in talents and aptitudes which resulted in classroom "classification" to differences in "attitudes, ideals, and habits" which lay outside the classroom and required remedial social programs. In each case, the school began to deal with differences in its constituency and in each the school was adapting to those differences with the new scientific apparatus provided "through the advances made in psychology . . . sociology, and social case work." In thus moving beyond the strictly academic sphere, the schools became first judges and then regulators. Of course, schools have always done this latently, but the visiting teacher program and many other social service functions did this with intent. Indeed, part of the visiting teacher's job was "to adjust home conditions whereby more favorable conditions will be attained in regard to schoolwork, conduct, attendance and interest." If necessary, "in extreme cases," visiting teachers could also "refer misconduct to juvenile court or society for the prevention of cruelty to children." There was always the fist behind the smiling face. Even when visiting teachers were themselves members of the community they served, as they frequently were, and spoke the language of the homes into which they came, they could at best humanize the intrusion that defined the very nature of their activity.[48]

Employing visiting teachers, like the addition of most other services, was an expansion of the ideal of schooling in the context of the need to educate an alien and often suspect population. Thus, New York's progressive school superintendent, William H. Maxwell, noted in his 1907 report that the most pressing educational need was for a Department of School Hygiene with doctors and nurses to administer physical examinations and make home visits to sick children. His reasons were explicit and to the point: "modern city life" and the need to "neutralize the evil effects of urban life upon children"; the need for "skilled physicians' advice in the treatment and training of children"; and finally because "The influence of such a department is needed, in addition to the influence of the teachers, to give to our enormous alien population new ideals and new habits in the rearing of children and to establish among them American standards of living."[49] The school as a conduit of influences for modifying the behavior and attitudes of the alien population was to move beyond instruction and become explicitly part of the process by which the foreign population would be modified, elevated, and reformed.

Individual differences, whether inborn as mental capacity was assumed to be, or learned as attitudes or bad health habits, profoundly altered the meaning of democratic education. No longer did it simply require that education be made available to everyone. It meant making education usable to individuals according to their needs and capacities. As a result, the schools assumed the role urged on them by progressive reformers and became intermediaries between the individual and society. At the same time, however, this required that the schools make judgments that were as often social as they were pedagogical. In this sense, the paradox of democratic education was that it often became far less democratic in fact than a less exacting educational ideology might have allowed. It would be a mistake to describe these developments as the consequence of fundamentally malicious motives—the contrary was frequently the case. Certainly, foreign children could use both the medical assistance of trained physicians and the added pedagogy of visiting teachers. Often indeed, the individuals most sympathetic to the needs and integrity of the immigrant community and the resources of immigrant culture, like Covello, were advocates of a closer coordination between the school and its surrounding community. The paradox of distinction in the service of democracy was expressed well by Frank Thompson, an unusually sympathetic observer of the immigrant: "It is to our credit that in our schools we have never made invidious comparisons with respect to the children of the immigrant; we have received them on the basis of equality. . . . Still, we cannot ignore the arguments for some sort of special educational provision for immigrant children. The motive is similar to that which has prompted us to make special provision for various kinds of atypical children. We wish in the schools to furnish an equality of educational opportunity; but we can no longer deny the fact of individual variation of powers and abilities, and the schools cannot bestow an equality of benefit through the same ministrations to all children."[50] As it came to be understood in the teens and twenties, equal educational opportunities required educational differentiation, remediation, and intrusion.

Although some of the programs incorporated in school expansion, like visiting teachers and medical inspection, had progressive roots and were specifically aimed at the reform of family environments and immigrant communities, others were directly a by-product of educational experience. In New York City, for example, the Board of Education had established 150 special classes for overage children by 1903. These classes continued beyond the normal summer recess. Even earlier, New York had established special intensive language classes for foreign children. In both cases, the classes succeeded in bringing students up to grade level, and thus permitted them to re-enter their regular classes as defined by their age.[51] The need for special instruction resulted in a variety of school adjustments and additions. A New York City circular of 1917 issued by the superintendent of schools and addressed to all high schools noted that "Each term the principals of our [teacher] training schools receive from the high schools a limited number of pupils who have some kind of speech defect, including that of foreign accent. These students are very unhappy when they are excluded from the training schools after a trial period in which they show no improvement. . . . After their rejection, they find fault with our system, which permits them to go through high school . . . with such defect. And they are right in their criticism." This kind of experience, serious enough to warrant a circular announcement, underscores several facts about the evolving school situation. First, a wide variety of students were progressing up the educational rungs. Secondly, the schools by not paying attention to the special needs of these students ultimately did them a disservice. Finally, the facts here called for an obvious remedy, namely auxiliary oral English classes.[52] While such English programs were clearly pedagogical, they also implied criticism of foreign accents and foreignisms of all kinds. Success in school, as the children of immigrants would find again and again, required a serious modification, if not an outright rejection, of the cultural bases of their identity, whether the schools specifically or directly challenged the culture of their homes or not.

IV

The results of the schools' efforts to meet the needs of their democratic constituency were nowhere more clearly demonstrated than in the attention given to education at the secondary level. As schooling grew in significance with the new century, the topic of public education in the high school moved to the center of educational discussion. Much of that discussion focused on curriculum differentiation and especially on the high schools' capacity to provide students with a variety of specialized skills. Educators explicitly connected this development to the prevailing ideology of "individualized instruction" and frequently to the specific requirements of the immigrant community. At Jordan High School in Los Angeles, for example, "the composition of the pupil population . . . affects notably the choices of the various curriculums. In the spring of 1931 almost exactly half the pupils were enrolled in the homemaking and industrial arts curriculum. . . . The large proportion of pupils enrolled in [a] non-academic curriculum is indicative . . . of the school's effort to direct pupils into types of work in which chances of success will be greatest for the individuals concerned"—hence the chef's course and the course in beauty culture. Blacks at Jordan were directed to courses designed to their futures as "cooks, waiters, restaurant helpers, or Pullman porters."[53]

The neighborhood serviced by Jordan High School may have been unusual in its specific mix of blacks, Asians, Mexicans, and European immigrants, but Jordan's educational efforts on their behalf were not. In discussing the complex ethnic composition of the Minneapolis schools, Riverda Jordan concluded that "the superintendent is failing in his responsibilities to the foreign constituents of his community if he does not seek to discover and apply such means as will most effectively serve their peculiar needs and requirements." This statement immediately followed a discussion of the Swedes' "undoubted gift for form, for mechanical skill," and the "different sort of training" needed to make the Jew a "most useful citizen in the mercantile and commercial pursuits for [which] his ancestry has pointed him." In the background of industrial arts curricula, as in that of the IQ and social services, lay quite specific perceptions about the needs and nature of the immigrant population. "Immigration has brought to our shores representatives of almost every race under the sun," Josephine Chase announced, "and with them certain racial skills and aptitudes which the schools should recognize and foster." Chase pointed with pride to New York's success at providing a range of vocational offerings and trade high schools which developed these capabilities.[54]

The problem of vocationalism in the high schools reveals not only the critical role of immigration in framing school issues and programs but also certain dimensions of the meaning of "individualized instruction" which relate it to the IQ in other ways as well. For what is troubling about the development of differentiated instruction "by the application of scientific and objective methods," to which educators like Riverda Jordan and others were committed, is not the wide variety of courses offered, some of them with a clearly vocational intent, but the degree to which the schools predetermined the limits of schooling for blacks and various immigrant groups according to judgments about native talents and probable future lives. Like the IQ with which vocationalism was intimately connected, these judgments were based not on the individual but on the group from which he or she came. Leonard Covello understood this immediately when an industrial or trade school was proposed for the Italian neighborhood where he lived. " 'An industrial high school,' I pointed out time and again at meetings, 'presumes to make trade workers of our boys. It suggests that the boys of East Harlem are not capable of doing academic work.' "[55]

Underlying all the discussions of high-school organization in the 1920s especially was a dilemma resulting from the complex of changes which had redefined the role of the schools in a specific historical circumstance. Between 1910 and 1930, American educators were inordinately impressed by the extraordinary growth of high-school attendance as the very fulfillment of democratic promise. At the same time, they studied and worried about the statistics on dropping out of high school. The number of high-school dropouts was as alarming as the snowballing number of high-school students was encouraging, and the situation produced a considerable literature, which, like the related studies of retardation, had a certain vogue throughout the teens and twenties. The studies were, in fact, part of the general infatuation with measurement, numbering, and the "science of education" which defined much of the educational discourse of the time. And the numbers, in good progressive fashion, became not only a way of seeing but also a measure of institutional failure that in turn defined the sphere for reform action.

The best known of these studies, *The Selective Nature of American Secondary Education* (1922) by George S. Counts, is still remembered today for documenting the close relationship between socioeconomic status and the probability of high-school attendance, since students from wealthier homes and from certain racial and ethnic groups were more likely to attend high school and much more likely to graduate than those from poorer and foreign backgrounds. Most of the studies came to similar conclusions. Thus, Mary Roberts discovered that most high-school dropouts left school for financial reasons, and that "a strikingly large percent . . . of the dropout pupils reported were of foreign-born parentage." Similarly, summarizing the data of several previous studies, Emily Palmer found that, "on the average, the families of pupils eliminated from school are on a low economic level." She also concluded that, "on the average, pupils who are eliminated from school make mental scores below the standard for their age." In her own investigation of students in Oakland, California, "the percentage of pupils whose parents were foreign-born is almost twice as large in the eliminated group." She found very large differences in the Terman group test scores between those who stayed and those who left school and concluded, "An examination of the relationship between the two factors of intelligence and nationality of parents shows that the occurrence of high scores with native white parentage and low scores with foreign white or other parentage is frequent enough to indicate some association between the two."[56]

All in all, the studies of high-school leaving provided a fairly detailed portrait of who stayed in school, who left, and why. In many cases, specific associations were made between mental tests and school retention, and always students were categorized by group. As these associations became a standard of analysis, they provided bountiful evidence that identified individuals' capacities and their potentials with the past experience of the group from which they came. It was a mere extension of analysis, in the context of a commitment to provide usable education, to guide students at school by the evidence drawn from the studies.

Some studies did more than profile socioeconomic, educational, and ethnic background. Edward Sackett, for example, used a questionnaire designed to elicit student views on what features of school would make them most likely to continue or stop their education. He found that students would leave school above all, "if next year the school would not train me for a job." When he asked what one single factor would persuade seniors to continue for another year, the most frequently chosen answer was, "If

I know that another year of any kind of school would get me a better job when I finally leave school."[57] Thus, all the findings seemed to point to the inescapable link between the students' practical concerns, the groups from which they came, and school retention.

The evidence on high-school dropping-out seemed to lead always to two related conclusions: The schools needed to deal with student maladjustment by providing more and better services like health and guidance programs, and schools needed to devise a richer variety of programs oriented to the practical needs and future work lives of their students.[58] Not only did the studies sometimes ignore their own data, which showed that students left school because of nonschool factors (above all, family economics) and not because school was not "interesting," but they also chose to interpret these handicaps as indications that students needed and wanted a more practical education. The schools, they thus insisted, could not overcome students' handicaps, but they could at least adapt to these handicaps. In this context, vocational education seemed to provide the necessary answer: it was practical; it seemed more immediately interesting (relevant); it seemed more suited to the mass of students who had no interest in or saw no purpose for Latin or trigonometry. No doubt, students, as Sackett discovered, and probably their parents as well, expected that more education would open up better job opportunities. At the same time, there is no reason to believe that schooling for better opportunities meant for them the specific training routines like those offered by Jordan High School. Nevertheless, educators often believed they were the same thing, confusing the desire for more adequate general skills with the kind of narrow restrictive education embodied in chef's training or beauty culture.

Some evidence for the fact that vocational curricula were not in fact fulfilling students' needs or expectations comes from the drop-out statistics themselves. Enrollment in vocational courses in no way prevented dropping out. On the contrary, students enrolled in vocational tracks were more likely to drop out than those in college preparatory courses. According to one study of commercial education in New York City (1911–12), "Every school reports a larger percentage of loss of commercial pupils than the city average or the school average."[59] Evidence of this kind is extremely difficult to evaluate because the constellation of factors—low socioeconomic status, foreign background, financial need, and low IQ—that were related to dropping out were also associated with enrollment in vocational courses. Moreover, it is by no means easy to discover just what students expected to derive from further schooling or what they meant by "preparation for a job." In 1911, Joseph King Van Denberg conducted a large-scale and impressive investigation of the background of New York City high-school students and the reasons they left high school. To a surprising degree, high-school students were the children of artisans, tradesmen, contractors, and petty officials, not only of clerks or professionals. They were likely to be the youngest in their families, and while native-white parents were proportionately best represented, a goodly number of immigrant children were attending high school. Among these, Irish daughters preparing to teach and Jewish sons preparing for professions were already prominent. All in all, Van Denberg found that high-school attendance was "a grand struggle upward," and reflected aspirations for "escape from manual labor," for lower-middle-class youth and their families who lived in poor and cheap surroundings and could afford to send only one of their children to high school.[60]

By 1930, when 60 percent of all high-school-age students were in school, the motives for attendance were certainly more varied. At the same time, there is no reason to assume that the belief in mobility through education had changed much. Sports activities and clubs did not keep students in school, although these helped to keep students involved and promoted loyalty and social identification.[61] High-school students went to school and stayed because they hoped this would lead to better lives. This does not mean that a better life did not often require practical skills, but that education was based on aspiration, practical aspiration to be sure, but aspiration for improvement and a desire not to stay in place. Once aspiration was arbitrarily limited by the schools, as was often the case with tracking and vocationalism, the schools cheapened their greatest asset.

Educators were not blind to the limitations of their vocational proposals. One student of commercial education observed that businessmen hoped to hire people with general, not special, skills and noted, "We cannot escape the conclusion that the non-commercial schools have a larger influence in the sum total of business than do the special schools and it is an open question whether or not the general school is not giving at present more appropriate training for the major business needs." He nevertheless recommended improvement and greater specialization in commercial courses, not their elimination. Similarly, in addressing the National Education Association in 1919, one speaker first waxed euphoric about the possibilities for democracy opened up by differentiated instruction and then cautioned against a too early channeling which would close, not open, opportunities for those directed into vocational tracks: "The use of the vocational schools, the fostering of the practical subjects which enlist the interests of the pupils carry with them serious dangers which threaten the realization of our aims. A president of a railroad recently declared that he found it exceedingly difficult to secure men even from the best technical schools who could be counted on to be capable of rising to the higher positions in the service." Likewise Edward Sackett explained, "Better vocational training means not more learning of specific, mechanical routines, but a sounder foundation in the fundamental skills, attitudes, and knowledge essential for getting a beginning job in some broad vocational field and growing in it."[62]

It may be that educators foundered in providing a solid vocational foundation for most students because they lacked a clear grasp of what vocationalism in education really required, confusing simple manual instruction with a richer and broader vocational preparation. The high school, after all, remained overwhelmingly an academic institution in which students who intended to go to college had status and priority. Alexander Inglis once observed that the attempt to "adapt instruction to varying capacities" had "been met in part by utilizing as salvage departments the various non-academic subjects, successively making the commercial department, the manual-training departments, and the practical-arts departments the educational wastebaskets of the high school." Vocational curricula usually concentrated on manual or simple clerical skills and watered-down or abbreviated academic subjects. The skills were frequently outdated before they were incorporated in the schools, and the academic instruction was usually second rate.[63] And yet, there was a kind of stubborn insistence on the need for appropriate and "progressive" vocational education that cannot simply be explained by institutional or pedagogical ineptitude. Indeed, this insistence, despite the attendant criticisms from the beginning about the lack of creativity and imagination in vocational programs, suggests a far more significant reason for the failures of these programs.

Historians who have written about vocational education have suggested reasons for the commitment to such training. Sol Cohen concluded that the industrial education movement resulted from a sense that immigrants were moving too fast and too far, challenging middle-class native groups for status, money, and power. But this does not seem sufficient as an explanation, if only because the children of most immigrant groups tended to be less successful than native whites in the schools. Marvin Lazerson has argued that manual education resulted from profound fears about the nature of work in industrial society and the need for the schools to train for that character and those values no longer available through artisan employment. Certainly, this was part of the progressive framework and fundamental to the attitude of reformers. At the same time, this commitment cannot explain why less theoretical or reform-minded educators held on so tenaciously and well into the highly industrialized twentieth century.[64] It also fails to explain the place of vocationalism in the high schools rather than in the elementary grades.

The belief in vocationalism is, however, related to another and simpler fear having to do with containment rather than either mobility or reform. All the studies in the first three decades of the twentieth century pointed to the inability of the schools to keep their charges, not to their rapid progress and ascent. Retardation, the statistics on high-school drop-outs, truancy, all pointed to maladjustment and suggested that American youth, especially the children of the slums and ghettos, were not being kept in school long enough. As Caswell and Campbell noted in their book on curriculum development, the child-labor laws and the new educational requirements in the business world were raising the age of entrance into responsible work with a resultingly large problem of youth unemployment: "The possibilities of several years of idleness during a period of life so important in developing character is appalling in the possibilities for developing traits inimical to the pursuance later of worthwhile activities and for developing qualities that are definitely anti-social." As a result, "the tendency for the age to increase at which young men and women become gainfully employed increased the demand and need for education."[65]

The search for practical, interesting forms of education for those who would soon be working but not soon enough responsible resulted from the increasing custodial function which high schools began to assume in the twenties and thirties. Better to have adolescents at school, engaged in any kind of activity, than out on the streets idle. For this reason, educators insisted on the need for vocational courses. Educators assumed that vocational courses were not only more practical, they were also *easier*. Unlike John Dewey who saw manual and mental activity as intimately related, most schoolmen and women saw manual activities as a substitute for mental work. This of course explains why those with lower IQs were pressed into nonacademic "vocational" tracks. Since the pressure to keep children at school increasingly affected the high-school age groups, and educators did not believe that the new masses were smart enough to benefit from traditional academic subjects, they could do one of two things—dilute the academic subjects and lower standards or provide convenient substitutes which included some academic courses but concentrated on manual skills. Vocational education curricula were one result, and later, general education curricula oriented to duller students whose aspirations were assumed to be limited anyway became another. Thus, in New York City high schools as early as 1911–12, about one-third of the population was enrolled in commercial tracks or in the two special commercial high schools. Of these students, one investigator observed, "The usual testimony of the principals of the general high schools is to the effect that commercial pupils constitute the less desirable element of the school." Academically less talented, frequently of foreign parentage, and usually from poor homes, vocational students were indeed less desirable. Moreover, this was commonly and popularly perceived from the first. Covello observed "The stigma attached to an industrial high school! The psychological effect upon the pupils and the community. Sure, people say on the outside 'The proper school for them dumb immigrants. They don't deserve any better.' "[66]

The schools, proud of their growth, prouder yet of the enrollments in high schools, were committed to keeping the democratic mass at school, even the "less desirable element." To some degree they had no choice. As Josephine Chase explained, "Compulsory laws, made to benefit large numbers of pupils who might otherwise be too hastily forced into industry, have forced into the high school pupils of little aptitude for 'book learning' to whom the work of the secondary schools . . . presents great difficulties."[67] The progressives had of course turned to the schools in their efforts to protect the young from industry and to provide them with a proper start in life. But, unlike the progressives whose aim had been reform of the schools and of their new populations, the schools hoped only to grow and to contain as best they could those they were forced to serve. By the 1920s and '30s, this became the objective of all the efforts at program differentiation, as it was for the emphasis on social life and extracurricular activities. Thus, the emphasis on vocational education was very much a result of the democratization of schooling, but it was less the result of adapting the schools to the genuine needs of its democratic constituency than of a priori perceptions about the new constituency. In this sense, as in others, vocational education was related to IQ because both emerged from the perceptions and needs defined by the immigrant presence, not only in the society, but in the schools as well.

Vocationalism, like IQ, allowed the schools to appear progressive, since it seemed to be a clear adaptation of the schools to the needs of the individual and of the society. Thus Oakland, California, in 1930 was described "as having a thoroughly modern school system" because it had "junior high schools, testing programs, and provision for accelerated, retarded, and other atypical groups" and because of its "vocational classes and other special curricula."[68] IQ testing, vocational education, age-specific schools, and tracking all were part of the modern school system and part of what educators usually meant by "progressive." Unlike Dewey's progressivism, however, strict age-grouping, systematic tracking, and vocationalism resulted less from an appreciation of the variety of individual talents or accomplishments, or dedication to releasing these potentials for social progress, than from the needs of schools to operate efficiently and to fulfill their obligations to all those who were entrusted to their keeping. As one study of school retardation appropriately concluded: "The sincere effort which has been made by many public-school systems to adapt the school programs of studies to the children enrolled, by improving classification, curriculum, and teaching methods has, no doubt, decreased the amount of retardation and increased the holding power of the school. . . . In fact, the changes which have come about in the school system were necessitated by the presence in the schools of pupils who were compelled by law to attend, yet who could not adapt themselves to the program of the school as then organized. *In progressive school systems the differentiation in curriculum has become so great that completion of sixth grade or eighth grade means different things in different cities, or even in different parts of a single city, or in different sections of the same grade."*[69]

By the 1920s, the enrollment of all students in the elementary grades was no longer enough. Instead, the emphasis shifted to the high school which now represented the fulfillment of democratic promise and became the appropriate arena for supervision. In an address to the National Education Association, aptly titled, "The Reorganization of the High School for the Service of Democracy," John L. Tildsley made clear that the high schools' great accomplishment had been the opportunity extended to immigrants to partake in the promise of American life and that further development in this direction remained their great task: "The high school must take *every* boy and girl from 13 to 18 years of age, *appraise* his ability, and discover his aptitudes by a series of *intelligence tests, group* him with others of approximately equal ability and then *assign* him subjects of study and give him a training which shall so enlist his interest and develop in him a *habit of success* not of failure. . . . The habit of failure can be eradicated only as we really see each pupil as he is and deal with him as an individual." Only the successful remained at school, and therefore the schools' task was to make each student a success by appraising his ability, grouping him with others like himself and assigning him only such subjects as he could master. Rather than determining how each student could best be instructed in what all should know, the ideology of individual differences encouraged schools to seek ways in which students could be occupied for longer periods by learning only as much as was easy for them to acquire. The ideology of individual differences had shifted the burden from the schools to the students. The determination of IQ was the central foundation for this process, since it provided the basis and primary form of definition. As Tildsley concluded, "These intelligence tests give greater promise than any other single factor for the remaking of our secondary education to be an instrumentality of service for democracy."[70] Democracy in education had thus come to mean hardly more than higher enrollments.

Addressing a much wider audience, Alexander Inglis, one of the most prominent and intelligent proponents of the new "democratic" high school, also explained the new meaning of the high school and the sources of its changes to readers of the *New Republic* in 1923. He began with an illu-

minating assertion, "The history of secondary education in America . . .
has been marked by two tendencies which are reciprocally related—the
expansion of its curriculum and the attraction of larger proportions of the
population, drawn from successively lower social, economic and intellectual levels. As the curriculum was expanded, larger portions of the population were enrolled and pupils of different types were attracted." Then to
bring the point home with crystal clarity, Inglis noted, "The increased enrollment in the secondary schools indicates not only that we have tapped
successively lower social and economic levels, but also that we have tapped
lower levels of native capacity and acquired ability." The social and intellectual merge here very easily and, I think, intentionally, since Inglis could
expect his readers to assume a certain coincidence between lower social
status and lower intelligence. Inglis's lesson did not stop here, however,
for he raised the significant democratic issue as this was understood by
educators in the 1920s. Even as presently organized, the high schools, according to Inglis, still demanded "intelligence probably not lower than five
points above average ability." "Can anyone hold that the selective factor
should be such as to exclude more than one half of our children from
reasonably successful accomplishment in English, foreign languages, mathematics, science and history?" Having posed the question in its most democratic form as an issue of open opportunities, Inglis then makes an important mistake by introducing what in any cultural context other than
the American would be a non sequitur. "It is not on record that the French
language is reserved in France for those only who have an intelligence
above the average."[71] Of course, the French neither reserved their language for the intellectually gifted nor opened their *lycées* to all comers.
But in the United States "English" was a code word used to describe the
urgency of schooling for those who were outsiders, poor, and assumed to
be less intelligent. Even the high school, Inglis made clear in the best democratic prose, had to be remade in their image.

Inglis was important as a proponent of the new democratic high school.
He is important to us for the significant expression he gave to the unconscious, or at least culturally loaded, meanings that went into the re-creation
of secondary education in the twentieth century. The specific "democratic"
problem that Inglis had to address was one with which his audience as
well as school officials were familiar. The "democratic" conclusions and
their implications were understood just as well.

In an important essay on progressive education, Richard Hofstadter described the substitution of a concern with life adjustment for a commitment to learning as anti-intellectual.[72] If this was the result of developments in American education in the early twentieth century, it was less a
by-product of a too exuberant democratic enthusiasm, as Hofstadter sug-

gests, than of an inadequate faith in democracy. In the full flush of victory at the opening of the schools to a new democratic constituency, educators restricted the access of that constituency to the education for which they had come by predefining them as not fully able to benefit from its best resources.

V

It is ironic, though not hypocritical, that this education should have been described as progressive. The beliefs of the progressives were often ambiguous but also more complex and more optimistic than the educational forms in which they issued. Above all, reformers had looked to education as a means for political revitalization and to schools as a fundamental force for change. If the progressives believed in order, they saw order as the foundation for democratic responsibility and for effective individual action. Their faith in science, though often simplistic, was fundamentally tied to a commitment to amelioration. The schools however took over the progressive prose without the urgent vision. Educators and administrators sincerely believed they were adapting to the people's needs and opening the schools to the masses, but they did so in ways that lost the sense of what democracy was all about. Order was translated into containment and science into a means for organizational efficiency as progressivism was transformed from a challenge into a convenience.

By the Great Depression, "equal educational opportunity," as a description of fact and an ideal, was a phrase common in educational parlance. But its meaning was inevitably linked to the belief in "individual differences" that would define for each pupil the degree to which he or she could benefit from the schooling now presumed to be generally available to everyone from the primary school through the high school. Individual differences provided the natural mechanism that allowed equal education to operate efficiently in the context of the cultural understandings of how the population differed in needs and abilities and of the requirements of school organization. The two concepts were intended to describe a school system that was at once pedagogically enlightened, progressively democratic, and protected the individual from the tyranny of sameness threatened by mass education.

The linking of these two concepts was seriously misleading. The commitment to individual differences had indeed permitted twentieth-century schools to adapt to a changing constituency, but it had also provided the foundations for systematic differentiations in curricula, in performance, in expectations, and ultimately in social rewards. Moreover, whatever genuine solicitude progressives and educators felt for the children of the disadvantaged who were now present in the schools in such large numbers, their concern was too often accompanied by a patronizing face and a gloved fist. This does not mean that some, perhaps most, of the children of immigrants and native whites alike did not benefit from various school-centered programs, but that the schools were caught up in the straightjacket of a form of thought and organization which democratic as it often seemed, and sometimes was, had confused an aspiration to educate all the people equally with the need to provide schooling for the population as efficiently as possible. To this, the particular historical juncture contributed mightily, as the cultural problem of a pluralistic population and the tendency to define and order that population hierarchically infused the assumptions and programs of the schools. In being committed to systematic expansion and deeply embedded in their broader cultural context, the schools had taken shortcuts which too often lacked faith in that population's ability to be educated and placed too much faith in scientific nostrums which defined the limits of education rather than in methods for providing better education. In all, despite their loud proclamations of faith in American education and their pride in its accomplishments, educators found ways to define, limit, and circumscribe their province and the promise of their schools.

Educators had defined a differentiated schooling as in the best interest of their students. But while their perceptions of student variation were inseparable from their perception of different talents and their views of the groups from which students came, they did not expect that the schools would encourage group ties. On the contrary, the schools would, for all intents and purposes, replace former group associations with new American loyalties while individualizing talents and aptitudes. That, after all, was what Inglis had meant when he refused to reserve the American language only to those who were most talented. In this fashion the schools would be both unified and diverse, creating a pluralism whose sources were lodged in individual aptitudes trained and contained in the unity of American school culture. But diversity is a tricky reality, and in the context of the expanded version of the democratic high schools created in the 1920s, its expression was not so easily contained or redirected.

3

"Americanizing" the High Schools: New York in the 1930s and '40s

We, the graduates, pictured on the pages following, stand united in our determination to continue "Life, Liberty, and the Pursuit of Happiness." These are the ramparts we love and fight for and for which, if need be, offer our lives and our fortunes.

What we really want to say can be best expressed by this excerpt from that very stirring song, the "Ballad for Americans." [by John Latouche and Earl Robinson]

. . . I'm the everybody who's nobody, I'm the nobody who's everybody
Are you an American?
Am I an American?
I'm just an Irish, Negro, Jewish, Italian,
French and English, Spanish, Russian,
Chinese, Polish, Scotch, Hungarian,
Litvak, Swedish, Finnish, Canadian,
Greek and Turk, and Czech and double Czech American!
Holy mackerel!
And that ain't all. I was baptized Baptist, Methodist,
Congregationalist, Lutheran, Roman Catholic,
Orthodox Jewish, Presbyterian, Seventh-Day Adventist,
Mormon, Quaker, Christian Scientist—and lots more.
You sure are something.
Our country's strong, our country's young
And her greatest songs are still unsung.

"We . . . The Graduates," High School of Commerce,
This is America, June 1943

By the 1930s the children and grandchildren of the progressive-era migrations were filling the classrooms of American cities. This was not unprecedented; drawn or driven, the children of the masses had been attending school in America since the nineteenth century. New, however, was the degree to which the children of immigrants were moving into the higher educational reaches, attending high schools and even colleges. In the nineteenth century, high schools had occupied an ambiguous place in American education—sometimes merely feeders for colleges and universities, sometimes academies in their own right, often no more than stopping places for females not quite certain what else to do. Still, they were always an educational attainment, "high" schools, and they embodied a level of achievement that is difficult for Americans in the last decades of the twentieth century to appreciate. As late as 1926, Niccolà Sacco still believed them to be bastions of privilege: "The capitalist class, they don't want our child to go to high school or college or to Harvard College . . . they don't want the working class educated; they want the working class to be low at all times, be underfoot and not be up with the head." Leonard Covello, another Italian immigrant, made the same connection between high school and college but with a very different intention. In Italy, he observed, poor children never went beyond the fifth grade. "But here in America we began to understand . . . that there was a chance that another world existed beyond the tenement in which we lived and that it was just possible to reach out into that world and one day become part of it. The possibility of going to high school, maybe even to college, opened the vista of another life to us."[1]

As we have seen, educators had begun to look differently at high schools. Linking them up with the elementary schools, not the colleges, they invested them with the peculiar socializing influence appropriate to adolescence. The high schools became the last institutional link in the socio-developmental process by which educators increasingly organized their perceptions of schooling. "The period of adolescence," Elwood Cubberly noted, "we now realize is a period of the utmost significance for the school."[2] Students went to high school because they were of a certain age, and age, not the fact of learning trigonometry or Latin (subjects of "high" intellectual seriousness), made the schools higher-learning stations.

Most immigrants probably thought of the high schools in the more traditional and noble fashion of Covello and Sacco. Evoking memories of what it meant to send a child to a *gymnasium* in the old country, sending a son or daughter to high school became a mark of present accomplishment and future expectation. Coinciding in time with the more aggressive posture that immigrants assumed between the wars, especially during Franklin Roosevelt's presidency, the attendance and graduation of Jews, Italians, Slavs, and others from high schools became the echo of a new insistence on a social place in America, as well as the result of quite pragmatic efforts by educators to keep children in school longer. Not all groups took advantage of high schools to the same degree, but by the 1930s, and certainly by the 1940s, the graduation from high school of the progeny of the great early twentieth-century immigrations marked the arrival of a new common-school era. In contrast to its predecessor, the common-school era of the early twentieth century concerned adolescents, not children, and in large cities, it replaced the pious air of Protestant respectability with a complex cosmopolitanism.[3]

In cities like New York and Chicago, the high schools, like the neighborhoods in which they flourished, became ethnic enclaves. This fact had two distinct consequences. On the one hand, the children and grandchildren of immigrants had arrived at the threshold of American success. From the progressive perspective, the individual was offered the keys to the kingdom while the kingdom itself would be secured. On the other hand, the very success of numbers portended a new social environment—an environment that was as much ethnic as it was traditionally American. In this context, the high school as a fundamental agency of socialization became both more and different than its planners had anticipated.

I

The school was, of course, the great institution of assimilation. This was one of its primary purposes and there is little doubt that it succeeded in introducing the children of immigrants to the culture of twentieth-century America. But assimilation is a complicated process. It is most likely not a single process at all but a way of summarizing a range of individual and group experiences. School officials were not so naive as to believe that assimilation meant simply replacing one set of cultural habits with another, in the manner of Henry Ford's famous pot, but they had to operate on the assumption that a shrewdly designed curriculum and a school environment oriented toward effective socialization would permit rapid absorption of a heterogeneous population into the American mainstream. This was, in fact, one of the beliefs behind the emphasis on the developmental significance of adolescence.[4] Educators eagerly looked to the special aptitudes and clannishness of adolescents as a force for social assimilation and, with this in mind, tried to construct a broadly conceived school program.

Integral to that new program by the 1920s and '30s was a range of activities in which students, although under adult auspices, exercised their own forms of self-direction in social, civic, athletic, and academic affairs. Through these extracurricular activities, Earle Rugg noted in a special issue of the *Yearbook of the National Society for the Study of Education,* "The school may well make itself the laboratory for training pupils for efficient citizenship."[5] Educators recognized, but did not sufficiently credit, the degree to which the extracurricular activities depended upon student self-direction. For while they certainly helped to introduce students to the meaning of social life in America, they did so in ways that school officials never entirely controlled and of which educational theorists would not have approved.

Various informal school activities had existed at the fringes of the academic curriculum since the nineteenth century, but it was not until after World War I that educators made a concerted effort to align them with the expanded concerns and "progressive" pedagogy of the modern school.[6] "Largely within the past decade, and wholly within the past two," Elwood Cubberly noted in 1931, "an entirely new interest in the extra-curricular activities of youth has been taken by the school. In part this change in attitude has been caused by the new disciplinary problems brought to the school through the recent great popularization of secondary education." Like the development of an appropriate curriculum, the opening up of education to a newly democratized constituency with its special "disciplinary problems" went hand-in-hand with the attempt to design an educational regime to enlist the specific aptitudes and differences found among individuals. Charles R. Foster, an associate superintendent of schools in Pittsburgh, noted, "Children differ in ability, aptitude, sex, probable career . . . social status, environment, traditions, habits of work, race, nationality, age, health, . . . and in numerous other ways such as to make it imperative that our secondary schools provide not only a differentiated curriculum . . . but also such forms of extracurricular activities as may utilize the socializing, integrating factors important in establishing a common basis of feelings, aspirations and ideals, essential in a democracy."[7]

The themes underlying extracurricular planning thus defined the educational issues of the period—institutional expansion and a new view of schooling as socialization, a democratic invocation of the significance of individual differences in aptitudes and talent, and the injunction that schooling assist in the re-creation of national community. When Leonard V. Koos polled school principals about their reasons for sponsoring extracurricular activities, he discovered that "socialization," "training for social cooperation," "experience of group life," "improved disciplinary situation and improved school spirit," "training for leadership in a democracy," and "recognition of adolescent nature," were the most frequently given.[8]

Finally, the extracurricular activities were meant to express and uphold a commitment to what one writer called "equalization of opportunity." "The one place where democratic ideals and objectives may function in a natural matrix is in the conduct of the extra-curricular activities. Whether a student is notably dull, studious, clever, rich, poor, handsome, or ugly he should have an equal opportunity to be a member of a school organization which ought under all circumstances to be organized upon a basis of democratic society." Leonard Koos said the same thing more compactly: "The scope and plan of operation should be such as to encourage participation by all students, with membership in all organizations equally open to all."[9] With the new attention to the pedagogical imperative of differences and in the context of a manifestly differentiated curriculum, IQ, and vocationalism, the extracurricular activities became the repository for the old common school ideal.

The fit between adolescent socialization, citizenship training, and democratic opportunity, while conceptually neat, was not always perfect in experience, and many of the educators were concerned to suggest remedies for problems encountered in actual day-to-day administration. The aim was always as tightly integrated a school regime as possible. "It has seemed to me," Elbert Fretwell explained, "that the secondary school should plan consciously its whole life. . . ." At the same time, Fretwell and others understood that too much control would vitiate the purpose of extracurricular activities by sapping their vitality as effective learning experiences. Too much control would undermine their usefulness as theaters for training in genuine citizenship, initiative, and voluntary cooperation. In practice, the challenge was always one of careful planning and judicious control, but not too much repressive adult direction; activities effectively integrated into the curriculum but sufficiently different from regular studies to utilize the abundant individual talents and latent interests of students. The activities had to be fun as well as work. "There is," Fretwell noted, "the exploration of the pupil himself, his interest, his curiosities, his abilities. . . . There must be joy, zest, active, positive, creative activity, and a faith that right is might and that it will prevail." Foster too expressed the usual progressive appreciation of the creativity of adolescence: "Instead of frowning, as in olden days, upon the desire of the young to act upon their own initiative, we have learned that only upon these varied instincts can be laid the surest basis for healthy growth. Self-reliance, honesty, perserverence, and respect for the rights of others are needed. The school democracy must be animated by the spirit of cooperation, the spirit of freely working together for the positive good of the whole."[10]

The activities were thus both an obvious and tricky realm for educational efforts. It is clear, moreover, that despite some general theoretical consensus, schools differed in their treatment of the activities, especially in the degree to which they were directed by adults and integrated into the curriculum.[11] Obsessed as educators were with the "science" of education, the literature of the period is replete with questionnaires about extracurricular clubs. The questionnaire, reflecting the contemporary expectations for the activities, asked long lists of questions about the availability of activities, the extent of participation, the educational values of activities, and the coordination of activities with the regular school program. They never asked what students thought of the activities or what they meant to them. Fretwell, in good progressive fashion, proclaimed that "Pupils are citizens here and now, with rights, duties, privileges, and obligations. If they are to grow into still better citizens, ideas about citizens may be helpful, but satisfying practice is absolutely fundamental."[12] And yet, despite the students' presumed citizenship and the fundamental training in thoughtful individualism that the activities were intended to provide, educators assumed that the activities became for students what educators meant them to be.

Practice was of course at the heart of progressive pedagogy, and without adopting all its values, one can accord recognition to the truth of that claim. What then did the practice of extracurricular participation express about schooling that educators might have missed by not asking students? Obviously it is not possible entirely to rectify their oversight. But it is possible to get some sense of what the activities meant to students by observing patterns of participation. Like the questionnaires of the 1920s and '30s, this profile of ethnic patterns in extracurricular participation is informed by a specific intention. Questions always provide the skeletal structure of answers, and since I am asking about the role of ethnicity in determining the choices students made among the clubs, the answers will be framed around ethnicity. It is important to remember that ethnic concerns and associations were not the only meaning students gave to clubs and activities. I have, of course, been arguing all along that the immigrant presence was crucial to the manner in which the school operated and to the theories of schooling that emerged in the twentieth century. But the immigrant presence in the schools was more than a stimulant to educational theory and school organization. It was a dynamic element that often operated apart from the interests of teachers, principals, and theorists. Indeed, the legitimation given to extracurricular activities in the 1920s and '30s had some very significant, unintended consequences. By providing students with just enough leeway to exercise choices and with enough room for "social cooperation," the extracurricular activities became a significant arena for the continuing expression of ethnic association among the second and third generations. This was not a force opposed to assimilation, but part of the way assimilation was experienced in high schools. While educators had hoped to put the special stamp of the schools upon the social order of the twentieth century, they had not quite foreseen the degree to which this would be both true and not what they had expected.

II

The following discussion is based on the experience of 15,000 seniors in seven New York City high schools between 1931 and 1947 as recorded in high-school yearbooks (see Appendix 1). Specifically, I have examined the ethnic variation in the extracurricular participation of these seniors on the basis of a tabulation of all extracurricular clubs listed by students in selected years. The ethnicity, not nativity, of each student was determined by the student's name. This kind of identification is by its nature inexact, because names are not necessarily ethnically transparent and because of the frequent overlap among names—for example, between Jews and Germans; among the Irish, native whites, and Scots; etc.[13] First names often provided an additional way to differentiate between some groups. Finally, the category of undecided ethnicity allowed me to discard those students whose names were just too ambiguous. The seeming precision of numbers should not be allowed to obscure the imprecision of the method, and most of the figures and conclusions should be read as only near approximations to reality.

New York was in no way a representative environment, but it was the preeminent immigrant city and an investigation of the city's schools is especially illuminating. Students in New York came from a very wide variety of nationalities and racial groups, but only six have been selected for analysis—native white, Irish, German, Italian, Jewish, and black. The schools themselves varied. A school like Seward Park High School, for example, contained a large majority of students from a single ethnic group, while New Utrecht High School was dominated by two ethnic groups. In other schools, there was a complex mix of groups and important ethnic changes over time. Two schools, Bay Ridge and the High School of Commerce, were unisex schools, the former exclusively for women and the student body of the latter throughout most of the 1930s and '40s made up entirely of men. These differences mattered. Individual and group experiences often depended on the specific social, economic, and demographic characteristics of each school. Wherever possible, therefore, I have based my conclusions not only on the overall pattern but also on the behavior of groups within specific schools, and I have defined ethnic tendencies only when there were compelling similarities among several schools. (See Appendix 2 for tables detailing the evidence for the following discussion.)

For the high school graduates of the thirties and forties, extracurricular activities had become a regular part of school experience.[14] Four-fifths of all students had participated in some club or activity, although the extent of participation varied from a high of 99 percent at Bay Ridge High School to a low of 56 percent at Theodore Roosevelt, and women were every-

TABLE 1. Participation in Some Activity by Sex and Ethnicity.
(Each box signifies overrepresentation in designated school)

	Men						Women					
	1	2	3	4	5	6	1	2	3	4	5	7
Jewish	■	■	—	■	■	■	■	■	■	■	■	■
Italian	—	—	—	—	—	—	—	—	—	—	—	■
Black	—	■	■	■	a	—	—	—	■	—	—	■
Irish	—	—	b	■	■	—	—	—	—	■	■	■
German	—	—	■	■	—	—	—	—	—	■	—	—
Native	■	—	■	■	—	■	■	■	—	■	■	—

Note: School Key

1. George Washington High School
2. Evander Childs High School
3. Seward Park High School
4. Theodore Roosevelt High School

5. New Utrecht High School
6. High School of Commerce
7. Bay Ridge High School

a. No blacks at New Utrecht High School
b. Not enough Irish at Seward Park
   to be meaningful

where more likely to participate than men. Certain groups were more active than others and engaged in a wider range of activities. Jews and native-white students were the most active. (Native-white students will hereafter be referred to simply as "natives.") In the entire population as well as in each of the schools, except Seward Park, Jewish men were disproportionately represented in extracurricular clubs, and even in Seward Park, they were, for all intents and purposes, represented at par. Jewish women were always overrepresented. In other words, the participation of Jewish men and women was almost always well above what would have been expected considering their proportion of the school population alone. Natives, both men and women, were also very active, but not as active as the Jews. Italians and Germans, on the other hand, engaged less frequently in activities and were underrepresented in most schools. The participation of blacks and Irish varied among the schools but was neither as low as that of the Italians and Germans nor as high as that of the Jews and natives. Table 1 summarizes the experiences of the various groups.

A student's ethnicity clearly influenced his or her chances of participating in the extracurricular world of high school. More significantly, ethnicity had a powerful effect on the kind of activity a student was likely to elect. Sometimes the ethnic variations depended on a specific school environment, but at other times there were uniformities across school lines that suggest strong ethnic preferences. For the purpose of analysis, I have divided student activities into twenty-three categories.[15] Although these did not exhaust the range of activities available to students in all schools, they were generally comprehensive of the most important activities in which students engaged.[16] A few of the clubs turned out to be extraordinary ethnic differentiators, while others were weaker as indicators of ethnic interest or association. As a whole, however, they reveal the strategic role of ethnicity in determining student choices among extracurricular offerings and help to define the nature of social life in the high schools.

Members of some ethnic groups rarely or never joined in certain activities. Irish men almost never joined science clubs and rarely participated in the orchestra or in dramatics. Jews and blacks participated in religious clubs very infrequently. Blacks were the most consistently absent from a wide range of activities.[17] They almost never participated in dramatic or science clubs. No black woman or man was ever elected president of the senior class or the student body; no black was ever editor in chief of the newspaper. The fact that blacks were the group most frequently absent from a range of activities suggests a strong exclusionary bias against them. Blacks joined only certain activities and almost never others. No other group was absent in so many categories across schools. Blacks who were eager to participate in school activities, and three-quarters of blacks did participate, chose carefully and judiciously, consistently sidestepping activities in which they either had no interest or were clearly not welcome.

In one activity, black men were dominant in an unparalleled way. In every school with a meaningful population of black men, they were over-represented on the track team. Indeed, track was the activity most consistently associated with blacks—almost one-third of all black men in the entire population were on the team. At George Washington High School, where blacks were scarcely 5 percent of the male population, eleven of forty-four track men were black. At Seward Park, where black men were slightly more than 1 percent of the male population, three of thirteen runners were black, and these three were part of a population of only thirty-one black men. Half of all black men at Evander Childs (six of twelve) and one-sixth (ten of sixty) of all black men at the High School of Commerce were on the track team. These were small numbers, because the number participating in track was small and because the number of black seniors was small, but they were stunningly clear. The chances of a black man electing track as one, and possibly his only, high-school activity were very great.

The same concentration of black men was not evident in any other sport. Only in track did black men make such an extraordinary showing. Although black men were also active in basketball, they almost never played football and rarely participated in other sports. These sports were dominated by natives, Italians, Germans, and, to a lesser degree, the Irish. Jews, like blacks, were not drawn to football. When Jews participated in sports, they strongly favored basketball. Italians showed the opposite tendency, choosing basketball only rarely but strongly inclined toward football. Natives participated most heavily in football and "other sports," but very infrequently in track.

The pattern in sports choices is clear and illuminating. Members of different ethnic groups made definable distinctions among the sports offered at school. In many ways, they divided the sports among themselves as each group chose several of the sports categories and bypassed the others. We can only guess at how these preferences were established. Track, a highly individualistic sport that required little team cooperation or body contact, may have served as an ideal outlet for blacks against whom discrimination would preclude strong group involvement. Native men may have been particularly drawn to football with its collegiate aura. It is possible that the example of some sports hero, like Jesse Owens or Red Grange, helped to orient different groups to sports in a selective manner. Once the preferences were set, however, they most likely defined a clear status and prestige hierarchy that in turn differentially attracted members of various ethnic groups.

The strong symbolic meanings that divided men among the sports apparently did not affect women. Sports never played the ethnically differentiating role for women that they did for men. Women too tossed a basketball, and many were involved in "other sports," but there was less ethnic clustering and few marked group preferences or patterns. In the end, this may indicate that women simply did not invest sports with the social meaning that men often attached to these activities.

In general, ethnic patterns were less sharply etched for women. Nevertheless, certain preferences were notable. Jewish women tended to elect literary activities of all kinds across schools. In the sample as a whole, only Jewish and native women were significantly involved in the category of "other publications," which included the myriad literary activities other than the newspaper and senior yearbook. In every school in which there was a meaningful number of students involved in these publications, Jewish women were disproportionately active. Similarly, wherever women were editors of the student newspaper, they were almost certain to be Jewish. At Theodore Roosevelt, all three female editors were Jewish; at George Washington, both women were Jewish; at New Utrecht, the one female editor was Jewish; and at Evander Childs, two of the three women editors were Jews. Only at Bay Ridge High School, where Jewish women were a very small group in a completely female school, was there a much broader ethnic distribution. It is perhaps significant that at coed schools where women were forced to compete with men for the editorial posts, Jewish women were almost invariably selected. Whether women in other ethnic groups were reluctant to compete with men or whether men tended to stereotype women ethnically, Jewish women alone of all female groups held these posts consistently and widely. Overall, Jewish women were 60 percent of all female editors, but only 45 percent of the female population, a disproportion that would have been far greater if the special case of Bay Ridge were excluded (see Table 2).

This special interest in literary activities by Jewish women also found expression in participation on newspaper staffs. German and native women too were quite active on the newspapers, but Italian, black, and Irish women were consistently underrepresented, as they were on the "other publications."[18] In the 1930s and '40s in New York, literary activities for women took on a distinctly ethnic cast in the high schools. Jewish men were active in literary activities as well but were far less conspicuous and far less consistently involved than Jewish women.[19]

The yearbook staffs seem to have been an exception to the ethnic participation patterns in literary activities. Jewish women were not as conspicuous on the staffs of yearbooks as they were in other literary endeavors, while Irish and native women were far more heavily represented in this than in other literary activities. Native women, and especially native men, tended to be disproportionately active on yearbook staffs, far more active than any other group including Jews. Irish men also expressed a unique interest in the yearbook, quite unlike their usual reticence to join publication activities. A good part of the reason for these differences resulted from the fact that the yearbook had less to do with literary than with the social interests of students and played a strategic role in senior class activities. As the recorder of senior events and the publicity vehicle of dominant senior personalities, the yearbook was a political tool, a purveyor of status rather than aesthetics.

In fact, student politics drew upon a different constituency than literary activities. Overall, Irish men were the most disproportionately represented among presidents of the senior class and student body. Eleven percent of all male presidents were Irish, although the Irish made up only a little more than 4 percent of the male seniors. Native men were also very active, and German men, usually least active of all, appear to have been especially drawn to student politics. Native and German men were nearly twice as likely to be presidents as was warranted by their proportion of the male population. Jewish men, on the other hand, were somewhat underrepresented among student presidents, and Italians had fewer than one-half of the presidents warranted by their numbers. No black ever achieved the coveted position as principal student representative.

TABLE 2. Participation of Jewish High-School Women in Literary Activities by School and Activity

School	Total # Jewish women	% of women who are Jewish	Activity[a]			
			Editor	Other news	Other publications	Yearbook
George Washington	(525)	40.4%	(2) 100.0%	(19) 38.3%	(4) 80.0%	(9) 20.0%
Evander Childs	(858)	46.8	(2) 66.7	(23) 57.5	(20) 58.8	(31) 57.4
Seward Park	(770)	75.9	—	(29) 85.3	(25) 86.2	(20) 76.9
New Utrecht	(1026)	60.7	(1) 100.0	(73) 79.3	(33) 73.3	(55) 76.4
Theodore Roosevelt	(575)	43.4	(3) 100.0	(21) 63.7	—	(41) 50.0
Bay Ridge	(88)	6.4	(1) 16.7	(8) 13.3	(4) 18.2	(15) 9.7
All Schools	(3842)	44.8	(9) 60.0	(173) 56.0	(86) 63.2	(171) 39.5

[a] In each column, the number in parentheses represents the number of Jewish women in the activity in each school during four years of my sample. The percentage is the proportion of all women in each activity who were Jewish.

The extraordinary showing of the Irish is somewhat misleading. In fact, it was the natives, not the Irish, who dominated presidential offices. Whenever natives composed a significant part of the school population, they disproportionately controlled presidential offices, except at the High School of Commerce. It was the fact of three Irish presidents at Commerce, without any native presidents, that exaggerated the Irish presence. Elsewhere, while the Irish were very active in politics, as evidenced by their frequent election to "other political" offices, natives dominated the highest offices. At George Washington, four of five presidents were natives, although natives were only one in five senior men. At Evander Childs, where natives composed one-sixth of the male population, two of six presidents were from that group. At Theodore Roosevelt, one-half of all male presidents were native, but they were less than one-fifth of the male population. If we think of the president as a symbol of aspiration, a kind of beau ideal, the conspicuous position of natives among presidents becomes more explicable and significant. Only at Seward Park and New Utrecht, in each of which Jews were more than 50 percent of the population, were Jewish men consistently chosen. All the male presidents at Seward Park were Jews, and at New Utrecht, located in a heavily ethnic neighborhood which grew from the outmigration of Jews and Italians from the Lower East Side, four of the five male presidents were Jewish. In these two schools, the president was of course a direct expression of electoral realities, since Jews were a clear majority of voters. But the office was also more than that. If the presidency was symbolic, as I have suggested, then the pattern at these schools indicates an alternative social environment and another ideal. Both Seward Park and New Utrecht were heavily ethnic schools, with Jews being the dominant ethnic group. Within an overwhelmingly ethnic setting, Jews set their own standard of success, as natives did in schools that were middle class and where status was defined by them.

When we move from politics as an expression of prestige and a symbol of aspirations to politics as an endeavor—from the presidency to "other political" offices—the picture changes. The Irish remain very heavily involved in the lesser offices. Fifteen percent of all Irish men held some political office, compared with slightly more than 10 percent of the Jews, natives, and Germans. Irish men, not especially active in clubs generally, tended to gravitate toward political activities in high school, choosing political office over many other kinds of endeavors. Irish women were likewise more frequently elected to minor political office than either Jewish or native women. Neither Italians nor blacks were represented up to their proportion of the population.

Politics was an arena in which students were forced to seek and to get peer approval. Native men seem especially to have benefitted from this, at least at the highest political level in schools not dominated by a large Jewish population. One other measure of popularity and esteem was contained in the category "celebrity status," which was not an activity but an expression of school prominence. A celebrity could be the man or woman chosen most likely to succeed, prettiest, handsomest, best athlete, best musician, etc. Although some of these designations suggest special talent, they all depended finally on prominence in school affairs and required peer approval.

Among female celebrities, Irish and native women were far ahead of women from all other groups, while Jewish women were frequently selected at only one-half of their proportion in individual schools and were represented at just 66 percent of their population overall. This was the case despite the strong involvement of Jewish women in a wide range of extracurricular clubs and activities. Black women were never chosen. Among men, natives found the most approval with two times the proportion of celebrities as their population warranted. Jewish men did better than Jewish women but lagged behind natives. Italian men, like Italian women, were favored in only one school, George Washington; while Irish men, unlike Irish women, were uniformly underrepresented. Two black men made the list.

The discrepancies between male and female celebrities, most obvious between Jewish and Irish men and women, are revealing. Men appear to have been accorded celebrity status for their achievements, such as sports, politics, and editorships, which explains the fairly good showing of Jews and even the special instances of black success. Women appear to have been differently evaluated; often, given the specified categories, on measures of beauty, grace, popularity. Irish, native, and, to some degree, German women, not Jews, Italians, or blacks most consistently embodied idealized versions of these attributes. In other words, if this designation was anything more than a quirky and humorous yearbook game, Jewish men appear to have better approximated peer criteria of success than Jewish women. Irish and German women did far better approximating a female ideal. But, overall, native men and women were the most popular. (Table 3 compares the celebrity status of four ethnic groups.)

The achievement of Jewish men in the extracurricular realm was impressive, but they were especially prominent in academically related activities. Most conspicuously, Jewish men dominated the science clubs. Only native men also expressed a disproportionate interest in science clubs (see Table 4). This pattern in the sciences was repeated with only small variations in the category of "other academic clubs." Jewish men were the most active overall and in five of six schools. Native men participated substan-

TABLE 3. Celebrity Status by Sex, School, and Ethnicity in Percentages

	Jewish		Italian		Irish		Native	
Men	Population	Celebrities	Population	Celebrities	Population	Celebrities	Population	Celebrities
George Washington	39.2	40.9	4.0	4.5	4.8	0.0	21.0	45.5
Evander Childs	39.7	50.0	19.5	0.0	4.8	0.0	16.7	0.0
Seward Park*	—	—	—	—	—	—	—	—
New Utrecht	55.9	73.7	24.2	21.1	1.2	0.0	7.8	0.0
Theodore Roosevelt	32.9	31.8	23.3	13.6	5.3	4.5	17.3	45.5
Commerce	20.2	29.3	17.8	14.6	11.7	2.4	18.6	29.3
Total	44.9	40.4	17.4	12.8	4.4	1.8	13.4	29.3
Women								
George Washington	40.4	27.3	4.3	9.1	2.8	22.7	20.1	18.2
Evander Childs	46.8	14.3	15.1	0.0	1.5	0.0	20.0	57.1
Seward Park*	—	—	—	—	—	—	—	—
New Utrecht	60.7	47.4	15.7	10.5	0.7	0.0	9.9	31.6
Theodore Roosevelt	43.4	20.0	20.5	5.0	2.9	10.0	14.5	30.0
Bay Ridge*	—	—	—	—	—	—	—	—
Total	44.8	29.4	15.9	7.3	2.7	10.3	16.4	29.4

* At these schools, celebrities were not indicated in the yearbook.

TABLE 4. Participation by Men in Science, Other Academic Clubs, and Arista by Ethnicity, Across all Schools

	Chemistry	Physics	Other sciences	Other academic	Arista	Percent of All Men
Jewish	(46) 54.8%	(30) 60.0%	(58) 52.7%	(288) 52.7%	(273) 56.4%	44.9%
Italian	(6) 7.1	(3) 6.0	(11) 10.0	(81) 14.8	(59) 12.2	17.4
Black	(0) —	(0) —	(1) 0.9	(4) 0.7	(3) 0.6	2.0
Irish	(2) 2.4	(0) —	(0) —	(10) 1.8	(11) 2.3	4.4
German	(4) 4.8	(2) 4.0	(5) 4.5	(22) 4.0	(16) 3.3	5.3
Native	(14) 16.7	(8) 16.0	(17) 15.4	(72) 13.2	(62) 12.8	13.4

tially, while the Irish, Germans, and Italians were only weakly involved. Black men were least active everywhere.

The representation of men in Arista, the National Honor Society, brings this pattern home. Arista was not a voluntary activity; students were honored by election to Arista on the basis of academic record. But the parallel between academic standing and personal choice among the activities makes clear how cogent the club choices of high-school students could be. Only Jews were elected to the honor society disproportionately to their numbers in every school and by very wide margins.

Whatever it was that drew Jewish men toward science and other academic clubs—college and professional ambitions, cultural preferences, or possibly the safety of association with members of their own group—did not do so to nearly the same degree for their sisters. The number of women in the physics clubs was very small, but Jewish women failed to participate strongly in either physics or chemistry, and they showed only a weak interest in the "other science" clubs. Irish and native women showed a strong interest in chemistry and also made an impressive showing in the other sciences. Jewish women were also far less active than Jewish men in the "other academic" clubs. Italian women, on the other hand, made the strongest showing overall, and the special interest of Italian women in academic clubs, in sharp contrast with Italian men, requires some explanation. That explanation may lie in the selective attendance (and graduation) of Italian women in the 1930s and '40s. Unlike other groups, such as the Jews and the Irish who began to send female children to school much earlier and kept them there longer, Italian preconceptions about a woman's place and the limited expectation of women's ambitions[20] meant that only the most academically inclined and ambitious attended high school at all, and still fewer graduated. Those who did may have chosen academic clubs as a further expression of their seriousness of purpose and possibly even to legitimate their extracurricular participations to themselves and to their parents.

As important were the likely economic pressures on Italian men which prevented their participation. Italians were, next to the blacks, the poorest of the groups in New York during the 1930s and '40s.[21] Often required to work part-time and after school, young Italian men had little leisure time to engage in many extracurricular activities and probably had less time than their sisters who were more sheltered and quite likely were also from more economically privileged families. When forced to choose who among their children would graduate from high school, poor Italian families chose one of their sons. That is clear even from the larger percentage of male Italians than females in four of five of the coeducational schools. The high-school attendance of daughters was already a mark of improved social and economic position.

Even though Italian women were more conspicuous than Jewish women in the academic clubs, they were rarely as often admitted to the Arista rolls. Jewish women were everywhere the most frequently and most disproportionately elected to the honor society. Only at Theodore Roosevelt did Italian women gain admittance over their expected numbers, and at Roosevelt they did even better than Jewish women. Thus, while Jewish women, like Jewish men, were academic performers, they were unlike Jewish men in not being as conspicuously drawn to academic and science activities outside the classroom—Italian women were, despite their lower academic standing.

These differences alert us to an important fact. A discussion of ethnic preferences must not disintegrate into a series of stereotyped profiles of groups, nor to the related conclusion that club choices were an expression simply of ethnic culture.[22] Ethnic men and women often behaved differently. Native women were much more likely to join Jewish men in the chemistry club than Jewish women. Jewish women were not drawn to academic clubs to the same degree as Jewish men or Italian women. Ethnicity at school was often differently expressed by men and women and this suggests that whatever culture students brought to high school, it was shaped and refashioned in very specific ways.

In general, however, especially if we look at the schools individually, men and women from the same ethnic groups tended toward the same choices among the performance clubs—orchestra, glee club, and drama. There were exceptions, but the similarity in the choices of men and women *in each school* appears more striking than any marked consistency in ethnic choices across schools. This was so probably because dramatic and musical performances were social events as well as arenas for aesthetic expression and the resemblance between men's and women's choices within ethnic groups suggests that these clubs and activities provided social arenas for students. "Social activities" was an extremely vague category because it took in a large number of different activities, but it nevertheless provides some confirmation of this pattern. In social, unlike academic, clubs men and women of the same ethnic groups tended to make very similar decisions. At Theodore Roosevelt, Seward Park, and Evander Childs, male and female patterns ran a parallel course; men and women from the same group being either strongly or weakly involved. Only George Washington, of all the coeducational schools, proved to be an exception to this pattern. Social clubs, like dramatic and musical activities, provided the occasion for active and ethnically biased contacts between men and women.

The service category was the weakest ethnic differentiator, and not coincidentally, this was the activity least dependent on peer acceptance or approval. Service did not require a heavy commitment of time or a demonstration of strong interest. Usually rendered during a free period in the regular student schedule, work in the dean's office, on the projector squad, or any one of the myriad other services students performed was least peer intensive and peer dependent. It was in service activities that students tended to gain most of their extracurricular points. Since Jews engaged most extensively in extracurricular activities in general, it is not surprising that Jewish men and women were the most active in service in most schools, but every group of men and women was overrepresented in service in at least one school. Blacks often made quite a good showing in service, and this strongly confirms what might have been expected: in their desire to participate in school affairs, black men and women often chose just those activities that involved few group events, little teamwork, and few potentially exclusionary practices by other students.

The absence of a strong relationship between service activities and ethnicity places the other patterns into even sharper relief. Certain of these patterns are especially notable. Despite the generally high level of Jewish participation, Jews did not gravitate equally to all parts of the extracurricular network. This was especially clear in sports, in religion, and in politics, where Jewish men rarely made a conspicuous showing. Instead, Jewish men moved into literary activities, science and academic clubs, and service. The opposite tendency is evident among the Irish who were rarely engaged in scientific, literary, or academic clubs, except on yearbook staffs, but were very active in politics, religion, and many of the sports. Italians also chose religion, but much more rarely politics. They were much more selective among the sports, choosing football above all others. The Germans were less selective than the Irish, participating more broadly in literary and academic clubs without marked prominence, but like the Irish, they chose politics frequently. Black choices were the most limited of all—track, basketball, service, and to some extent the orchestra. Native men were least restricted in their choices and were very frequently overrepresented in a wide range of clubs and activities, but they were particularly conspicuous in the most prominent positions and those that were socially most strategic—the presidencies, editorships, and the yearbook staffs. And they were most often chosen as school celebrities.

The pattern among women was less sharp but still revealing. Jewish women were almost as active overall as Jewish men, but far less prominent among the celebrities. Very disproportionately involved in Arista, Jewish women bypassed the academic and science clubs to choose literary activities consistently. Native women also chose literary clubs, but not to nearly the same degree. When they did, native women, unlike Jewish women, emphasized the yearbook. Native women were also very active in the performance clubs and in social activities, which may explain their special prominence among the celebrities. Weakly involved in most areas, Italian women made a clear and specific decision to join academic clubs and religious activities. Irish women were far more dispersed among the activities than Irish men, but like Irish men, they chose politics and religion frequently. Especially unlike Irish men, they were active in the science clubs. Whatever the reason for these strong and clear expressions of preference, men and women from various ethnic groups made definable choices in selecting extracurricular activities, choices that describe a complex and busy social system in which ethnicity affected what students did and how they viewed each other.

It is important to remember that the ethnicity we can today trace with some difficulty was for young men and women visible, palpable, and meaningful. It helped them to define who they were, where they belonged in the extracurricular world, and where others were in that world. It not only set groups apart and provided individuals with effective networks of peers but also established a competing universe with hierarchies of power and status which provided tangible lessons in Americanization. This was, in many ways, the core of high-school assimilation, a process defined not merely by incorporation and cultural diffusion but also through the processes of differentiation, stratification, and group identification.

III

The patterns of ethnic group preference describe important differences in the experiences of students at school. But students' choices were always influenced by the specific school context. It is more difficult to describe individual schools numerically because the number of students in each activity was often too small to permit a statistically compelling analysis. Nevertheless, the differences among school environments are important to any substantial understanding of what ethnic choices meant. Therefore, even the small numbers are suggestive.

As environments in which the second and third generations learned about America, the schools and the activities provide an important focus for understanding the complex experience of assimilation. As such, the seven schools can be crudely described as illustrating three different paths to assimilation: schools where native patterns dominated, schools in which one ethnic group was especially powerful, and finally schools in which no one group seemed especially in control. This tripartite division only begins to suggest the intricacy of school life as it was lived. But it provides some basis for defining the nature of assimilation in New York in the 1930s and '40s.

Manhattan, symbolic center of New York's urban primacy, had surprisingly few comprehensive high schools in the period between the wars. School populations were often carved up along vocational lines. But two such schools, Seward Park High School and George Washington High School, illustrate the enormous range of the borough's social experiences. Dissimilar to each other in almost every other way, both nevertheless provided a social context within which blacks found it easier to participate in activities than elsewhere. In contrast, the experience of Jews in these schools was very different.

Situated at the very top of Manhattan Island, George Washington was located in a luxurious building in a prosperous and solidly middle-class neighborhood of up-to-date apartment buildings, although the school drew from a larger and somewhat more heterogeneous area.[23] Yearbook pictures documented the well-to-do appearance of students who were usually elegantly dressed, many of whom wore fashionable furs even during the depression thirties. Paul Robeson, Jr. (son of the famous American actor and singer) was among the 120 blacks who graduated from George Washington in this period, and black students, like students from other groups, were among the most economically privileged of their community. Seward Park, on the opposite end of the island, was unlike George Washington in almost every respect. Drawing its students from the tenements and alleys of New York's lower East Side, the classical American ghetto-slum, students at Seward Park could also depend on the similarity in the economic circumstances of their families whatever their ethnic origins. Poor Italians, Jews, Germans, and others attended Seward Park in the twenties and thirties, and gradually in the forties and after, they were joined by more and more blacks and Hispanics.

The most heavily Jewish of the seven schools (74 percent), Seward Park was in many ways a Jewish city. No other group had even 10 percent of the remaining population, although the Italians came close with 9 percent. Instead, at Seward Park, Germans, natives, Italians, and blacks contributed small spices to a homogeneous stock. (The Irish population was so tiny as to be inconsequential.) At Seward Park, black men played football as well as basketball and track, and they were more disproportionately active overall than either Jews or Italians. Moreover, blacks held minor political offices and were even duly represented in "social activities." The participation of blacks ought not to be exaggerated. They were scarcely, if at all, represented in a wide range of areas, especially literary activities, academic and science clubs, and drama and performance groups. But they seem to have been welcomed to an extent that was unusual in New York schools in the thirties and forties. Black women were even more widely involved in activities. They were proportionately the most active group of women. Unlike black men, who were totally absent from academic activities, sciences, and Arista, black women were represented to some degree in each of these. For black women certainly, Seward Park proved to be a hospitable environment for the expression of a broad range of interests and talents. Although the activities of black men were still typed, largely sports and service-oriented, black women appear to have been no more excluded or channeled than any other group.

The experience of blacks at Seward Park is illustrative of the mixed ethnic character of the activities. Most clubs were ethnically heterogeneous, although, of course, most were also overwhelmingly composed of Jews. Jews seem, in fact, to have participated less vigorously at Seward Park than in most other schools. Jews obviously dominated the activities in sheer number, but this did not exclude members of other groups from participation. At Seward Park, Jews were the host group and seemed to welcome other groups. The one exception was politics. Jews were politically very much in control at Seward Park. Their numerical superiority showed itself in this one activity which explicitly represented power and which required election by peers. In a Jewish school, Jews could depend on other Jews for very large voting majorities. All presidents, male or female, at Seward Park were Jewish. Of all male groups, only Jews were disproportionately represented in political office-holding.

It is worth thinking about the lackluster performance of Jews at Seward Park in the context of the record Jews made elsewhere. Bearing in mind that Seward Park, unlike other schools, catered to an overwhelmingly working-class population—and its poorest portion which had not yet made the trek to satellite areas of Brooklyn and the Bronx, it is still revealing that where the Jews were most at home, they were least competitive. Jews could assume their social acceptability (and political control) at Seward Park, and in that context they seem to have exerted themselves least. At the same time, their overwhelming presence did not exclude other groups, including blacks, from active participation in the extracurricular life of the school.

Jewish experience at George Washington was very different. Jews were disproportionately active in general and especially conspicuous in certain areas, orchestra, drama, "other political" offices, and the whole range of academic clubs. But the strenuous involvement which seems on the surface a demonstration that Jews had arrived may well have been the reverse. If Seward Park provides a yardstick of Jewish activity in a largely Jewish context, then, even given the large class differences, Jewish hyperactivity at George Washington may well suggest a kind of restlessness produced by a lack of manifest status and assured social position. Significantly, Jewish men were underrepresented as presidents and as editors of the newspaper, the most prominent positions a student could hold. In both, Jews took second place to native men who captured far more presidencies and editorships than warranted by their numbers. Unlike Seward Park, where Jews were the occupants of these positions but were not conspicuously active in most other activities, Jews were active strivers at George Washington, but they failed to capture the positions with the most power and prestige. Native men held these positions to a very marked degree.

Native men did well generally in most activities at George Washington, but their prominence was even more real than is apparent from a quick perusal of their overall level of participation, since they were especially strong in certain areas of strategic and visible importance—presidencies, editorships, yearbook staff, news staff, football—and they claimed almost one-half of all celebrities. The conspicuousness of native men in these areas, despite the fact that they were only half as large a group as the Jews, suggests a great deal about the relationship between ethnicity, prestige, and power at George Washington.

What Jewish men could not achieve at George Washington, Jewish women apparently could. Of all female groups, only Jewish women ever became either president or editor in chief of the newspaper. But, while Jewish women, like Jewish men, were broadly active, they lagged far behind native women in the significant areas of political office-holding and the news staff. In the academic areas (Arista, "other science," "other academic"), Jewish women tended to be way ahead of all other groups. This did not seem to enhance their popularity, since Jewish women were significantly low in their achievement of celebrity status (27.3 percent of celebrities, 40.4 percent of the population).

Jewish restlessness and achievement in extracurricular life as well as in academic activities seem to have been especially strong at George Washington. In part, this was a function of the largely middle-class composition of the population—it was full of students whose parents had already gained considerable economic success. This may also explain the strong showing made by Italians. Usually a quiescent group elsewhere, Italian men at George Washington were unusually active. Their choices were selective to be sure, but the heightened level of activity by Italian men is notable nevertheless and underscores the importance of the class structure of schools in affecting participation among certain groups. Italians completely, and uncharacteristically, ignored football at George Washington. Instead, they concentrated on social activities, the yearbook, orchestra, glee club, track, and "other sports." Italian men and even Italian women avoided the academic clubs where Jews were extremely dominant. The weak showing made by Italian women in academic clubs may possibly have reflected the very prominence of the Jews. Academics, formal and informal, at George Washington were even more than elsewhere an arena for Jewish achievement.

George Washington, like Seward Park, was a school in which all groups were actively involved in selected activities. At the same time and despite their very large portion of the population, Jewish men could not capture the most prestigious extracurricular slots. Both facts may be explained by George Washington's prosperous population which at once provided students from all groups with the leisure, money, and encouragement for extracurricular participation and accorded power and status to natives. It is revealing that native men, who were one-fifth of the population at George Washington, captured nearly one-half of the celebrity spots. George Washington illustrated an archetypical pattern of assimilation in which standards were set by natives who held the most visible campus offices and were selected as representatives of student values and ideals.

The experience of students at Seward Park is less easily definable. Most students at Seward Park were Jewish. That they were not unduly active in most clubs hardly affected the social environment of these activities or of the school. Jews were influential by their sheer numbers, and Jews did, of course, hold the important political offices and run the newspaper. The Jewish presence did not seem to dampen the enthusiasm of other groups for participation, but it did mean that Jews, not natives, set the standards for other Jews and possibly even for other groups. The behavior of the forty-one native men and forty-six native women (in four years) could not have meant much in a place like Seward Park. But Seward Park also represented a form of Americanization, although one in which it was possible for men and women to go through adolescence and graduate from high school without making significant contact with natives or members of other

ethnic groups either in class or out. As significantly, for non-Jews at Seward Park, Jews not natives defined the host society. For students at Seward Park, American urban culture and assimilation was a very different
experience than for those who attended George Washington. Both, of course,
were thoroughly exposed to American values and ideals in the classroom,
but neither the meaning of those values nor their practice in the context
of daily school experience was the same for students from the two schools.
Class obviously made an enormous difference, but the experiences of these
students cannot be easily reduced to class. They must instead be described
as alternative forms of acculturation that depended on a combination of
demographic and ethnic factors as well as class.

George Washington and Seward Park capture two different geographic
and economic corners of Manhattan. Evander Childs and New Utrecht
were suburban. Situated in the Bronx and Brooklyn respectively, these
schools serviced two of the many satellite immigrant communities growing
up all over the greater city in the 1920s and '30s as the second generation
fled from the older Manhattan ghettos. Largely white and lower-middle to
middle class, New Utrecht in Bensonhurst and Evander Childs in the Pelham Parkway section of the Bronx received many of their students from
the apartment buildings and two family homes developed by shrewd builders (many of these second generation Jews themselves) who had sensed the
aspiration of the second and third generations for both fresh air and prop-
erty.[24]

At Evander Childs, Jews and Italian newcomers (44 percent and 17 percent of the senior classes, respectively) met a large contingent of natives
(19 percent) in an area previously dominated by natives. Also present were
small groups of Germans, Irish, and an even smaller number of blacks.
New Utrecht was less complex and more Jewish. It had no blacks, few
Irish, and less than 10 percent natives. The Jews were a substantial majority with 58 percent of the population; the Italians the largest minority (20
percent).

The Jews were very active in both schools, but while they dominated at
New Utrecht, they were far less prominent at Evander Childs. Italians also
had different experiences at the two schools. At New Utrecht, Italians were
underrepresented in the social world of politics, dramatics, publications,
and the yearbook. Jews controlled these activities. Italian men tended to
cluster in the glee club, religious clubs, and football, and they showed an
unusual interest in academic clubs. They were joined in these activities by
Italian women who also joined religious clubs and academic clubs in very
disproportionate numbers and substituted "other sports" for the male interest in football. Despite the fact that "social activities" had too few participants at New Utrecht to be meaningful, the patterning of extracurricu-

lar clubs effectively describes a social world in which Jews exercised power and enjoyed prestige. Although Italian men, and especially Italian women, did join Jews in some activities, Italian and Jewish separation suggests both marked distinctions in choices and the probable exclusion of Italians from the most sensitive political and social areas. This conclusion is amplified by the fact that natives had less trouble joining Jews in politics, on the newspaper staff, and on the yearbook staff. Not surprisingly, Jewish men at New Utrecht dominated the celebrity categories. Even at New Utrecht, however, Jewish women were denied celebrity status commensurate to their numbers while native women, here as elsewhere, represented ideals of beauty and popularity.

Italians also showed no special prominence at Evander Childs, but they engaged more extensively in sensitive areas and were the most disproportionately active of all groups in the category of social activities. At Evander, Italian men also captured one of three editorships (one to a Jew, another to a native), and one of six presidencies. They were disproportionately represented on the yearbook staff. Blacks, on the other hand, were ranked at the bottom in almost every activity and were entirely absent from most. Other than in service, where 75 percent of all black men were involved, and in track, which absorbed one-half of all the black men, they were scarcely visible at all in the extracurricular world of Evander Childs. Physically absent from the New Utrecht population, blacks were effectively absent from the social world of Evander Childs's adolescents as well.[25] Suburbs within cities as well as outside cities have their racial boundaries.

In contrast, native men were in every part of Evander Childs's extracurricular world. They were especially active in all publications, the presidencies, other sports, football, Arista, other academic clubs, editorships, as well as social activities. In other words, the involvement of native men was both far-ranging and intense. Jews also fanned out into most activities, but they rarely dominated them as they did at New Utrecht. As was true at George Washington, Jews were bested by natives for the most prestigious posts, the presidencies and editorships. Even though there were more than two times as many Jewish as native men at Evander Childs, there were twice as many native presidents and as many native editors as Jewish editors. In "other political" offices, there were nine natives to only seven Jews. Native women did equally well in politics. Although Evander Childs had no female presidents, native women were the only group of women disproportionately represented in the "other political" category.

Jewish women at Evander took first place in a long list of positions and activities, including editors in chief, Arista, "other science" clubs, social activities, "other publications," and "other news."[26] As they had at George Washington, Jewish women did relatively better than Jewish men, and they did so in a similar social context—a middle-class school with a substantial native population. Native men tended to capture and hold strategic positions, and Jewish and native men appear to have been in continuous competition with their interests similarly focused. For whatever reason, while Jewish and native women were also active in similar activities, Jewish women were usually more active and more readily assumed prominent posts. Despite these achievements, native women overwhelmed Jews in celebrity status.

The differences in the experiences of Jews and Italians at Evander Childs and New Utrecht seem to have had less to do with the economics than with the demographics of the schools and their surrounding neighborhoods. Both were largely middle class, and neither was under special economic duress. But Evander Childs, like its Pelham Parkway neighborhood, was more recently developed and changing rapidly as it became increasingly populated by newer ethnic groups.[27] Between 1933 and 1945, the proportion of natives in the senior class at Evander was cut in half, from 26 percent to 14 percent, while the proportion of Italians more than doubled from 10 to 23 percent of the population. This growing Italian presence may help to explain the substantial participation of Italians. Sensing their developing role in the school, Italian men moved more smoothly into the social life of Evander Childs than they could or were allowed to at New Utrecht. In contrast, Italians were a constant minority at New Utrecht, with Jews a slim but clear majority. The other groups were too small to matter. In that context, Italians were an outgroup, and their status in the activities reflected that position. While Italians were active in a broad range of clubs, status and influence were exercised by the Jews who were heavily involved in most activities and controlled the newspaper, yearbook, and the presidencies. Seventy percent of all male celebrities at New Utrecht were Jewish. At Evander Childs, Jews had to compete with a large and active native group and a growing Italian population. In that context, Italians were not an outgroup, but only one of several minorities. This kind of complex and changing ethnic situation was also part of Americanization in the city's schools. It affected not only the experience of growing up in the neighborhoods but also the structure of social relationships in the schools. Young men and women from the city's ethnic groups often reacted as much to each other in their development as they did to any certain and stable native norm.

At Theodore Roosevelt High School in the Bronx, Jews experienced an even sharper set of constraints than at Evander Childs. An ethnically mixed school with a substantial Jewish minority, Theodore Roosevelt had very clearly defined patterns of ethnic participation. Despite its relatively small Irish population (4 percent), Roosevelt's location in the old Irish bailiwick of Fordham Road (directly across the street from Fordham University) meant that the extracurricular world at Theodore Roosevelt reflected the specific ethnic pressures of its location. During the 1920s, '30s, and '40s, parts of the old Irish neighborhood rapidly filled with newer immigrants, especially Jews and Italians, and the whole area was witness to the heightened friction between the Irish and Jews in the context of the depression and the special pressures of the city's political coalitions. Anti-Semitism became a familiar experience for Jewish youths who were frequently harassed by the Irish on the streets. Even the churches became embroiled in the controversies.[28]

At school too, Jewish men appear to have been on the defensive. At Theodore Roosevelt, Jewish men were far less prominent in the extracurricular realm than elsewhere. Again and again at Roosevelt, the small group of Irish men and the larger group of native men made a remarkable showing in the activities.[29] The strength of their combined influence may have intimidated Jews, or more likely, the Irish and natives in control of strategic areas of the extracurricular realm actively excluded Jews from participation. It is significant that so many Irish and a good many native men belonged to religious clubs at Roosevelt. Elsewhere, religious clubs were largely female preserves, but at Roosevelt, one-fifth of all the Irish men and one-tenth of all native men were members, as well as almost one-third of all Irish women. Italian membership, significantly, was very low. Religion may have become especially important at a school like Theodore Roosevelt as the Irish were forced to define themselves in the context of a growing group of Jews.

The Irish and natives were the most active groups in general, and both groups were especially prominent in politics, in "other political" offices, as well as in the presidencies. Native men were three times as likely to be presidents as was warranted by their population and twice as likely to hold other political offices. The yearbook staff, which as I have suggested had strong social and political possibilities, also had a large, disproportionate Irish and native presence. The Irish were particularly, and unusually, active on the newspaper staff as well as in dramatics clubs, a situation unlike that of most other schools, where Jews tended to be dominant in both these activities. These powerful areas, which often underwrote prominence in school affairs because they were highly visible, were strongly Irish and native at Roosevelt. Election to the presidency highlighted this pattern. Of six male presidents, three were native, one Irish, one Jewish, and one Italian. The Irish and natives together controlled two-thirds of the male presidencies but were scarcely one-fifth of the population.

Jewish men were active in some of these areas, but they were in each instance less active than the natives and the Irish and considerably less visible than elsewhere. Jews, Italians, and also the Germans tended to be strong in areas where the Irish and natives were conspicuously absent or only slightly interested—the orchestra, academic clubs, and the "other sciences." Indeed, Italian men made an unusually strong showing in academics at Theodore Roosevelt, taking first place in Arista as well as in "other academic" clubs. Italian women, like Italian men, showed real strength academically at Theodore Roosevelt. They took second place to German women in Arista and were also well represented in "other academic" clubs.

Although the pattern of ethnic exclusiveness and separation is not as clear for women as for men at Roosevelt as elsewhere, there remain strong indications of a prestige hierarchy in which the Irish and natives were dominant. The one woman elected president was native, and other political offices were disproportionately in the hands of Irish women. Jewish women differed from Jewish men in joining Germans and Irish in dramatics and not joining Jewish men in the orchestra. Above all, of the three women who became editor in chief of the newspaper, all were Jewish, and Jewish women were much more prominent than Jewish men on the news staff. The marked preference of Jewish women for literary activities is nowhere better illustrated than at Theodore Roosevelt where Jewish women, but not Jewish men, dominated most literary activities. Although not quite so consistently or strongly as men, there was also some tendency for Jewish, Italian, and German women to cluster in activities not interesting or important to Irish and native women. Overall, however, Jewish women were more active than Jewish men. Jewish men apparently, far more than Jewish women, felt the brunt of the power of the natives and Irish in the extracurricular world of Theodore Roosevelt.

Theodore Roosevelt was the school in which Jews, men especially, had the most difficult time. Unlike most other schools, the Irish were extremely prominent and in tandem with a large group of active native men effectively stymied the extracurricular ambitions of Jewish men.[30] This was in part an expression of demographic realities. Jews tended to be most prominent in extracurricular activities where they were a large part of the student population, as at New Utrecht. But this could not be the whole reason. Jews were an even smaller population group at the High School of Commerce which had even more Irish than Roosevelt. But at Commerce, Jews were the most active group and held some of the most coveted positions. A larger set of neighborhood issues, of which the specific demography of the school was an expression, probably held the key to Jewish experience. High schools exist within the broader context of the neighborhoods they serve. In the 1930s and '40s, they reflected not only the economic realities of those locations but also their special social and cultural conflicts. At Roosevelt, the Irish made a much stronger showing in activities just as they were a powerful presence in the community. Whatever the high school's role in the assimilation of immigrant youth, schooling never operated in isolation from the other pressures young people experienced at home and in the streets.

We have thus far examined only the situation at schools that were co-educational and which reflected the specific social and economic realities of the communities they served. But New York was full of special schools of all kinds. Some of these were sexually restricted, and many were vocational and technical. Bay Ridge High School and the High School of Commerce were two such schools. Although quite different from each other, both were sexually exclusive. Bay Ridge was and remains a female school. Commerce went coeducational during the war but was a men's school throughout the thirties and most of the forties.[31]

Located in Brooklyn's southwest corner, Bay Ridge High School drew its students from a fairly wide geographic area. While it was noted for its academic excellence in the 1930s and '40s, Bay Ridge's most prominent feature was social rather than academic. As a woman's school, Bay Ridge was considered "safe," a factor of some consequence for parents, many of them first or second generation immigrants who hoped to protect their daughters from daily associations with men.[32] This seems to have been especially important to Italians, who sent large numbers of their female children to Bay Ridge, but other groups, like the Scandinavians, also responded to Bay Ridge's appeal. Bay Ridge was unlike the other seven schools, not only in being exclusively female, but because it was also largely non-Jewish. Jews composed only about 6 percent of the population. Italians with 31 percent and natives with 26 percent (the school was situated in a heavily native enclave) were the two largest population groups.

At Bay Ridge, Italian women participated more widely and actively in extracurricular activities than elsewhere, a fact that may explain why Italian women appear to have been reticent to participate in coeducational schools or at least restricted their participation largely to academic clubs since at Bay Ridge extracurricular activities did not result in coeducational socializing. At Bay Ridge, Italian women were unusually active in politics and drew one of four presidencies. Italian women also landed two of the editor positions; two others went to natives, and one each to an Irish and a Jewish woman.

At the same time and despite their small portion of the population, Jewish women were very active in a large number of activities and were especially prominent in literary activities. Indeed, Jewish women appear to have expressed their literary interests regardless of specific environment. Italian women, on the other hand, were not especially drawn to any of the literary activities—yearbook, news staff, or other publications.

Native women at Bay Ridge were also very active and were especially prominent in literary activities and the performance clubs. But it was the Italians, not the natives, who dominated the social activities. Part of the explanation for the conspicuousness of Italians in the social activities may lie in the unisex character of Bay Ridge and therefore the "safe" quality of these activities. But a good part of the reason may be that a club name can hide its real purpose. At Bay Ridge, Italian women, together with Irish women, dominated the social activities to a degree which reminds us of their strength in religious clubs elsewhere. Bay Ridge had no clubs that could be designated religious. It is certainly possible that at least some of the social clubs at Bay Ridge were religiously oriented and served this purpose for Catholic women.[33]

Bay Ridge was hospitable to all the ethnic groups. Indeed, its most prominent characteristic was the extraordinarily high, practically universal, participation of students in extracurricular activities. Despite some clear preferences among the activities, the differences among the ethnic groups seem less marked than the universal participation. This is well illustrated by the experience of blacks. The number of black women was tiny, only four, but they were involved in a surprisingly wide range of activities—in sharp contrast to the experience of the small numbers of black men at Evander Childs and Theodore Roosevelt. Bay Ridge appears to have provided a uniquely mixed environment. Ethnically heterogeneous, the school provided all groups with substantial access to the activities and encouraged an exposure of members of different groups to each other in the clubs. It is significant that Bay Ridge had only women students. Women, as we have seen, tended to demonstrate fewer sharply defined ethnic patterns. In addition, at Bay Ridge the social functions of the extracurricular clubs in the dating-and-rating games of adolescence were missing and therefore also some of the reasons for ethnic associations.

The High School of Commerce, like Bay Ridge, was a unisex school. It had an even more ethnically balanced population. Of the 1138 senior men, 20 percent were Jewish, 18 percent Italian, 5 percent black, 12 percent Irish, 7 percent German, and 19 percent native. Unlike Bay Ridge, Commerce was a vocational school, one of the many located in Manhattan. In fact, many seniors indicated that they planned to go to college and into the professions by the late forties, and some had specific plans to attend City College. Still, by emphasizing business, Commerce's population was more skewed than that of most high schools in this sample, and the school gave specific terminal training in commercial skills. The unusual proportion of Irish men and the substantial number of blacks may have been related to this fact. As we shall see, Catholic students unable to compete for places in Catholic high schools were often sent to public vocational schools (Chapter 6). Similarly, blacks may have found the security of a business education attractive. The proportion of blacks increased over time, from just 2 percent in 1933, to 6 percent in 1939, and finally to 13 percent in 1947. The Jewish population trend was in the reverse direction, declining from a high of 31 percent in 1933 to 15 percent in 1947. Jews were clearly finding the opportunities provided by a commercial high school less obvious, while blacks found them more attractive. One cannot discount the strong possibility that as blacks increasingly attended high school, they were directed into vocational programs like the one available at the High School of Commerce.

The extracurricular pattern at Commerce demonstrates and amplifies the tendency for Jews and natives to take the lead in student activities. Except for sports, the orchestra, and social activities, Jews were almost always the most active group. Half of all the editors were Jews, and two of the nine presidents, although both the Irish with three and the Germans with two did better proportionately. Most surprising was the Jewish absence from the science categories. The tiny number of science club members and the business orientation of commercial students are two of the possible reasons for this. Jewish students with clear academic and professional interests probably went elsewhere than to Commerce.

Among the other groups, the Irish showed considerable activity. In addition to their control of one-third of the presidencies, the Irish were heavily involved in "other political" offices, on the yearbook staff, in other news, and in all the sports. But the Irish did poorly on various measures of academic interest—Arista, "other academic" clubs, "other publications." Indeed, Irish interests and avoidances in general are well illustrated at Commerce. Wherever the Irish attended in any number, they concentrated in sports, politics, and on the yearbook staff, as well as in social and religious clubs. They rarely took an interest in the sciences or "other academic" clubs and usually made a poor showing in Arista. This pattern was related to the peculiar pattern of attendance of Irish Catholics at public high schools (see Chapter 6).

The Germans, like the Irish, were very selective in choosing activities at Commerce, as they were in most schools, and they tended to cluster very markedly. Lowest ranked of all the groups in general activity level and in service, they were prominent among presidents. Completely absent from basketball, they were especially active in "other sports." Germans ignored the orchestra, but chose the glee club and especially drama. For Germans at Commerce, this marked clustering seemed more significant than the choices themselves.

Natives joined Jews at Commerce to take the lead in campus activities. In most instances they trailed the Jews in the degree of participation, but they were more broadly active. Jews and natives were designated celebrities to about the same extent at Commerce, and this was unlike the usual native preeminence in schools with a substantial native population. Above all, natives were largely absent from the newspaper staff and had no editors and no presidents. In these categories of symbolic prestige, the absence of natives is curious.

Perhaps the most reasonable explanation for this anomaly lies in the different ambitions of students attending the High School of Commerce from those attending the usual comprehensive high school. Unlike most high schools, Commerce, at least in the thirties and early forties, had few students with college plans, and thus the extracurricular world did not serve as leverage for college entrance. This raises questions about the extracurricular patterns in general that deserve some attention. To what extent were the patterns of extracurricular participation largely determined by differential college goals among ethnic groups? Certainly, by the 1920s college extracurricular enthusiasms began to spill over into the high school, and those students most imitative of college patterns would be most readily drawn in. More significantly, as colleges imposed a variety of restrictive admissions policies, they began to evaluate students according to nonacademic or marginally academic criteria.[34] Among these, the demonstration of initiative or leadership potential or special talent as evidenced by extracurricular performance would capture the attention of admissions officers. Especially attractive therefore for the college-oriented were the plums of the extracurricular arena like editor in chief of the newspaper or a presidency. In this context, the special propensity for extracurricular participation of Jewish students and those of native background becomes more comprehensible. These students were most likely to have college plans or ambitions. Their prominence in the choice positions of the extracurricular world as well as their prominence on the Arista rolls substantiates this.

It would be a mistake, however, to attribute the complex patterns in extracurricular activities to selective college-going ambitions of different ethnic groups alone. These ambitions could intensify the pattern and might explain certain features of it but would be inadequate to explain the complex patterns we have been finding. Far more students participated in activities than could or would attend college. Moreover, the diversity in school experiences and the specificity of choices made by different ethnic groups cannot be understood by reference to college ambitions. One example will suffice. Although both editorships and presidencies were prominent and attractive positions, native men were much more likely to be presidents, while Jews were more frequently editors. Rather, the elaborate patterning of extracurricular participations in New York high schools must be understood at least in part as the effects of ethnic preference and evidence for the continuing significance of ethnic group association at school.

As the child of immigrants, Leonard Covello understood how much school life was a shared group experience. "Whatever problems we had at school or in the street, we never took up with our parents. These were our personal problems to be shared only by companions who knew and were conditioned by the same experience. How could parents understand? Parents belonged in one of the many separate watertight compartments of the many lives we lived in those days."[35] Certainly the experience of the children was unlike that of their parents, but they shared that life with others of their own group and high-school students still lived in a world strongly shaped by ethnic bonds. Even at school, where assimilation was an educational objective, among adolescents who could be expected to view their parents' old-fashioned world with disdain or pain, ethnicity was a palpable experience.

IV

The ethnic experiences of high-school students were not the same as those of their parents. But student experiences were deeply influenced by the specific features of their schools and their neighborhoods. A Jew at Theodore Roosevelt did not have the same experience as one at New Utrecht or at George Washington. An Italian at Evander Childs had different American experiences than an Italian at New Utrecht. The specific mix of ethnic groups, the neighborhood context, the size of the native population, as well as traditions specific to the school's history all influenced the nature of high-school extracurricular and social life.

The evidence also suggests that ethnicity often affected men and women differently. Two things are especially notable. First, women sometimes made different choices than their ethnic brothers. Italian women placed a heavy emphasis on academic clubs, a choice more rarely made by Italian men. Jewish women chose literary activities more consistently than Jewish men but hardly ever joined Jewish men in the science clubs. Irish women chose science clubs, while Irish men almost never made the same decision. These were strong variations, and they remind us that ethnicity, like culture in general, is not homogeneous but operates in a socially differentiated universe and may mean and imply different things to each sex. When confronted by the American school environment, the sons and daughters of immigrants responded to the opportunities offered by their new environment in different ways.

Second, ethnicity seems not to have been as consistently expressed by women as by men. That is to say, we can see the ethnic patterns across schools and within schools much more clearly if we look only at men. Male ethnics divided their activities more regularly among themselves, as each group emphasized different kinds of interests. Sports participation is a good and symbolic illustration of this. Certain ethnic women, such as the Jews and the Irish, seemed always the most active wherever they were present in number, while others participated less often and in less varied ways, but rarely was the sharply competitive and exclusionary world visible among male ethnics as clear for women. Bay Ridge, an all female school, illustrated this greater homogeneity well. Participation in extracurricular activities and successful competition especially may have had status resonances and possibly relationships to future goals that appear not to have influenced women to the same degree as men. Some of this was probably the result of differences in college-going plans between men and women. The Jews and the Irish, women with a strong orientation to teaching, engaged more broadly in extracurricular clubs in general, and this tends to support this conclusion. But women may have been more accepting of other groups and less exclusionary in general.

The class composition of a school population also mattered. Except for service activities, the ability of students to engage in many extracurricular activities was dependent on the time available to students after school hours. Some groups, like Italians and blacks, probably had less leisure because they were poorer. At a prosperous school, like George Washington, Italians and blacks were far more active across the board than at a lower-middle-class school like New Utrecht.

It is also important to remember that even when activities were ethnically stratified, students from different groups did meet. That mixing was most notable at a school like Bay Ridge where four years of the yearbook staff, for example, introduced fifteen Jews, thirty-one Italians, nine Irish, eight Germans, fifty-three natives and thirty women from other ethnic groups to each other. But it was true almost everywhere that, except for blacks, students from different groups elected to participate widely in a range of activities. Since I have made conclusions about ethnic participation on the basis of group disproportions, it is easy to overlook or discount the degree to which *individuals* from all the groups made contacts with those from other groups in the social life of the school. In fact, meeting individuals from other groups may have been an extremely powerful experience for fifteen-, sixteen-, and seventeen-year-olds, whose previous contacts with students from other backgrounds was probably slight. Even at Theodore Roosevelt, Jews met Irish, as well as Italians, Germans, and natives on the yearbook staff. These exchanges were defined by considerations of status, and even hostility, as well as shared interests and friendship, but that only meant that the activities reflected the larger realities of American society.

In this sense, the high schools and the social and extracurricular activities exposed students to various critical features of American social and civic life. This was certainly what theorists and administrators had in mind when they developed the activities as allies in socialization and Americanization. But Americanization and assimilation were never neat or uniform. At different schools, students began to grasp the complex features of the society differently. While they met students from other groups, they did so in ways that were mediated by ethnic bonds and various kinds of stratification. The prestige of natives, the ambition and drive toward success of Jews, and the exclusion of blacks were variously experienced, and these introduced students to the broad features of American urban life in which ethnicity was as much a part as voting, caucusing, and the ability to change one's name. The cliquishness and selection of friends, the preference among activities, and the inclination to attribute status, popularity, even beauty to some and not to others was a fundamental experience of adolescents in school and out. Those who theorized about the potential of the extracurricular activities in socialization were correct to this extent, although they could not have foreseen all of the consequences.

After all the complex and varied differences in students' experiences are considered, one is left with the sharp and clear impression of a high-school society divided along ethnic lines. On the simplest level, this meant that it was more likely for men and women of the same group to associate together in performance clubs and social activities at individual schools. Beyond that, groups became identified with different talents and characteristics. Some groups, notably the natives and Jews (and the Irish and Germans in politics and sports) usually captured the limelight and strategic posts. Certain activities, like science, track, and football for men and literary clubs for women, were disproportionately selected by members of some ethnic groups rather than others. Wherever females were appointed editors, they were almost certain to be Jews. In schools with a native population, males of this background were most likely to represent the school and the senior class. Religious clubs were almost exclusively composed of the Irish, Italians, and natives. The track team was likely to contain a good portion of the male black population. And almost everywhere, Jews were the academic achievers, a fact that must have been repeatedly brought home as the newspapers published Arista lists and as the long rosters of Jewish-sounding names were read in the common assemblies.

The data do not allow us to speculate with any real insight about the sources for the diversity among groups. It would be tempting, but unwarranted, to ascribe the differences to cultural traditions pure and simple.

Certainly the strong preference of the Irish and Italians for religious clubs, for example, points to the strength of Catholicism in even a secular setting. But it would be a mistake to reduce all the choices to inherited and tenacious traditions. It is also important to avoid a too easy stereotyping of the groups as they adapted to American circumstances, stereotypes that can be attached to the Jewish interest in science and the Irish fascination with politics, for example. In this context, it is useful to remember how frequently men and women from the same group differed, a difference especially notable between Italian women, who consistently joined academic clubs, and Italian men, who did not; but also clear in the choices registered by Jewish and Irish men and women. We simply do not know the degree to which these patterned choices were circumstantial, functional, or traditional. That is to say, it is possible that Jewish women initially chose literary activities merely because they were imitating what they understood to be native patterns rather than expressing in secular form the long-standing Jewish regard for the book hitherto largely confined to men. Italian men who chose football may have been doing just the same, based on their selective perceptions of native patterns and the factors underlying campus popularity. Once the patterns were established, members of ethnic groups may have been drawn because of associational rewards and because these choices themselves had become traditional within the school context rather than because of some clear-cut immigrant or ethnic preference. The regularity across schools suggests a strategic interaction between inherited traditions (possibly different for men and women), which shaped perceptions initially, functional patterns of adaptation, and forces of imitation in an environment which encouraged and rewarded certain kinds of imitation. Obviously the rewards were broad, because the choices made by various groups differed a great deal. If imitation of natives was important, as I believe it was given the extraordinary popularity of natives registered in celebrity status, the dispersal of natives among almost the full range of extracurricular activities suggests that different groups could and did attach themselves to different areas of the school social world in the process of assimilation. What the data describe is a complex and pluralistic range of ethnic expressions which point to patterns of second generation accommodation (i.e., ethnicity) rather than to some exact translation of immigrant traditions.

As significantly, the patterns in activities permit us to observe a deeply divided social universe. In fact, the channeling and clustering is far more consistent and more strongly etched than any special set of choices. Ethnicity provided an important form of differentiation and association in the schools in the thirties and forties, and it was a potent ingredient in the status and prestige hierarchy of most schools. Ethnicity seems to have endowed students in high school with a continuing source of identification within the mass culture and the impersonality of the schools.

The extracurricular activities were neither coextensive with the students' social world nor even a very large part of it. Their significance lies in revealing a very small fragment of an active, competitive culture which drew heavily on who students were before they came to the classroom and affected who they would become after they left. In connecting school activities to ethnicity, the data are significant far more for what they suggest than for what they say. The school, as it was envisaged by educators and often imagined by historians, was never as powerful an integrator, equalizer, or socializer as it has been portrayed. Important and powerful as it was, it was one of many institutions operating consecutively and concurrently on students' lives. The young brought to school as much as they took away, or rather, what they brought gave meaning to what they learned. Ironically, the high school not only did not destroy the preschool associations of students, it encouraged, supported, and thereby strengthened them.

The data also suggest something about the culture of schooling, by which I mean less the specific forms and mores of school life than the range of resources made available for personal and expressive development. It is hard, if not impossible, to reconstruct what difference belonging to one kind of club or another may have meant to the young women and men who proudly announced these affiliations. It is also difficult to say what influence these may have had in later life, and it would be foolish to assume that all those who belonged to science clubs went on to careers in science and medicine or that football players worked on the docks. Yet one cannot ignore altogether the patterned ethnic variations uncovered in this study. Clearly some students in the 1930s and '40s had more cultural options open to them, more avenues for self expression, than others. We can catch only a fleeting glimpse of this from the data, but it is a lasting impression. A black student, in even the best of circumstances, found a more impoverished and restricted culture at school than a white student, especially if the latter was of native descent. A Jew, in most places, found a richer set of contexts for personal development and more social approval than an Italian. In other words, the culture of the school, like the culture of which it was part, operated selectively on the groups who entered the schools, and it accorded success more readily to some than to others.

To say this is not, however, to argue that some groups were more or better assimilated by the schools than others—for example, that Jews were more assimilated than Italians or Germans, or any variation on that theme. The widespread absence of blacks in many activities may be something of an exception, since it appears to have registered some sort of exclusion. But to stress better or lesser assimilation is, I think, to misunderstand ethnicity and certainly to ignore the findings of this study. It is important to avoid the progressive fallacy that is especially pervasive in educational history and also frequently the wolf among the lambs in an analysis of immigration. In education, we are all more or less progressive, since we assume that doing well academically is a positive objective, and therefore anything that inhibits this—race, sex, class, ethnicity—is an impediment to progress. The virtue of looking outside the classroom is that in charting the significance of these factors on activities they become not obstacles to achievement but indicators of legitimate differences less easily defined as good or bad. As such, neither the extent of activity nor any one kind of activity can be taken as a measure of assimilation. Instead, the choices among the activities indicate different expressions of assimilation and represent varying strategies of assimilation.

By the time they graduated from high school after twelve years of schooling, second and third generation men and women had absorbed great quantities of American culture, beliefs, habits, and attitudes. They absorbed them in the classroom and from each other outside the classroom. But the experiences of these New York City youths illustrate just how complicated assimilation was. For the almost gothic structure of extracurricular participation must be understood as the fine tracery of evolving ethnic (not immigrant) preferences and the continuing significance of ethnic group association within an assimilationist environment. Indeed, this kind of association was part of the process of assimilation as it was experienced historically at a particular juncture in the development of twentieth-century society.

The culture of the school provided students with different options and rewards in that process. For while the range of activities was wide, it was not infinite, and there were clear hierarchies of prestige. By reinforcing only some cultural or functionally adaptive patterns and not others, the schools helped to shape these patterns among the groups. Thus the high-school culture acted as a filter, letting through some qualities from among its diverse population and not others. In so doing, the school reinforced, even accented, these patterns. In that sense the very process of acculturation which the school successfully pursued in teaching students about American manners and values was acculturation with a significant by-product. In blunting the edges of the ethnic cultures from which the students came, it set certain aspects of those cultures in high relief—accentuating the Jewish drive in science and literature, Irish political acumen, or the competitive brilliance of black track and field athletes. It identified certain groups with particular talents and did so in a socially approving manner. It shaped how members of one group saw members of other groups. What this meant was that selected features of a complex ethnic culture were singled out as all right while others were either ignored or specifically disapproved. This re-created ethnicity at the same time as it incorporated various ethnic talents into the American mainstream. The schools thus provided different groups with rewards for remembering as well as for forgetting aspects of their past.

Viewed in this way, assimilation was not simply a one way process, and it cannot be easily separated into a set of staged progressions in which the progeny of immigrants become first culturally more and more like natives and then integrated into a society defined once and for all by a native majority. Even so finally textured a sociological theory as that of Milton Gordon misses the complex and dynamic cultural process of which assimilation was a part. In an ingenious theoretical insight, Gordon proposed to divide assimilation into two phases, an initial acculturation and subsequent assimilation.[36] But Gordon's hypothesis, by adopting an Anglo-conformist and largely middle-class perspective, might be useful for a school like George Washington but meaningless at Seward Park or even New Utrecht. Separating culture from society may be useful heuristically, but it has neither the feel of reality nor the sense of history. The high-school graduate of an immigrant family was not like his parents, as Covello and his friends knew well. That difference must be understood as part of the experience of assimilation which is not a clearly ascertainable end product but a process defined by the strains of cultural change and adaptation. Having participated with others like himself and unlike himself in the classroom and in the school, the high-school graduate had participated in a strategically assimilative environment. That this environment was colored and shaped by ethnicity meant that it participated in the complex realities of American urban culture. The student, the high school, and the culture of which the school was a part as well as an anticipation had been changed in the process.

II

Other People, Other Schools: Race, Sex, Religion, and American Education

4
New Day Coming:
The Federal Government
and Black Education
in the 1930s and '40s

We have seen a lifting of the horizons of our youth through increased educational and recreational opportunities. Such Federal programs as student aid under NYA, Adult Education under WPA and the vocational training available in CCC camps brought the light of training to thousands of Negroes whose own economic resources would have held them in darkness, ignorance and dependence.

MARY McLEOD BETHUNE (1939)[1]

Why, then, does the Army trouble itself with the problem of illiterates? Why not permit them to remain in the deferred status? . . . The fact is that deferment on account of illiteracy has aroused considerable resentment in certain sections of the country. There are towns where all the physically fit literate young men without children have gone, leaving the illiterates to engage in their usual occupation or to engage in idleness. To the American, this is *unjust*.

"Upgrading the Illiterate Registrant
For Use By the Army" (1943)[2]

Race had often been an issue in school discussions in the early twentieth century, but largely as this word was attached to immigrants. The special concerns of blacks had rarely been thrown into the pot of problems that schools were asked to solve. That, as much as anything else, suggests the degree to which black Americans were social outsiders. Although educators had begun to question the equal educability of some immigrants in the 1920s, the incorporation of the second generation by the schools, both academically and socially, had always been a central school objective, and the problem of the second generation underwrote the reconstruction of secondary education in the twentieth century. Blacks had not figured in these changes. Blacks, too, attended high schools in New York City and elsewhere, but their numbers were small, and they were treated by fellow students largely as they were treated by educational policy makers—as if they hardly existed. In fact, nationwide in the 1930s and '40s the level of

education achieved by blacks only infrequently brought them into the pur-
vue of the high school. This is not surprising, since before World War II
three-quarters of all black Americans lived in the South where educational
facilities were segregated by law, and very few high schools were established and maintained for black students. As a result, nationwide, black
educational attainment lagged behind that of second-generation immigrants by three to four years for men and two to four years for women.
As many Americans would discover during the war, many blacks were
totally or largely illiterate.

Blacks did not enter the sphere of educational discussion in an important way until the federal government began to exercise significant authority in educational matters during Franklin Roosevelt's presidency. Even
then, however, the problem of black education did not result in consciously designed policies. Rather, a new awareness of black educational
needs and their social implications came in unexpected ways and through
nontraditional institutions, as a result of the operations of relief organizations and through the armed forces. As the crises in the economy and in
defense exposed the special deficiencies of black educational preparation
and the social costs of those deficiencies, the federal government was forced
to act in limited ways that would have fundamental consequences both for
the future of black educational reform and for subsequent patterns of federal intervention in matters of schooling.

The opening of schools to immigrants had never resulted in a federal
policy or significantly disturbed national politics.[3] What schools had
achieved, as well as the manner in which they had been transformed in the
context of the masses of the second generation, was a local and state concern. The nationwide scope of the changes were monitored by professional
organizations of teachers, principals, and superintendents (increasingly
centered in the National Education Association) that developed with the
extraordinary expansion of schooling and the heightened self-awareness of
school professionals at all levels. Indeed, immigration had been fundamental to the professional development of education. In responding to the new
urban populations and the pedagogic challenges posed by the second generation, educators had fashioned the instruments and concepts of twentieth-
century education. Progressives had, of course, been keenly aware of the
political dimensions of education and had insisted on the necessary relationship between education and social reform. But neither progressives nor
educators who saw themselves above politics envisaged a social policy that
required active federal participation. On the contrary, federal intervention
or a national policy for educating minorities, or anyone else, was actively
despised by most educators as an unwarranted and illegitimate intrusion

beyond the bounds of the Constitution and in opposition to the democracy and efficiency believed to reside in local control.

When the federal government did finally direct its attention to the schools, it came at just that point when optimistic expansion and growth—the fuel of professionalization—could no longer be sustained. In the 1930s and '40s, in the midst of depression and war, the federal government shed an unflattering beam of light on the state of American education that profoundly challenged the heady self-congratulations of the educational community. As the federal government moved to remedy the educational deficiencies of Americans, it directed its attention to those who previously lay so far on the social periphery and so distant from traditional sources of power that they had hardly provoked notice or attention. In so doing, it created a new awareness of the problem of black education, not as a pedagogical concern, but as a social issue. The particular circumstances of federal intervention, centered as it was on issues associated with poverty and national security,[4] would indelibly mark future federal policy on education. So too, the depression, the New Deal, and the Second World War made the unequal burden of educational deprivation borne by blacks visible, problematic, and unforgettable.

I

It is tempting to see the new federal role in education during the New Deal as inevitable; another instance of the aggressive march of state influence into new social spheres for which the period is remembered. In fact, the New Deal's unprecedented educational programs were not the result of policy or of intention but usually of fortuitous and partial decisions whose purpose had little if anything to do with education. Looking back, we can see clear patterns, but at the time, educational policy was undefined, and almost all the new initiatives disappeared after the economic emergency gave way to war. At the same time, the particular association created during the New Deal between poverty and educational deprivation and the specific discovery of black exclusion from educational progress left indelible mental blueprints which by the 1960s created the new genre of educational understanding which underlay a deliberate set of federal activities.

On the eve of the Great Depression, Herbert Hoover appointed a National Advisory Committee on Education whose report pinpoints the pre–New Deal understanding of the role of the federal government in education. The report issued by the committee in 1931 provides an unusually good framework for understanding the departures the New Deal would shortly initiate and the revised understanding to which those changes would lead. In keeping with the self-congratulatory spirit of three decades of school expansion, the committee proclaimed the American system of schooling as "without a peer." "In responsiveness to popular sovereignty, in adaptability to varying need and aspiration, and in richness of experimentation conducive to flexibility and to progress, our management of public schools is without a peer. Certainly no national system of public schools managed in a highly centralized spirit shows such substantial democratic qualities." The American system of education was great because it was democratic, and it was democratic because it was responsive to local needs and free of central control and direction. Even in the twenties, however, the increased financial needs of schools propelled educators to make demands for the resources of the central government, and Hoover's committee, in line with those demands, urged that federal funds be used for educational purposes. In large part, this resulted from the recognition by committee members, who were overwhelmingly educators themselves, that school funds were unequally distributed regionally and among the states. Thus, while the Hoover committee was thinking in national terms, indeed hoping to equalize national resources to smooth out differences within the national system of education, its members expected to prevent centralized control and initiative and to sharply circumscribe the federal role. "The American people are justified in using their federal tax system to give financial aid to education in the States, provided they do this in a manner that does not delegate to the Federal Government *any* control of the social purpose and specific processes of education."[5]

The call for federal tax dollars was new and would become louder as the depression proceeded to flatten local sources for school financing, but the Hoover committee stood by the firm belief in the superiority of a decentralized educational structure in which the federal government played no role in policy or programs. The Hoover committee expressed the overwhelming consensus of opinion of those who were thinking seriously about educational issues by proclaiming its sacred faith in local control. The report took great pains to describe at length, and often with sentimental flourishes, the traditional roots of American localism in education and the fundamental contribution of localism to democracy and citizenship. It also clearly distinguished between legitimate and illegitimate precedents for federal aid to education, rejecting what it saw as the "growing trend toward federal centralization," contained in legislation like the Morrill and Smith-Hughes Acts.[6]

While the National Advisory Committee on Education attempted a nervous balance between exclusive local control and federal "cooperation" in finances, a far less audible plea was relegated to the last few pages of the report. Here a minority report raised an issue which had been largely ignored by the majority. Issued by three presidents of Negro institutions of higher education, the minority report asked that the federal government assume the "moral obligation which binds a central government to exercise special solicitude for disadvantaged minorities." While carefully worded to ask for assistance "in full accord with the principle of State autonomy," the plea by Presidents John W. Davis, Mordecai W. Johnson, and Robert R. Moton for "some definite increase in the per capita amounts and in the percentages of State support made available for Negro education," was, in fact, a challenge to the report as a whole.[7] Where the majority had located educational inequities in state and local resources, Davis, Johnson, and Moton introduced a wholly different kind of educational distinction into the discussion, a distinction which concerned not the providers of education but the recipients. The majority had gloried in the democracy of American education and hoped to remedy its incidental deficiencies, spots of local poverty, unequal state resources, and geographically unequal economic development. The black members of the committee were proposing that Americans were being unequally educated not only because some localities could not afford the best education but also because groups and individuals were being excluded from its equal benefits. In that context, the federal government had an obligation to provide directed assistance. Davis, Johnson, and Moton made clear that in the case of blacks, the historical experience of the limitations of local action and the unequal state of black education, apart from regional sources of poverty, demanded federal consideration.

The minority report represented a fundamental challenge to the traditional perceptions upon which the report was organized. The majority had largely ignored the issue of minority education and had raised the matter of black education only to dismiss it as one of the "perplexing problems" whose solution "might appear to be hastened by the Federal Government" but which was, in fact, only an "imperfection" resulting from "the political, economic, physical and social conditions often surrounding them [blacks]." Private charity, not directed government action, would solve the problem of black status and with it the issue of educational disadvantage. This view not only protected the democratic character of American schooling by ignoring the issue of segregation but also, and as significantly, saw educational advancement for blacks as the product of and not the stimulus to social change. Black Americans, the majority report noted, had already made an "impressive advance" and would continue to do so.[8] Throughout the twenties, schools had been adapting to their varied populations, providing new programs, and tailoring curricula to the perceived needs of students. Equal opportunity meant largely the opportunity to get the schooling one needed. It was in this light that the National Advisory Committee on Education saw the issue of black schooling. Black education would develop as the needs of black people developed. Schooling was not a force for social reform, and equal opportunity for blacks in education could not substitute for or produce social equality.

Throughout the 1930s, the National Education Association (NEA), the most powerful, vocal, and largest professional association, would stand by the views of the Hoover committee. Emergency conditions would make federal dollars seem all the more necessary as local school resources diminished or dried up, but the NEA never altered its fundamental commitment to unconditional tax dollars, and the various NEA investigative committees, educational coalitions, and its strategic Educational Policies Committee stood firm on this traditional line.[9] The educational profession had gained its self-identity during the expansion of American education and in the absence of federal involvement. School reform, not national reform, had been the issue for educators, and that had been unsponsored by the government. It was in that period with its obvious self-satisfying successes and growing professional power that the NEA formed its views of school progress and of how that progress would naturally continue to evolve.

Moreover, NEA officials, like the majority of Hoover's National Advisory Committee on Education, had no special interest in or sensitivity to the educational problems of blacks. When John Sexon, Chairman of the Educational Policies Committee, addressed the National Advisory Committee on the Education of Negroes, for example, he bluntly began his talk by exclaiming, "I find it difficult to discuss the subject you have asked me to talk upon. I find it difficult to think of the problems of Negroes as being any different from the problems of any other race." He went on to note that in speaking privately "to one member of your race . . . I had this to say, 'I don't know enough about it to be prejudiced about it.' And this person said 'Well you have missed something.' " Sexon's address demonstrated his lack of knowledge, and he proceeded as if he were talking to a white audience, bemoaning the depression's effect on the former opportunities of youth: "Making a living is the big problem of today. . . . they [youth] are learning to accept positions now which are not pointing to the Presidency of the United States or some other great position."[10] Sexon had clearly missed something about black education. Whatever relevance a romantic nostalgia had for the education of the majority of white youth, it had none at all for blacks. That position, with its yearning for the good old days, continued to define the NEA's views on the role of the federal government. Certainly, some educators knew of and were concerned about the differences between blacks and whites, but as an organization of educators, the NEA did not voice their views.

Hoover had not entirely ignored blacks. The National Advisory Committee on Education was graced by the presence of three black educators, and beyond that, Hoover had appointed Ambrose Caliver as a senior specialist on Negro education to the Office of Education. Thus, by the time Franklin Roosevelt came to office, the educational status of blacks, though still a peripheral issue, was already perceptibly on the horizon. But as long as the federal role in education was traditionally defined, as long as blacks were an addendum to a report, or black schooling was relegated to separate volumes of investigation, the problem of black education could not serve as a spur to federal policy.[11] The new understanding that Davis, Johnson, and Moton had introduced needed to be brought from the periphery to the center, where it could redefine the potential role the federal government could play, before any real change could take place. That change required an entirely different set of perceptions than those adopted by the Hoover committee and those that underlay NEA policy throughout the 1930s. It required that educational problems, not successes, take center stage and that the federal government refashion itself into an educational advocate with special responsibility for those slipping through the existent educational net. It moreover required that the federal government not allocate money across the board, as educators demanded, but direct funds selectively to individuals or groups.

There is no little irony in the fact that the New Deal did precisely this. Franklin Roosevelt was no advocate of federal educational activity. He was on record as opposing federal aid to education, and the initial and onetime emergency allocation in 1934–35 to faltering school districts could best be described as stingy in the context of the massive need. Certainly, the New Deal had no educational policy at the outset, and in light of the aggressively expanding administrative apparatus in Washington in the 1930s, the Office of Education remained an insignificant backwater, largely ignored, frequently chastized, and usually despised as a third-rate organization. Shuttled between different agencies in the 1930s and '40s, the Office of Education was hardly part of the dynamic changes introduced in Washington by the New Deal.[12] Even Caliver, whose presence as black advocate might have been expected to elicit some business for this putative center of national educational life, was largely relegated to developing statistics and organizing meetings. Surveys and conferences were the raison d'être of the Office of Education, but for blacks, these held little promise of change. As one black educator noted at a meeting of the Advisory Committee on the Education of Negroes over which Caliver, as usual, presided: "We have been surveyed to death in Arkansas, we know what we are going to find."[13] Serving more as a representative of the Office of Education to blacks than the reverse, Caliver was clearly hamstrung by the limitations of the office where he was a subordinate in an organization with little understanding of black needs. Moreover, Commissioner John Studebaker often showed remarkable ignorance of the particular concerns of blacks. In 1935, he and others from the Office of Education met with members of the Advisory Committee on the Education of Negroes to assure its members that the Office of Education would try to obtain federal financial support without federal controls for Southern states, only to be told by his audience that "the members of the group generally felt that what they do want is Federal Control." It was not the Office of Education with its traditional attitudes that produced the New Deal's fundamental educational initiative, and it was certainly not the office, despite Caliver's best intentions, which would bring black Americans the help they sought from the federal government.[14]

The New Deal entered the educational arena through the back door, as it were, not as an agent of education, but as a dispenser of relief. Throughout the thirties, the Roosevelt administration never overtly questioned the local basis of educational policy or the autonomy of the states in decisions about schooling and did not set out to establish a federal responsibility for education. Instead, in the course of its relief efforts, the New Deal developed educational programs and facilities that paralleled those of traditional educational institutions. Those programs were federally administered and controlled but did not technically interfere with or challenge local and state control over education. In devising and administering relief programs, the federal government not only became an active participant in all phases of social life, including education, but also uncovered basic inequities, inefficiencies, and "perplexing problems" that had been dormant or taboo subjects. In the end, the Roosevelt administration injected the federal government into the educational arena in such a way that it both exposed educational failures and defined their redress as a federal responsibility.

Roosevelt and the relief administrators most immediately involved—Harry Hopkins and Harold Ickes—responded to the school emergency of the depression not by assisting the schools as organizations but by assisting school people and school plants. They did this through a mixed bag of work relief programs, work-study schemes, supplementary social work enterprises, and public works construction and repair projects, organized and administered through FDR's alphabet-soup agencies—the Public Works Administration (PWA), Works Progress Administration (WPA), Federal Emergency Relief Administration (FERA), Civilian Conservation Corps (CCC), and the National Youth Administration (NYA). These separately run agencies, relying heavily on discretionary administrative policies whose purpose was to provide maximum individual relief, were coordinated with a variety of federal departments but almost never responsible to the Office of Education. Thus, to speak of the New Deal's educational activities is both to describe a massive program of improvements—school construction and repair, teacher employment, courses in literacy and naturalization, vocational training and rehabilitation, nursery schools, correspondence courses, educational radio programs, and subventions to high-school and college students—and to describe no educational policy at all. In most cases, (the NYA was in part an exception to this) education was a by-product of work relief, and the educational content and purpose were defined in the course of the agencies' activities by the need to find appropriate employment for teachers, carpenters, masons, students, nurses, and unskilled laborers.

Since many of its educational endeavors were unfocused, the New Deal often discovered its educational commitments in the process of program administration. When the CCC, the most popular of the New Deal work projects, got under way, the aim was to provide out-of-work youth from relief families with immediate employment in conservation work. The expectation was that CCC recruits would pick up what they needed to learn in the process. Despite resistance from CCC director Robert Fechner, it soon became clear that explicit instruction, not only in the technical aspects of conservation but also in basic literacy, was often urgently needed. Additionally, as the officials of the CCC sought to occupy and stimulate camp enrollees in their nonworking hours, they turned to education in subjects such as Latin, mathematics, and history, as well as in vocational skills and literacy.[15] At first these activities were entirely voluntary, but the moral pressure on enrollees to occupy their time usefully made the educational supplements almost as basic to CCC activities as the work regime.

By 1938–39, more than 90 percent of the members of the corps were enrolled in some instruction, averaging four hours per week. Two-thirds of these enrollees were in job-related classes, but one-third were in strictly academic classes. An educational adviser had early been attached to each CCC camp, and it is clear that the camps, by utilizing various local resources, helped to educate thousands of young men, providing many with basic literacy and remedial instruction and some with welcome advanced education. When it extended the life of the CCC in 1937, Congress formalized the educational activities of the CCC by providing each camp with a school building and by increasing specifically educational appropriations. By 1941, credit for educational work completed in CCC camps was provided by forty-seven states and the District of Columbia. The CCC had certainly become the center of a federally administered educational enterprise, but the camps were run by the War Department, with personnel and responsibilities shared with the Departments of Agriculture, Labor, and

Interior and, to only a limited degree, the United States Office of Education.[16]

The National Youth Administration, while more focused in its goals, was even more administratively fragmented. Established in 1935 as an autonomous division of the WPA, the NYA had a clear objective: to permit students in secondary schools and colleges to continue their education by providing them with part-time, often on-campus, jobs as clerks, janitors, and research assistants or jobs on construction projects, on playgrounds, and in nursery schools. The NYA also provided work relief with a prevocational objective to unemployed, out-of-school youths of school age. In 1938, the NYA channeled its grants through 26,751 colleges and secondary schools and reached 368,921 students. At the height of its activities, in 1935–36, NYA provided almost half a million students with various kinds of work-based financial assistance.[17] Organizationally autonomous, though nominally under the WPA, the NYA, according to the 1938 Advisory Committee Staff Study, "has in principle worked in close cooperation with local, State, and other Federal governmental agencies and numerous non-governmental agencies." This close cooperation with traditional school authorities was strongly disputed by some educators who felt they had, in fact, been ignored in both the organization and administration of the NYA. Moreover, until 1940 the Office of Education had no role in its organization or operations.[18] Despite administrative complexity and professional hostility, the NYA was sufficiently able to enlist the cooperation of local school officials to become one of the New Deal's most successful programs, popular with students and the public and effective in terms of New Deal policies whose principal objective was to keep young people out of the labor market.

In addition to the CCC and NYA, the only exclusively youth-oriented programs, the New Deal also provided various educational programs through the WPA. These included worker education, nursery schools, vocational retraining, and parent education. In all these programs, the federal government's stated objective was simply to provide relief funds. It chose personnel on the basis of relief needs but left program content to various professional groups and state departments of education. "Under the Works Progress Administration the emergency education program is conducted on a State basis. This practice derives from the principle of operation underlying all Works Progress Administration policies, which assumes that the determination of the nature and content of the program is essentially a State and local government responsibility." In short, according to the advisory committee that issued this statement, the federal government had no intention of determining educational content. Indeed, it had no policy concerning education.[19]

This was no doubt what Roosevelt wished to believe and was probably initially also true. A glance at the WPA projects makes clear that the programs were carefully designed to provide educational offerings that did not conspicuously compete with traditional school programs or to compensate for cuts made necessary by the economic emergency. In fact, however, this was a less than candid assessment of the impact and consequences, if not the intent, of New Deal educational endeavors. First, the New Deal programs, ad hoc and administratively derived as they often were, made statements about the role of education in American economic life. The programs were all work-coordinated; that is, education in the CCC camps, student supports, and various supplemental programs sponsored by the WPA were heavily job-related. In the process of administering relief, New Deal programs uncovered not only massive illiteracy but also a population with outdated and inadequate skills. The relief projects became actively involved in underwriting a practical vocationalism and helped to define this as a deeply educational issue and responsibility. In so doing, the projects helped to emphasize the value of education in job terms and as essential to economic opportunity in America.

Secondly, the educational programs of the New Deal were aimed at the poor. As the Advisory Committee Staff Study on the WPA explained, "Here, perhaps lies its greatest contribution and its strength. An educational offering of major significance has been made available to the poor and the needy. . . . That there was and is a demand for the services rendered is manifest in the persistence and growth of enrollments. The people can learn; the people want to learn; the people intend to learn. What the regular educational agencies have failed to provide the people have found—in a relief program."[20] The point was clear: education for all was a possibility and an imperative. Only the inattention of traditional educational institutions had failed to awaken people to their legitimate educational needs and to service the needs of all the people. The New Deal programs were at once an implicit criticism of established educational offerings and a demonstration that the federal government could do what established agencies had failed to do.

The criticism implied by an educational agenda for the poor meant more than an extension of education to those previously ignored. The New Deal programs encouraged an awareness of how poverty often underlay inequalities in educational attainment. Before the 1930s, equal educational opportunity was more often a catch phrase for providing people with only as much education as they could use than it was a platform for eliminating inequalities in access to education. But New Deal programs and especially the NYA subsidies provided a challenge to this perspective. As Harry Hopkins made clear in an informal address to NYA state administrators:

Well, I think we have started something. It seems to me that what we are
starting is this: that anyone who has capacities should be in college and
should get a higher education, and that he is going to get it irrespective
of his economic status. That is the crux of the thing, to decide once and
for all that this business of getting an education and going to law school
and medical school and dental school and going to college is not to be
confined to the people who have an economic status at home that permits
them to do it. . . . All this about anyone being able to go to school who
wants to go to school is sheer nonsense and always has been, in my opinion. I grant you there are a few exceptional students who can do it, but
the great majority of people cannot; and anyone who knows anything
about this game at all knows that in the good old days of '28 and '29
tens of thousands of young people were leaving school to go to work for
no other reason than that they were poor. They were quite capable of
going to college, far more so than some of us in this room.[21]

Hopkins was not alone in this challenge to the educational status quo,
nor in his slap at college-bred egos. Fresh from his experiences as chief of
the NYA, Aubrey Williams came to the very center of elite education,
Harvard University, to give the coveted Inglis Lecture in 1940. His address
was on vocational education, but he made clear that "I am not talking
now of vocational training. In fact, I think there has been an overemphasis
in the past on a strict division between vocational and academic education." He proceeded to condemn the idea that education was the preserve
of the gentleman and training that of the laborer. "I do not believe that
our democracy can afford to provide less educational opportunity for any
of our people. On the contrary, I think it should provide more adequately
so that children in all parts of the country, from all races, and all economic
groups, may have the best we know to give them." To do this, Williams
insisted, required that "we find a way to extend to them the opportunity
to work during the period of their schooling."

NYA experience helped to frame Williams's perceptions about practical
solutions to educational inequalities. That experience also resulted in
something more than practical ideas, however. "I can imagine no worse
menace to our democratic tradition than the development of a hereditary
system whereby only the children of the well-to-do might enter the professions and the children of the now less honored should remain permanently
bound to follow in their parents' footsteps." Williams did not offer a uto-
pia in which all would become doctors and lawyers. Instead, "it follows
inevitably that if there is to be opportunity for children of unskilled, manual workers to move into the skilled and intellectual pursuits there must
be an acceptance of the fact that there is no tragedy when the child of a
professional person becomes a factory worker." One can only wonder about
his audience's response to a proposal that would perhaps turn their own
offspring into auto workers and janitors. After this solid slice of educa-

tional radicalism, Williams proceeded to deny, in good New Deal fashion, that the federal government in its programs had intruded in any way in educational policy: "We have purposely left the actual direction of the program in the hands of school people themselves. . . . it has eliminated any possibility of an effort, or even the appearance of an effort, on the part of the federal agency to interfere with the sacred area of educational policy traditionally reserved to the state and local authorities."[22]

Williams was not being entirely disingenuous. The New Deal had left the schoolrooms to the educators. But it had not only provided education outside the classroom, it had also revised the meaning of democratic education at its most basic level. The federal government had made education available to those who had previously been ignored. This political redefinition had once again made education more than a question of classroom learning; it turned education into a vehicle of reform, indeed into a potential instrument of social reconstruction. Whether the source of this radical understanding was the eye-opening experience of a long economic depression during which from one-quarter to one-third of a normally hard-working population was unemployed, or whether the depression and the Roosevelt administration provided a haven for the expression of radical ideas that could not have been voiced in such high places before, the New Deal provided a context in which a new view emerged of the role the federal government could and should play in making education available for all. That view adopted a vision of freely available education that was vastly different than what had been previously understood as the American tradition of public schooling. The New Deal programs exposed not only educational deficiencies but also the social conditions that explained them. In this context, the federal government became responsible for education as part of its newfound obligation to eliminate gross inequalities and social deprivations of all kinds. Once again, education became part of a much larger national picture, too large in fact to remain exclusively in the jurisdiction of the states or in the care of those professionals whose concerns were largely pedagogical.

Finally, the New Deal's educational programs exposed and were attentive to the educational needs of black Americans in a wholly unprecedented way. Much of this attention was the result simply of the discovery of black poverty—a poverty long borne but deeply exacerbated by the depression. But a good part of it was more pointed and explicit as various New Deal officials and members of the black community seized on the opportunity implicit in new federally directed programs and sympathetic government personnel.

Roosevelt initially had no plans or policies to deal with the special needs of black Americans in education or in anything else, and his closest advisors were, as Nancy Weiss has shown, often racist and hostile to black entreaties. By the mid-thirties, however, often through the intercession of his wife Eleanor and in response to the aggressive advocacy of Mary McLeod Bethune, president of the National Council of Negro Women (appointed to serve as head of the Negro Affairs Office in the National Youth Administration by Harold Ickes), as well as by the activist head of the NYA, Aubrey Williams, Roosevelt began to take note of and make provision for the needs of blacks. While blacks had received far less than their fair share of relief in the early phases of the New Deal, they began to be employed in larger numbers on construction projects and in other relief programs by mid-decade.[23] More significantly for our purposes, black schools and colleges received significant federal appropriations, some of them specifically earmarked for Negro colleges in the South. Blacks responded enthusiastically to New Deal offerings. Turning eagerly to the many opportunities for instruction offered through the WPA, they benefitted especially from skilled manpower programs and literacy classes. One hundred thousand black adults were reported to have learned to read and write because of the WPA program.[24]

The experience of blacks with the NYA and CCC is especially instructive because it reveals something of the manner in which New Deal programs operated and the possibilities of federally sponsored programs. At NYA, Williams, attacked as a "nigger lover," saw progress in black people's educational and economic status as one of his top priorities. And Bethune put her considerable energies and keen eye for advancing black interests to work. NYA regulations specifically forbade discrimination in student selection and paid black students exactly what was paid to whites for doing their jobs. As the Final Report of the Negro Affairs Office of the NYA made clear: "The relative extent to which Negro youth shared in the student work program is indicative of the 'equality of opportunity' policy of the agency. . . . the number of young men and women of any minority racial group given aid shall not represent a smaller proportion to the total number aided than the ratio which this racial group bears to the total population of the *school district or state.*" By the time the NYA was dismantled, 300,000 black youth had participated in its varied programs.[25] The NYA also had a special fund to aid "eligible [black] graduate students who cannot be cared for within the quota for graduate aid of a particular institution, after it has made a just allocation for Negro graduates from its regular quota." Set aside for use by blacks only and carefully cultivated by Bethune, this fund was specifically aimed at overcoming the lack of opportunity for professional education for blacks to meet the great need for professionals within the black community.[26] The special fund benefitted over 4000 black graduate and college students.

At the CCC camps, blacks who had initially been vastly underenrolled despite stated Labor Department policy barring discrimination were by mid-decade enrolled near their 10 percent quota, and 200,000 black men eventually participated. At the same time, the CCC program, useful as it was for individuals, had less to commend it as an advance for black equality since blacks were sequestered in segregated camps where educational advisers, but not other supervisory personnel, were black. Black CCC units constantly provoked local opposition, and according to one student of the camps, "in response to any slight pressure CCC camps for Negro enrollees were cancelled or moved."[27] Black leaders consistently supported the camps despite their shortcomings but also expressed their dismay at "the practice of establishing separate CCC camps in states where there is no legislation prohibiting interracial groups."[28]

This paradox—an apparently aggressive program to provide blacks with their due and a program that continued traditional social policies—was thoroughly in line with the New Deal's record in general. The explanation has as much to do with the fragmented way the New Deal programs were organized and run as with Democratic party politics, with its strong southern base, to which the ambivalent policy toward blacks is usually attributed. Since each agency had broad discretionary power, individuals with strong commitments, like Williams and Bethune at NYA, could make special provisions for blacks without forcing a general administrative policy position that would have antagonized the southern bloc crucial to Roosevelt's congressional coalition. Because the agencies provided a wide berth for discretion, positive leadership as well as stand-pat policies were possible. The CCC, run by a War Department accustomed to segregated units and headed by a conservative, found it difficult to give blacks even their due; the NYA, run by Williams and Ickes, both sympathetic to blacks, sought to do more.

A good part of the achievement at NYA on behalf of blacks resulted from Mary Bethune's incessant efforts to defend and extend black interests. In the annual report for the Division of Negro Affairs in 1937, she defined the responsibilities of her office: "To my mind the only reason for the existence of a Division of Negro Affairs and for such a Report is the recognition of the special nature of the problems and difficulties faced by a minority group of twelve million who seek integration into the American program. We believe intensely in the adherence of the National Youth Administration to the Democratic principle of integrating the members of minority groups as completely as possible into the warp and woof of its program." The report went on to list the considerable achievements of the previous year: "The results, generally speaking, have been highly gratifying and have profoundly influenced the promotion of educational opportunity for Negro youth in all sections of the country." Especially noteworthy was the assistance provided to high-school and college students. Of the special graduate fund she observed, it "made it possible for the few universities for Negroes to have strong graduate groups. This fund represented a veritable God-send to Negro graduate students and schools." In Bethune's usual style, the report was hardheaded and hortatory. It used statistics not just to inform (as was the usual case at the Office of Education) but also to encourage directed federal involvement.[29]

Little wonder then that when black educators gathered to discuss federal relief programs, they were eager for more than flat grants to states and sought the very controls and strings that white professionals abhored and the Office of Education disdained. "No developments have occurred since the last conference to cause this Committee to change from the attitude expressed at that time on the matter of federal vs. state control of work projects and relief administrations," members of the Second National Conference on Problems of the Negro and Negro Youth concluded. "On the contrary, experience has shown us that relief administrators in many states and municipalities either lack sympathetic understanding of the problems of minority groups, or are desirous of preventing Negroes from equitably sharing the benefits of relief programs. Under these conditions, it is apparent that federal control of relief funds furnishes the most practical safeguard for the protection of minority groups."[30] Less wonder still that black educators and leaders made strenuous, and at NYA often successful, efforts to have blacks represented on state advisory committees which distributed funds and ran programs on the local level. Williams had promised such appointments, and Bethune worked to see that promise fulfilled. According to the Final Report of the Division of Negro Affairs, "From the period 1935–43, State Supervisors of Negro Affairs functioned in twenty-seven states, and when in 1942 the program was administered on a regional basis, Regional Negro Affairs Representatives served in nine of the eleven regional structures."[31]

Bethune and the Office of Negro Affairs could rightly claim credit for the major advances for black youth as the NYA, in Bethune's words, "tunneled its way into the rural and urban conditions of our country, awakening and inspiring thousands and thousands of youths, opening doors of opportunity." Bethune was always careful to share that credit with the various local and state officials. As she put it in addressing one conference of college and NYA officials, "I am expressing my gratitude. . . . Whatever has happened to the Negro in the forty-eight states, we are responsible. We together have worked." Nevertheless, Bethune did not pretend to an unwarranted modesty: "I have worked and fought with my sleeves rolled up night and day." And she acknowledged her influence with Williams,

"They pretty much felt up there that whatever I wanted . . . let her have what she wants."[32] Williams was not her only resource, however. "I have taken the time to go into the President's office when nobody but the President and God and myself were present, and to pour into his ears the cries, the needs, the desires and the possibilities of the masses of Negroes, shut up in Texas, in Florida, in Mississippi, in South Carolina, and in many other states." While Bethune pled her case in Washington, she kept an eagle eye on what was happening at the local level, and she warned one group of NYA state representatives, "Don't think I don't know when your Negro program is not adequate. Don't think that I don't know when there is just a little makeshift over here and over there. I do know. I am not being fooled at all."[33]

Bethune's broad reach up toward the president and down toward local officials certainly facilitated black progress, and the success of NYA programs for blacks would not have taken place without it. But that success was made possible by the very structure of the relief program, by its ad hoc and experimental nature, and by the fact that it operated apart from more traditional views of education and their professional defenders. Thus, at one point Bethune admonished black college administrators not to be overly suspicious of their students nor stingy when applying relief standards. "Bend backwards," she urged, to include all forms of potential talent, even athletic ability, in their assessments of academic merit: "I never see a child upon a street without thinking, 'That might have been Mary McLeod Bethune.' Sometimes you say you will not give a preacher's child help. You cannot always tell by outside appearances. Let us continue to dig deeply. I want this Committee to think. . . . Sit down to your desks and create things. Make up jobs and put them into motion. Get as many students as you can give jobs to or make jobs for."[34]

The NYA in Georgia had an especially good and broad program for educational assistance to blacks. With eleven blacks employed at the state level and a sympathetic state administrator, D. B. Lasseter, the Georgia NYA put innovative programs—like an educational forum for blacks—into effect. The forum provided a symposium format with lectures open to the public on a wide range of subjects: "Negro Health," "The Negro and the Church," as well as more general academic and public affairs addresses, like "The British Empire," "World Peace," and "Citizenship and Voting."[35] Georgia also promoted informational conferences and studies pertaining to black life, and state NYA publications provided a variety of practical guides to vocational fields for black youth. In a more traditional vein, Georgia had a highly successful school program that assisted even black youth who had never reached high school (as many blacks had not) but "had dropped out of school at the second and third grade level, and

many [who] had never attended school." "With the passing of the National Youth Administration," the final report of the Georgia NYA aptly
observed, "there passed one of the greatest friends of Negro youth in
Georgia."[36]

Georgia's efforts on behalf of black youth were appropriately greeted
with enthusiasm and gratitude by students, principals, and others involved
in the programs. C. L. Harper, principal of Booker Washington High School
in Atlanta, noted that "It is impossible to estimate the value that NYA aid
has been to students in a large public high school such as Booker Washington in Atlanta. Hundreds of our best students would not have been able
to continue their high school education without these grants. . . . These
grants have created confidence and hope in the hearts of these youth and
brought success where failure threatened. . . . It is to be hoped that the
NYA aid may be continued [so] that the door of opportunity may remain
open to thousands of American youth who, otherwise, would not be able
to continue their preparation in school." D. D. Hubert of Morehouse College expressed similar sentiments: "Had it not been for this assistance,
more than one-third of our student body would have found it impossible
to remain in school. . . . A continuation of this kind of help is not only
desirable, but an urgent necessity." And the president of Morris Brown
College in Atlanta added that "My only regret is that the financial consideration granted students was not large enough to accommodate more deserving students."[37]

In line with Bethune's advice that they think hard about finding things
for students to do, Georgia school officials and NYA administrators set
black students to work on a range of activities to earn their stipends—in
nursery schools, churches and Ys, libraries, and laboratories. They gardened for the poor, collected historical data, and created music bibliographies. Georgia also had an effective resident training program where students studied half the day and worked on group projects the other half.
The school was entirely vocational in emphasis and housed 421 black youths
who engaged in agricultural, trade, craft, and homemaking projects.[38]

While Georgia's programs were especially effective and alert to black needs
on all levels, black students throughout the South were given new opportunities to attend or remain in school or to learn new skills. During the
decade, black high-school enrollments rose 126 percent. "The National
Youth Administration," the final report of the NYA proclaimed, "in giving
youth a fair share of its benefits, blazed a new trail of federal procedure—
it gave Negro youth their first real slice of American democracy." NYA
publications were, of course, eager to present the administration's programs in the best light possible, but student responses clearly demonstrate
the degree to which something new had taken place. "I am a poor boy

with no one to help me to go to school," one student noted, "the NYA has made it possible for me to remain in school." Another offered the suggestion that "The NYA should continue always, because it is so much help to the students that are receiving it."[39]

Beyond the specific programs and enterprises of the NYA, the CCC, and the WPA lay a significant change in black self-awareness, aggressiveness, and visibility. The black appointments at the national and local level and the large and vocal conferences which not only applauded New Deal programs but also issued complaints about insufficient efforts and made demands for fuller opportunities provided blacks with a political education in the broadest sense. "These conferences," the Final Report of the Office of Negro Affairs explained, "represented the first time in the history of the government that Negroes from many fields were called together to discuss and make suggestions for problems affecting Negroes and Negro youth, and to gain information on how federal agencies served the needs these problems represented."[40] These experiences inaugurated a new spirit of expectation and propelled blacks toward demands for justice. They also provided blacks on all levels—students, teachers, principals, as well as leaders in the states and in Washington—with lessons on how the federal government worked and how blacks could benefit from its operations. The New Deal programs, in their redefinition of federal responsibility and the new meaning attached to equal educational opportunity, as well as in their posture of receptivity to black demands and interests, provided blacks with new hopes as well as new opportunities; with new confidence as well as new tools.

Above all, the New Deal provided blacks with new visibility. "We are people with a grievance, and we have been invited to the very capital of the Nation and requested to express that grievance," Charles S. Johnson observed in an address to the most successful of the organized conferences, whose roster of attendance was a who's who of leading black Americans. "If there is anybody here today or yesterday or the day before who has not expressed his opinions, it was his own fault. . . . I don't know when I have felt freer than I have here in the Department of Labor under the Administration of this meeting." Eleanor Roosevelt was an invited speaker at this Second National Conference on the Problems of the Negro and Negro Youth (as she was at the first), and the conference report was addressed and delivered directly to Franklin Roosevelt. As Mary Bethune observed in her letter of transmittal to Roosevelt of the recommendations of the first such meeting, "We feel now that this is the one time in the history of our race that the Negroes of America have felt free to reduce to writing their problems and their plans for meeting them with the knowledge of sympathetic understanding and interpretations."[41] Both conferences had, not coincidentally, been centered on the problems of black youth and were heavily based on an assessment of New Deal programs. The educational concerns of blacks had come out of the shadows, to which they had been relegated in the past, into the light of federal attention. In good part, because New Deal programs had delivered in the area of education, education became once more a pivotal issue for blacks in their pursuit of justice and new opportunities. For black Americans, the New Deal had helped to confirm the belief, initiated during Reconstruction and present in tempered tones in the 1931 National Advisory Committee, that salvation might lie with the Lord, but educational opportunity would come at the hands of the federal government.

II

The New Deal relief programs did not survive America's entrance into the war. By 1940 the NYA, which had been most innovative and effective for blacks, was being rapidly transformed into a war-industries training program increasingly under the auspices of the Office of Education, and the CCC was being linked to military training. Black leaders at conferences repeatedly asserted their interest in the continuance of federal, school-related programs, but those aspirations were not to be realized. FDR's program had been ad hoc and emergency related, and his own visions had not gone beyond that. Bethune had, at least once, demonstrated her profound understanding that a solid institutional setting for the new experiments was needed, and she sought to "cooperate with Dr. Caliver" because "it seems to me very important to work with some permanent agency of the government in an endeavor to build a permanent youth program."[42] The Office of Education was hostile to any such effort, and its own traditional concerns had remained largely unaffected by New Deal experience. Most educators and their most prominent organization, the NEA, found the New Deal programs repugnant to democratic traditions and, because FDR had largely ignored them in his efforts, a threat to their power and control over the direction of American education. But the New Deal, in operating apart from the institutions with which educators identified, had not really prepared the soil for any future growth. This was in part because the framers of the New Deal had operated without a sense of the future, and in part because their bold new efforts could not have succeeded if they had been forced to compromise with existing institutions and those whose own power was defined through them. The New Deal programs had been just that, a set of uncoordinated programs without explicit policy directives. Beyond that, Roosevelt's own conservative bias predisposed him to look forward to a return to the normal functioning of traditional American institutions once the economy recovered.

As we have seen, New Deal programs could reach out to blacks in unprecedented ways because they were not bound to traditional institutions with their vested professional interests and antique conceptions of the role of the federal government. The programs were effective because they were directed from Washington where, in some agencies at least, broadening black opportunity mattered. Aubrey Williams made this plain in a statement to the Chicago Urban League in 1936, "It is only by having a national administration . . . that it has been possible to break down and overcome . . . attitudes and provide a program in which all men are treated as equals . . . their need and not their birth nor their color the only criterion for their treatment."[43] Whether it would have been possible for the federal government to overcome the biases and self-interest of the profession and its allies in the state departments of education and the United States Office of Education in order truly to institutionalize their programs is a matter of speculation. What is clear is that the New Deal had raised hopes in the black community that it could not satisfy.

Moreover, because they acquiesced in segregation, New Deal efforts on behalf of black education had probably reached their limit. Those efforts, at their best, can be summed up in the phrase, "separate and equal." This sounds incongruous to our post-1954 ears, but it had meaning in light of the manifestly deprived condition in which black schools were kept in segregated states by state appropriations—a condition plainly revealed by New Deal investigations. Nevertheless, the retention of segregated schools in the context of a developing ideology of equality meant that New Deal activities were both ultimately limited and fundamentally distinguishable from the issues that define equal opportunity today. Because of segregation and because of a failure broadly to defend black interests as such, the Roosevelt administration could neither articulate a thoroughgoing policy of educational equality nor establish goals for black education that could move substantially beyond the remedial advances achieved by relief efforts.

The changes in conception created by New Deal experience, as well as its policy limitations, are tellingly revealed in the report issued by Roosevelt's own Advisory Committee on Education in 1938. The composition of the 1938 committee was very different than Hoover's. Significantly, educators were now in the minority, their places taken by a kind of Rooseveltian coalition among labor, government, agriculture, and industry.[44] This personnel profile anticipated the new, more comprehensive view of education as a necessary part of a functioning society that the report would adopt. The committee report and its twenty-one staff studies were based on an exhaustive set of investigations geared to defining the legal precedents for federal aid and possible financial aid formulas. These were conducted by a staff of ninety-nine researchers and advisers. A number of staff reports summarized the educational results and implications of several New Deal relief agencies, thus at once evaluating and legitimating these new educational endeavors. Black concerns were no longer relegated to an appendix but were fundamental to the report as a whole. The Second National Conference on the Problem of Negroes and Negro Youth, which met after the report was issued, applauded "the technique of the President's Advisory Committee on Education in including the Negro as an integral part of the report rather than in a separate minority statement, and in stipulating definite legislative guarantees for equitable Negro participation in the expenditure of federal funds for education."[45]

From the outset, the committee report adopted a broad perspective on education, noting that the schools had become the central socializing agency in modern urban society, eclipsing community, church, and family. According to the report, children needed and the schools had to provide new social and welfare services that would assume the burden of socialization once carried by an integrated network of family and community agencies. The Hoover committee had described the schools as part of the richly functioning life of small communities and an extension of local democracy. Roosevelt's committee proposed that education provide a means for bringing that democracy about. The report took note of the many new social services provided through various agencies of the New Deal and declared, "The committee is convinced that the Federal Government must continue and expand its efforts to improve and enlarge the social services, including education, and that it must exercise a large measure of constructive national leadership, because in no other agency can representative national leadership be vested." The benefits of localism as the primary context for democratic schooling had given way to a new imperative for national goals for the education of all America's children: "If the educational programs of local communities and States could and would accomplish all of the purposes that are vital to the nation as a whole, the Federal Government would not need to participate in education. Past Federal participation in education has been necessitated by the fact that local programs never have been adequate to accomplish all vital national purposes." Education, newly revitalized, could become a force for social change. "Education can be made a force to equalize the conditions of men. It is no less true that it may be a force to create class, race, and sectional distinctions." Finally, the report not only urged that education become a force for greater social justice but also proposed that education was an entitlement and that the federal government had an obligation to provide and protect the legitimate rights of its citizens to an education. "The American people are committed to the principle that all of the children of this country, regardless of economic status, race, or place of residence, are entitled to an equitable opportunity to obtain a suitable education. . . . The principle has never been fully realized in practice. There is now no reason why it cannot be, and it is time that it should be."[46]

And yet, the report shows the strains of the mixed New Deal experience whose innovations were at once radical and limited. Those limitations make the report a less than completely convincing document of the possibilities of federal leadership in education. In the end, the report short-circuits its radical new vision by concluding not that the schools were inadequate but that their financial structures were inefficient: "The major reason for the great inequality in educational opportunity is the manner in which financial support is provided for the public schools." "If every locality were equally provided with taxable resources in education, there would be little need for Federal participation in the financial support of education."[47] The spirit, they implied, was willing, but the purse was weak.

In fact, this conclusion flew in the face of some of the evidence, especially that provided by Doxey Wilkerson in a detailed study of the state of black education in segregated school systems. The inequalities in facilities, the disparities in funding and teachers' salaries, the blatant discrimination against blacks and black schools in segregated states could not be defined as good faith inefficiently underwritten. And the New Deal programs had uncovered the special needs of blacks in ways that could not be ignored if the federal government were to exercise real leadership. The response of the committee, however, was not to call for special aid for black education or for new federally administered programs; it was to make each of the elements of the federal funding program (divided by goals such as teacher education, adult education, vocational education, apprenticeship training, etc.) contingent on "an equitable distribution of the Federal grants between white and negro schools."[48] This proviso was repeated throughout the recommendations made by the committee which, by prohibiting a reduction in state and local funding when federal funds were received, further protected black schools. The recommendations of the committee were much more far-reaching and much more specific than those of the 1931 report, defining a host of target areas for appropriations in addition to the general fund and making each of them contingent on equitable allocations to black schools. It is in the context of this much expanded view of federal obligation that the statements about equal opportunity for blacks must be placed. Blacks were to get their fair share of each of the allocated funds, but the report did not call for equal education for blacks (which was obvious from its acceptance of segregated schools) and not even for equal though separate facilities. The demand was restricted to equal distribution of federal funds and a maintenance of contemporary levels of state and local appropriations. In other words, the federal government was not to correct the fundamental inequalities, but it would assist in improving education for blacks commensurate with the improvements offered to education in general.

The New Deal's mixed legacy for blacks is nowhere clearer than in this report, which was so much a product of New Deal experience. The Roosevelt administration raised the issue of inequality to national consciousness and made it central to any federal aid to education, but it never challenged the traditional institutional matrix within which this inequality functioned. The New Deal had not questioned segregated schools, as it had not challenged segregated CCC units. Certainly this was based, in part, on political considerations, since Roosevelt had always to act with a careful eye to the support of southern Democrats. But the limitations of New Deal activity also resulted from the pragmatic manner in which New Deal perspectives had evolved and the fact that the New Deal experience had generated principles without policies, goals without long-term implementing procedures. Its goals for education as part of an enlarged commitment to social welfare were large, but its procedures were limited to measures such as an "equitable distribution" of federal funds. In one instance, the report suggested how New Deal experiences could be institutionalized. In urging that the CCC and NYA be retained and newly coordinated in a National Youth Services Administration to be run as a separate agency under a new department of health, education and welfare, the committee hoped to turn the special lessons of the New Deal into an effective government instrument. Ultimately, however, the report of Roosevelt's advisory committee fell back on the good will and cooperation of local school districts whose segregationist and discriminatory policies had been the source of inequities in the first place.

The advisory committee report was never adopted by either Roosevelt or his immediate successors. The experiments in education were effectively over when one by one between 1939 and 1943 the depression agencies were reduced and disbanded. The war, the accompanying renewal of prosperity, an increasingly recalcitrant Congress, and the strengthening opposition of the NEA and the Office of Education to programs operating outside of "normal" channels together turned the New Deal programs into temporary experiments of an emerging welfare state. But, while New Deal education had come and gone, it left a significant legacy to blacks. Having raised their sense of the potential for federal activity in education, it had also brought the peculiar facts of black schooling into the limelight of public attention where they would serve as a reminder of federal responsibility.

Ironically, the Office of Education, whose activities had been largely confined to collecting statistics and calling conferences, played a part in that process. Well before Roosevelt's advisory committee documented the poverty of southern schools, Caliver had been conscientiously accumulating the statistics for a basic indictment of segregationist policies. Some of those statistics did more than accumulate dust on the shelves of government agencies. Throughout the New Deal, pamphlets, leaflets, radio, and news releases were bringing the realities of black education to public attention. One of those pamphlets graphically presented the facts in "black and white"—and did so with wit and insight. After showing the enormous discrepancies between per pupil allocations, building capitalization, teacher salaries, length of term, and other features of the schooling of black and white children in segregated schools, the pamphlet made its point: "If we assume the democratic principle of equal educational opportunity for all children, it would appear that it takes seven times as much to teach a white child as a Negro. As Booker Washington used to say, it is too great a compliment to the Negro to suppose he can learn seven times as easily as his white neighbor." The pamphlet and its message would have been inconceivable before the depression and the New Deal.[49]

III

The implications of black educational disadvantage were exposed with great clarity in the brutal light of war. New Deal educational enterprises disappeared with the war, but the educational needs of Americans did not similarly vanish. On the contrary, the total effort required by American entrance into world war made the education of the nation a strategic concern and the special handicaps of blacks a national dilemma.

During the war, the federal government, through the Department of War and the various arms and services, entered the educational field with force and determination. Federal funds underwrote a massive instructional apparatus in which vast numbers of Americans learned to become radio men, engineers, mechanics, airplane pilots, and medics (among other things), and an equally large number were trained in skilled civilian tasks. The government also sent 235,000 especially capable young men to 350 colleges and universities across the country in its effort to provide American forces with specialized skills.[50] If the New Deal began to define education as a matter of equity, the war made it an urgent necessity, not for some time in the future, not as part of a long-term reform agenda, but as a

matter of immediate social survival. The war put an emphasis on immediate payoffs that made the issue of justice seem less important than the
training of the best and most easily educable. But it also brought to the
army and into the glare of nervous publicity the peculiar educational disadvantages of black men who, as potential soldiers and a necessary part
of the nation's war effort, were simply unusable because they were either
badly educated or entirely illiterate.

The complex and problematic experience of blacks in the armed forces
has been the subject of considerable attention, and it is not my intention
to repeat that story here.[51] My concern is with the more limited and specific issue of how the War Department and especially the army were forced
to devise a program of instruction whose primary, though unstated, objective was to provide remedial literacy training for black soldiers. The army's educational program for illiterates was an unexpected by-product of
its larger strategic objectives, and it was far more single-minded and pursued even more pragmatically than New Deal programs. Nevertheless, that
program was effective in both illuminating and correcting what traditional
schools had ignored.

From the beginning of war mobilization, but especially after America's
full-scale entry into armed conflict, the black soldier presented the army
with special problems.[52] The number of blacks rejected and classified 4-F
was about twice that of whites, and the difference was almost entirely
"attributable to the Negroes' relative failure to meet minimum educational
requirements," since fewer blacks than whites were rejected on physical
grounds (with the exception of venereal infection). "Moreover, the proportion inducted was substantially lower among Southern Negroes than
among urban Negroes."[53] Since inductions were local and selective service
set quotas for localities, the high level of illiteracy among southern blacks
would have a particular significance in that region as whites were inducted
and blacks left behind, especially as call-ups were accelerated and larger
and larger pools of potential recruits tapped.

Rejections kept black participants below their population proportion
throughout the first two years of the draft, despite executive orders man-
dating proportionate participation. Beyond that, the "low quality" of black
draftees meant that most black soldiers wound up in supportive, unskilled
service commands rather than in combat roles. This problem was further
exacerbated by the dearth of adequately trained black officers. In the words
of a postwar confidential study of black performance during the war, "The
most important non-military factors affecting the aptitude of the Negro
for war were the low educational level and the meager administrative and
technical experience of the overwhelming majority of Negro trainees."[54]
During the Second World War, the army made its selections and assign-

TABLE 5. Educational Attainments of Blacks and Whites in the Army,
1941–45

	Black	White
High school—graduates	17%	41%
High school—non-graduate	26%	29%
Grade school—graduate and non-graduate	57%	30%

Source: "The Training of Negro Troops," Study #36, Historical Section, Army
Ground Forces, 1946, p. ii.

ments largely on the basis of a battery of mental tests which classified
recruits into five categories of mental alertness and training potential. I will
have occasion to return to these tests and the meaning attached to them
later. For now, it is sufficient to realize that the vast majority of black
draftees were, not surprisingly, classified very low in these examinations,
overwhelmingly in the bottom two categories. Because the army was
throughout the war a segregated organization, the difficulties of the individual black soldier were translated and magnified into a problem of black
units. "About 80 percent of the enlisted personnel of typical colored units
were in classes IV and V of the Army General Classification Test (AGCT),
as against 30–40 percent in white units." One postwar secret report made
the discrepancies between white and black educational preparation especially vivid when it set the figures for educational attainment of black recruits "with some schooling" alongside those of whites (see Table 5). These
figures, in fact, exaggerated the degree of schooling among black soldiers
by including only those "with some schooling," and by treating grade-
school graduates and nongraduates as a single entity. In addition, as noted
by the report, "These figures understate the educational handicaps of the
Negroes because they do not take into consideration such factors as shorter
school terms, inferior teachers, and less adequate facilities. The disparity
in educational opportunities was greatest in the South whence came three
out of every four colored registrants."[55] Those educational handicaps, the
report noted, were directly related to the low AGCT scores of black soldiers. Table 6 gives the distribution of those scores for the period January
1, 1943 to June 30, 1943.[56]

Precisely because the army was organized into separate black and white
units and low-scoring black soldiers could not be dispersed and scattered
thinly among higher-scoring white soldiers, the glaring deficiencies became
prominent and infinitely troublesome. Committed to the view that "it was
not clearly the province or the responsibility of the Army Ground Force to
change the pattern of society," Lieutenant General Lesley J. McNair, the
commanding officer of the Army Ground Forces, took "a strictly military

TABLE 6. AGCT Scores for Black and White Soldiers in the U.S. Army
(January 1, 1943 to June 30, 1943)

AGCT Grade	White		Black	
	Number	*%*	*Number*	*%*
I	102,143	6.4	419	0.2
II	480,330	30.1	5991	3.4
III	532,215	33.5	23,402	13.5
IV	413,006	25.9	83,104	47.8
V	65,818	4.1	61,023	35.1
Total	1,593,512	100.0	173,939	100.1

Source: "Command of Negro Troops," War Department Pamphlet No. 20–6, February 29, 1944.

and therefore pragmatic view," of the problem of "making effective soldiers of the material given him."[57] Not about to reform society through an attack on segregation, the army was forced to take short-term measures whose objective was to raise the quality of black performance. In the long run, the army's problems and its pragmatic remedies resulted in more than the education of vast numbers of black soldiers. They also illuminated the inequalities in black educational preparation and helped to set the strategic stage for future pragmatic decisions, including desegregation of the armed forces. The consequences for combat success of the educational handicaps of blacks exposed during World War II contributed to the agenda for army desegregation, a decision with far-ranging social implications.

In the meantime, faced with the poverty of black preparation, the army had to choose among a number of practical solutions, including partial integration of combat units, rejection of perhaps one-half of all black registrants, or restriction of blacks almost entirely to support functions and menial jobs in service commands. None of these was politically acceptable. The most direct and least objectionable alternative was for the army to try to make up for the deficiencies with which it was presented. The army finally moved in this direction, but very reluctantly and only after the political situation, the strategic necessities of a manpower crisis, and the magnitude of the educational problem forced its hand.

The army made no provision for remedial literacy training when recruitment began in 1940. Selective service policy was to reject all registrants who could not demonstrate reading and writing skills at the fourth-grade level. According to selective service director John Hershey, "American education was so general that surely there were no persons who had not attained the equivalent of a fourth-grade education." The army was nevertheless aware that the fourth-grade standard might provoke adverse reaction in the Fourth and Eighth Corps Areas which covered the southern region from which most blacks could be expected to be inducted.[58] By mid-1941, army regulations and selective service policy began to allow recruitment of illiterates at the level of 10 percent of all black and 10 percent of all white registrants on any one day. The army also began to establish remedial literacy centers as the need arose at various replacement training centers all over the country. Most of these were attached to the services of supply to which less desirable men were normally sent. Hershey also changed his mind and his tune. In September 1941, he said, "We believe that no man should be permitted to avoid military training when his only ground for deferment is a remedial one," and he called on the War Department to "cosponsor with this headquarters a project that has as its objective the correction of illiteracy." He added ominously that "some communities are very definitely feeling the effects of deferment due to illiteracy."[59]

The War Department waffled back and forth on its policies concerning the drafting of illiterates during the first two years of recruitment, first reducing the number of illiterates drafted to 5 percent of inductions on any one day and then once again returning to the figure of 10 percent of black and 10 percent of white registrants. The army's parallel literacy training program remained unfocused and undirected. Various voluntary, preinduction literacy programs were also in operation. These were run cooperatively between selective service offices and either local school districts or branch WPAs. While effective, they were rarely able to reach large enough numbers of illiterates to be sufficiently useful for the expanding needs of the army.[60]

By the summer of 1942, as the American war machine went into high gear, a conference called to examine the issue of recruitment estimated that there were approximately 433,000 illiterates in the recruitable age group (20–44) of whom almost two-thirds (247,000) were black. The figure was dramatically revised upward to 900,000 by Dean William Russell of Columbia's Teachers College who was called in as a special consultant. In January 1943, Russell issued a report based on an analysis of selective service records. By eliminating men with children, those over thirty-eight, and those with subnormal intelligence, Russell concluded "with confidence" that there were 500,000 illiterate but otherwise draftable men and that this estimate was "conservative."[61] The report and its recommendations for an immediate program of postinduction remedial instruction came in the midst of a new recruitment campaign and growing social and political difficulties that included well-publicized and violent racial disturbances within black units and between black soldiers and white civilians. Southerners were also becoming increasingly vociferous in their expressions of outrage at black deferments. As Senator Theodore Bilbo of Mississippi testified at a Congressional hearing in the fall of 1942, "In my State, with a population of one half Negro and one half white . . . the system that you are using now has resulted in taking all the whites to meet the quota and leaving the great majority of the Negroes at home. . . . I [am] anxious that you develop the reservoir of the illiterate class . . . so that there would be an equal distribution." By 1943 also, black resentment at the lack of combat opportunities was beginning to mount with the accompanying problem of low morale in black units, and the concentration of less desirable, often black, illiterates in service commands was creating additional difficulties.[62]

In this context, the army committed itself to a major effort to upgrade the education of black recruits by instituting a revised plan for sifting and selecting enlisted men accompanied by a large, new instructional program. As officially established in June 1943, that program was aimed at illiterates, the more capable grade V men, and non-English speaking recruits.[63] It is important to remember that the Special Training Units, as they were called, were not organized exclusively for blacks. On the contrary, they were to provide instruction in basic literacy for all salvageable recruits whose "mental capacities" were judged adequate for military training. Army directives concerning the Special Training Units never admitted that the program was aimed at solving special racial difficulties. Nevertheless, the most pressing issue facing the army was not salvaging any and all potential soldiers, although this was obviously desirable, but how to salvage and raise the level of black recruits specifically, and the initial expectation was that about two-thirds of the members of the newly reorganized Special Training Units would be black. In fact, that estimate proved incorrect, as the lifting of literacy restrictions revealed the degree to which white as well as black Americans had failed to be schooled to even a minimal level of competency. Moreover, selective service continued to reject blacks who appeared at induction stations to a much greater degree than they rejected whites.[64] Still, it is clear that the program was instituted with a special view to black service performance. A candid report on personnel utilization, commissioned after the war, made explicit what was usually left unsaid: "Experience indicated that [combat] units could not absorb men classified in Grade V in excess of 15 percent without material reduction in efficiency. Despite this, from March 1941 to December 1942, 8 1/2 percent of whites and over 49 percent of Negroes inducted were in Grade V on the basis of the Army General Classification Test. The percentage of grade Vs in white units was small enough to present no particular problem, but the percentage of grade Vs in Negro units was so high as to be serious. The War Department could not afford to send Negro units overseas in proper ratio to white units if the former were inferior to the latter, because shipping had to be reserved for the best units. This in the end would have resulted in a denial to the Negro race of its fair share of battle honors as well as losses. . . . Because of this situation, it was decided to make mental capacity, rather than literacy, the criterion for induction of *all* personnel. This offered a means of developing better Negro units and enabled them to be employed more profitably overseas." In light of federal policy, selective service regulations which required that "the strength of Negroes in the armed forces . . . be maintained on a general basis of their share of the population," and Roosevelt's insistence that blacks be used proportionately in all parts of the service, the problem of black education had finally forced the army into a thoroughgoing strategy of remedial instruction.[65]

In June 1943, all previous limits on the induction of illiterates were lifted. In line with this new policy, the army instituted a massive new effort which involved sophisticated examination procedures and follow-ups aimed at distinguishing the uneducated from the truly deficient.[66] The Russell report had recommended that remedial training be compulsory but under civilian control and administered as a sub-basic training program. The army rejected this proposal and made the literacy program part of its military operations, to be run, supervised, and manned by military personnel in tandem with military drill. Between June 1, 1943 and September 30, 1945, during which time all potentially educable illiterates were accepted into the army, over 217,000 men defined as strictly illiterate were inducted. Almost one-half of these were black. The addition of men with salvageable scores on the AGCT, those from the top two-fifths of Category V, brought the total in Special Training Units to well over 300,000.[67] Despite the army's new elaborate testing procedure, a good many, if not all, of those recruited from Category V AGCT were probably illiterate rather than mentally deficient. "Undoubtedly," one study noted, "some who failed were truly mentally retarded. But both theory and fact indicate that most of those who fail have simply been deprived of reasonable educational and cultural opportunities."[68] In all, 139,054 of a total of 321,049 black inductees were in one of these two groups, either illiterate or illiterate and/or Category V on the AGCT. The great majority of these were in the Fourth and Eighth Service Commands (before 1942, known as Fourth and Eighth Corps Areas), that is to say from the southern or southwestern districts, one half of them from the Fourth, deep South region, alone. Although the proportion of blacks who were illiterate was higher than the proportion of whites who were illiterate in every service command, the discrepancy in the Fourth and Eighth was the widest. These had proportionately more white illiterates as well, but while 11.7 percent of white inductees in the Fourth Service Command, for example, were illiterate, a whopping 45.3 percent of the black inductees were illiterate. In addition to those strictly defined as illiterate, 43.3 percent of all black inductees during this period, as opposed to 6.6 percent of white inductees, were classified as either illiterate or Grade V. This was true for 16.1 percent of all whites in the Fourth Service Command and 61.1 percent of all blacks.[69] It was toward these men, black and white, that the army now turned its attention, as the earlier makeshift policy was transformed in the early summer of 1943 into an integrated and efficient effort to teach American soldiers to read and write.

After June 1943, the army's efforts became clear, well-organized, and centrally supervised as the Special Training Units were "transformed into an efficient school system." The central command kept close watch over all aspects of instruction, including supplying satisfactory instructional materials. That system, like the army, was racially segregated.[70] Twenty-four units were organized and consolidated from the previously scattered 239 units, six each in the Fourth and Eighth Service Commands. All the Special Training Units were attached to reception centers where soldiers were first organized. The Special Training Units enlisted the efforts of 5,291 army instructors, 1,271 of them black. A few civilians were also used, and their numbers grew as the war progressed in order to release military personnel for more immediate war duties. By June 1944, there were 3,861 civilian instructors "carefully trained in army methods and regularly supervised by responsible military personnel." Most of the teachers, both military and civilian, were well suited to their tasks and extremely well-educated. The Special Training Units varied in size from 309 trainees in Fort Devens, Massachusetts, to 3,809 men at Fort Benning, Georgia, and despite some unevenness of performance and procedure, the units were usually rated successful or excellent in the regular and repeated inspections made from June 1943 to September 1945. The overwhelming majority of men were educated to perform in reading and writing at the fourth-grade level, and over 250,000 men left the Special Training Units to assume regular army service.[71] Among these men were illiterate black and white soldiers, literate but non-English speaking Chinese, Hispanics, and Italians, as well as illiterate Native Americans.[72]

Pamphlets of instruction for teachers and texts used in the Special Training Units underscored the goals and objectives of schooling in the units. These were modest: the ability to read signs, understand command language, write letters home, speak clearly, to calculate sufficiently to use a clock and a timetable, and to participate in timed drill. The methods were, however, often innovative and sophisticated, and they incorporated the most advanced techniques for adult education available. In addition, recruits were to learn something about American war objectives in the broadest civic terms. Instructional techniques, though sometimes individualized, emphasized those group activities most necessary for success in army life. Group games and devices were, in the words of one pamphlet outlining procedure, "especially worthwhile. . . . They help to establish confidence, gratification and success in activities involving other persons. . . . Moreover, such exercises often provide a means of developing skill in cooperative participation. Well-planned activities of these kinds will not only make learning a pleasurable and worthwhile experience, they will contribute, also, to the practice of teamwork which, in the Army, is the pervading ideal." [73] Members of Special Training Units stayed in school for a maximum of twelve weeks, or in exceptional cases sixteen weeks, and were transferred to regular service if they achieved the fourth-grade standard before that time, which many did. Failures were dismissed with an honorable discharge. [74]

In evaluating the success of the program, it is important not to exaggerate how much schooling the army undertook and how much soldiers actually learned. Although army regulations about proficiency at the fourth-grade level were taken seriously and attempts to circumvent the purposes were denounced and promptly stopped, the fourth-grade standard was not very high, and soldiers were not quickly schooled to great command of literate skills. Nevertheless, the testimony of soldiers and observers about the program was immensely enthusiastic, as much an indication of how little had previously been done for these necessary components of the war machine, as of the efforts now being expended on their behalf. [75] One graduate of a Special Training Unit announced, "I am proud of what I learned in school. . . . I am out here and I made Corporal first and now I have made Seargent. Tell all of the Boys the more they learn there, the Better it will be for them at the next Camp. . . . I sure do appreciate what you taught me down there, it is helping me in many ways." [76] When questioned after the war, soldiers testified to the benefits they received. One noted, "Yes, the classes help me in lots of ways and I all so know of boys that came there that couldn't write their name did before we left there." Another testified, "If I hadn't gone to school I wouldn't have been able to write home." Most of the complaints were directed toward the brevity of instruction. "Yes, they helped me to read and write a little But not much because I didn't get enough. I liked the classes I would like to get more of them if possible." One former soldier was more specific, "I believe it would have been better if the School would have last all way through training." [77]

The army's instructional program had an effect back home as soldiers were encouraged to write letters. One mother wrote: "I thank you all Because My child did not know nothing. . . . I did not have the time to send him to school I did not have no husband I raised him from a Baby By my self." Another mother, not able to write herself, conveyed her thoughts through her husband, "Mother was so proud to get your letter, to think you could write a letter yourself. I will always keep it as a remembrance. God bless the man that taught you." Some relatives back home may not have been quite as charmed by the new skills and boldness that literacy brought: "I ask my wife for a divorce and I think I will get it. I appreciate what you done for me. I take a shoar every day and shave every day. I am proud of my uniform. . . . I am not a messup any more."[78]

Although the army was never really satisfied with the military performance of graduates of Special Training Units, and "the consensus was that the Army's investment in this group was not repaid," evidence gathered after the war by Eli Ginzberg and Douglas Bray leads to different conclusions. Only 10 percent of the students were actual failures in school, and their military performance, judged by various criteria, was adequate, as good as that of comparable soldiers who had not required instruction in literacy.[79] Certainly, the brief period of instruction could not overcome the vast deficiencies common among black soldiers, and blacks were never able to catch up to whites in AGCT scores or to develop all the various specialized skills which would have made them equal to whites in the many capacities required by the army. Ulysses Lee, who wrote the most comprehensive study of blacks in the World War II army, observed, "Negro troops often had to be instructed . . . in the bare rudiments of existence in a machine age." Whatever the payoff to the army, however, the schooling provided to over 384,000 illiterates during the entire period of the war (and to another 35,000 in the navy), of whom about 45 percent were black, had a positive effect on the soldiers themselves. One soldier was very specific: "Now since I learned to read and write I can keep my farm record and do all my writing." Another put it more simply: "I wood not have been any good at all because I could not write my name before I went in the army . . . it learn me how to write and spell Just as I wont to no more about Those Things—which I did not have a chance to get in my groing up days."[80]

Many others also wanted to know more and to make up for the deficiencies of their "growing up days." In a sample study conducted in 1951, Ginzberg and Bray found that one-half of the graduates of Special Training Units sought further instruction after the war under the GI Bill. Blacks were equal to whites in their postwar pursuit of education. That pursuit for blacks far more often than for whites was an attempt to move in new directions and away from the enterprises of prewar days. Thus, while white men sought farm training, blacks hoped to put their educational benefits to use in providing them with nonagricultural skills. Southern blacks were the most eager of all to move beyond the bounds of their former lives. "The group that had the highest percentage entering further training was the Southern Negroes. At the opposite extreme were the Northern whites. About 70 percent of the Southern Negroes took additional training, whereas this was true of only 35 percent of the Northern whites." For southern blacks, so long deprived of opportunities, even the minimum training offered by the army raised expectations and desires. One soldier specified this new interest: "Yes the class did help me. I enjoyed them. at that time i had begin to see where i needed more Education, and I put everything I had in it. . . . I started to Perry Business School, January 3, 1950, and I finished 20th Century Accounting and Clerical course. and then I got a chance to take a G.E.D. test and passed it. I then got a High School Equivalency Certificate."[81]

Not all blacks had entered the army as illiterates, and certainly most did not subsequently receive high-school equivalency certificates. Nevertheless, almost one-half of all blacks who entered the army had been subliterate, and almost all of these left their stint with new knowledge and a much expanded sense of postwar possibilities. The federal government had once again, willy-nilly, taken a hand in black schooling and had once again demonstrated its potential, not only for black soldiers but also to the white world.

The significance of the schooling provided black soldiers in the Special Training Units was larger than the instruction it provided. The undertaking required that the army and the federal government adopt a new perspective on their own screening examination, a perspective with important, socially revisionist implications. Before the summer of 1943, the army had administered the AGCT mental aptitude examination to all potential recruits. The findings on black-white differences were immense. As the army removed barriers to the induction of illiterate recruits and began seriously to educate them as well as men in Category V of the AGCT, it had to admit that the tests registered achievement and that test performance incorported educational circumstances as much as they reflected some pure native ability or intellectual aptitude. It continued to use, indeed to depend, on these examinations, but it did so with a new view which informed all internal publications as well as public relations instruments. Thus, in a booklet directed to commanding officers of Negro troops (1944), the subject came up very early in the discussion. "The very material differences between white and Negro soldiers in terms of knowledge and skills important to the Army is illustrated by their comparative performance on the Army General Classification Test. The test . . . is a roughly accurate measure of what the new soldier knows, what skills he commands, and of his aptitude in solving problems. *It is not a test of inborn intelligence.*

Assignment to class IV or V on the basis of a solid AGCT score is not to be accepted as evidence that a man is unteachable, but it does indicate that his training requires extra patience, skill, and understanding on the part of the instructor."[82]

Another booklet, issued by the Army Service Forces, went to great lengths to discuss the educational implications of differences in black and white scores and took special care to question the testing conclusions of the First World War. Of black performance on those earlier tests, it asserted, "when the conclusions with regard to the Negro-white differences were questioned . . . it was shown that Negro and white men of the same school grade had not been provided with equal educational opportunity. . . . group comparisons on the basis of intelligence tests are valid only when backgrounds of equal opportunity have been assured." It concluded, further, that the army's general classification tests "were not developed to measure a man's native intelligence and they should not be interpreted in that way." The pamphlet also contained a long and elaborate discussion of the unequal opportunities for blacks in education, a discussion that included many of the findings on disadvantage and discrimination collected by Caliver and the Office of Education and various New Deal agencies throughout the thirties. Among other conclusions, it noted that "Competent scholars in the field of racial differences are almost unanimous in the opinion that race 'superiority' and 'inferiority' have not been demonstrated despite the existence of clearly defined and tested differences between individuals within every race." "It is agreed also that most of the differences revealed by intelligence tests and other devices can be accounted for in terms of differences in opportunity and background. . . . It is true that in certain areas Negroes show relative inferiority, but this is not necessarily inherent inferiority. There is always the deadening differences of inequality of background and opportunity."[83] Where subnormal mentality was the bombshell of the First World War draft, illiteracy and unequal educational opportunity became the equivalent shocker of the Second. But while mental deficiency, defined strictly as a genetic trait, was an irremediable threat to the nation, illiteracy was both remediable and a telling commentary on the much vaunted tradition of American democratic schooling.

Although these army pamphlets were intended for army personnel and marked for restricted access, this revolution in the meaning of mental tests and their revelations about unequal opportunity quickly spread beyond the confines of army life. In 1944, a pamphlet based on a conference on the postwar education of blacks, hosted by Ambrose Caliver, noted that the differences between black and white scores were "not due wholly to inborn differences of *native* intelligence. They reflect both inborn and acquired capacities. Parents, ancestors, and the race have contributed something" but, "by the time a man comes into the army his ability to learn what the army has to teach him is largely determined by . . . many environmental factors which have modified for better or for worse the *native* intelligence with which he was originally endowed."[84]

The effort to salvage the teachable black recruit thus meant an acceptance by the army of the limitations of its tests as merely expedient instruments of selection for practical army needs. The test results were useful and necessary but deeply tainted by the educational and other experiences of recruits. "Many Negroes make low scores on the Army General Classification Test. This is consistent with previous studies which show that Negroes do poorly on this type of test which stresses academic achievement," an army pamphlet on instruction in Special Training Units noted. "This is probably a further confirmation of the fact that many Negroes have not had sufficient academic training. However, Negroes appreciate the opportunity to learn. They are eager students in the Special Training Units. . . . With adequate opportunities for academic learning, their usefulness will increase considerably."[85]

By the end of the war, the army had put itself on record as supporting the teachability of blacks. It had furthermore acknowledged that while intelligence and aptitude were important to its technological and highly differentiated operations, test results should not be confused with definitions of individual intelligence and certainly not be associated with racial distinctions in intelligence. These views, officially adopted and publicized, were critical for blacks and whites alike. While the former might turn to them with relief and the latter dismiss them and cling to other less enlightened opinion, official views had registered a profound redirection between the two wars. Whatever alarm the Second World War rang about black attainments and performance, the army had carefully made its official charges against black education and their unequal and inadequate schools, not against blacks themselves.[86]

The army also used the evidence of black performance in the Special Training Units as documentary evidence for this new position. Of the black trainees between June and December 1943, for example, only 9.3 percent were discharged as unteachable, while 90.3 percent were assigned to regular army units. This record bested that of whites in Special Training Units, since 15.5 percent of white recruits were discharged. Blacks also took no more time to reach the required fourth-grade level than whites did. "The colored group as a whole, tends to make better progress than the white group," one report during the war concluded.[87] "The results," of this training, a training pamphlet observed, do "much to refute the theory of innate Negro inferiority." "The fact that Negroes have succeeded fully as well as white trainees in Special Training Units suggests significant implications for the further training and utilization of Negroes in the Army. It certainly tends to show the Negro needs only the chance to learn."[88]

The army had given blacks a chance to learn. It was no more than that. Recruits were not taught more than they needed to know, and they were taught in segregated classes, largely by army personnel. Efficiency not justice had been the purpose. Yet, limited and narrow as that instruction was, it was more of a chance and a more equal chance than most of these soldiers had been given before. The army's program had been a response to three sometimes conflicting and not altogether strategic demands: the desire for more and better manpower; public resentment at the under-utilization of physically fit but uneducated personnel—largely black; and executive policy requiring that blacks be proportionately represented in the war effort. The army had never operated in a strictly technological manner—using only those who most efficiently fit its high-powered war machine. In response, the army began to educate blacks, and just as that decision was never strictly a military one, so the results had consequences beyond education for the military.

H. D. Bond, president of Fort Valley State College, a black school in Georgia, visited a Special Training Unit at Fort Benning and was boundlessly optimistic about the possibilities this schooling provided for black soldiers. In a letter to the adjutant general, he wrote, "My observation is that the work there constitutes one of the most astonishing contributions to the theory and practice of educational method on record, in any historical period, or in any country. . . . This truly wonderful work was called again to my attention by a visit to the Reception Center and the Special Training Units, recently made by a group of Negro public school teachers from Columbus, Georgia. . . . This visit was the best education these teachers had ever received. The *main* point of this letter is, respectfully, to inquire as to the possibility of giving wider public notice to the devices, methods, and the philosophy of education developed by the Special Training Units. . . . What is being done at Fort Benning has significance, not only for the education of Negroes, but for the education of people, children, and adults, everywhere in this country and all over the world."[89] By the time Bond wrote his letter, the war and its experiment in special education was coming to an end, but the message Bond had found in the Special Training Units would be broadcast for some time to come.

The most immediate result was in the army itself. Although the army stopped inducting illiterates in 1945, it continued a variety of experiments to elevate the skills of black soldiers. In 1946, an educational program organized at Fort Benning for the black 25th Combat Regiment sought to raise the educational level of black soldiers to the eighth-grade level. A wider effort was instituted in the Far East command for "all soldiers lacking the equivalent of a fifth grade education." But the most radical and innovative attempt was made in 1947 among black soldiers stationed in postwar Germany. In that program, according to Morris MacGregor who has written about army integration, "thousands of black soldiers [were] examined, counselled, and trained." Convinced that "a program could be devised to raise the status of the black soldier," the officers in charge, including Marcus Ray, a black lieutenant colonel, hoped to "lay the foundation for a command-wide educational program for all black units." The experiment they started in Grafenwohr Training Center was adjudged successful in light of the "improved military bearing and efficiency of black trainees and the subsequent impressive performance of two new infantry battalions." The results of the experiment led to the establishment of a permanent training center at Kitzingen Air Base through which all black troops were to be rotated. Each black soldier was "required to participate in the educational program until he passed the general educational development test for high school level or until he clearly demonstrated that he could not profit from further instruction." The program was a success and at its peak in 1950 trained "more than 62 percent of all Negroes in the command." Finally, the experiment allowed men to raise their average AGCT scores by twenty points. According to MacGregor, the program provoked jealousy among white soldiers "who claimed that the educational opportunities offered Negroes discriminated against them," but spokesmen for the army justified it because "Negroes on the whole had received fewer educational opportunities in the United States." In its attempts to produce better black soldiers and in light of its assessment of the deficiencies underlying black performance in World War II, the army had at last produced a thoroughgoing and successful program aimed specifically and exclusively at blacks. The army revealed both how education affected performance and how these deficiencies could be remedied. For an organization whose own goals were always limited to military payoffs, the army had made quite a considerable contribution to black schooling—one which perhaps, in MacGregor's words, "although rarely so recognized . . . qualified it as one of the nation's major social engineers."[90]

The word about the army experience in World War II that Bond hoped would go forth also went beyond the walls of army headquarters and soldiers' barracks. After the war, the Special Training Units served as the basis for discussions on adult education programs, like those proposed by the American Council on Education.[91] Even more grandly, the educational experiences of the Special Training Units informed the materials and policy initiatives of the United States Office of Education, which after the war directed its efforts at providing both educational information about adult education and repeatedly sponsored attempts to enlist federal funds for adult education programs. Not surprisingly, Ambrose Caliver, who had so long resided at the Office of Education as an expert on black affairs, presided over these efforts and amassed the supporting documentation for the educational deficiencies of so many Americans—especially but not exclusively black—which the war had made manifest. Caliver was at the Office of Education for three decades, and his various assignments, from collecting separate statistics on black schools to promoting adult education programs from which blacks would benefit, in many ways reflect upon the story I have tried to tell in these pages. Always marginal to the real action, he nevertheless lived through and participated in a transformation in which blacks moved from the periphery of national concerns about education to the center of concerns about the role of education for national security. But the early 1950s, during which time the Office of Education tried to stimulate Congress to appropriate funds for adult educational endeavors that would compensate for the deficiencies of American schools, was not a good time for these efforts. Americans were certainly worried about education and its role in manpower development as a national security issue, but the unequal distribution of educational deficiency among adults seemed less significant than other school related issues, especially America's competitive technological position in a world once more threatened by larger and future wars.[92]

During the war, the army, unlike the New Deal programs, had provided no arena for the expression of radical ideas about justice. Yet, the practical experiments of the war and of army life offered significant lessons beyond the very sizable instruction it provided black soldiers. Those lessons are best grasped in the form of a question: How, in the context of a national life dependent on literacy and specialized skills, could the federal government afford to ignore the educational needs of an enormous group of its most disadvantaged citizens, a group almost entirely defined by race? All through the 1930s and '40s the answer to that question lay beyond the locked door of official segregationist policy that FDR had managed to avoid confronting in peace and in war. The Roosevelt years were always limited by that fact. And yet, without them the question itself might never have been asked.

IV

"It seems strange," Eli Ginzberg and Douglas Bray wrote in 1953 on the eve of the Supreme Court's decisions on school desegregation, "that the serious shortcomings inherent in the population revealed by these examinations [AGCT] had gone unnoticed in previous years, or if noticed, had failed to lead to remedial action. . . . The United States has long been recognized as one of the richest countries in the world as well as one of the most democratic. One reflection of this economic well-being, and democratic orientation, has been the emphasis that has been placed for many generations on education, particularly free education for every boy and girl in the country. Yet at the outbreak of World War II more than 4 million men in the labor force had less than five years of schooling; about 1.5 million were totally illiterate." "Although the American effort to provide an ever-higher level of education for the mass of the population was unique and largely successful," they continued, "relatively little attention was paid to conditions which indicated that in certain parts of the country and particularly with reference to certain groups, there was little or no participation of the local population in the expanding educational efforts that characterized the country at large."[93] For Ginzberg and Bray, social planners and policy analysts, American disregard for the uneducated and the undereducated—disproportionately black Americans—was both strange and irresponsible. From a historical point of view, it was neither strange nor unexpected. Indeed, the discovery of this large and geographically concentrated minority came only as national concerns came to the forefront of attention, largely the result of depression and war. Education had never been more than marginally a national concern, and it was not directly a federal concern even during the New Deal and the war. Educational matters had emerged incidentally, unexpectedly, and unintentionally as a national economic crisis and national security considerations eclipsed the visions of a more parochial past. It was only then that the failures of American education became prominent and replaced the euphoria of educators about democratic progress. And it was only then that the black child in the American midst and the uneducated black adult he would soon become became visible. Neither Roosevelt nor the commanding officers of the armed forces had set out to find the black American; neither had they set out to educate him. They had, in fact, in good pragmatic fashion proceeded to do both as the occasion called for it. In so doing, they once more linked education to social reform and inaugurated a tradition of federal responsibility for black education that encouraged blacks to look for federal redress. That tradition would strongly inform, perhaps even define, the federal thrust in education for the rest of the century.[94]

In 1965, in the midst of yet another war, Lyndon Johnson would rediscover the New Deal, poverty, and the black child. And he would do so, in lingo appropriate to the time, by waging a war on poverty from the battlements of the schools.

5

The Female Paradox: Higher Education for Women, 1945–63

The paradox of women who are educated like men and can do most of the things men do, but are still taught to prefer marriage to any other way of life, causes most of the confusion that exists for women today.

MARGARET MEAD[1]

One of the paradoxes which plague thoughtful persons who are actively interested in matching educational patterns to the special needs of women in today's world is that an overwhelming majority of the young women in whose behalf so much pioneering and soul searching has gone forward appear to be unwilling to believe or be interested in the variety of roles that changing times offer and require of them. . . . they drop out of college and rush into early marriages and announce firmly that they have no interest beyond home and family.

Commission on the Education of Women (1958)[2]

A well-adjusted individual who has majored in Greek archaeology, in my opinion, is much better prepared for marriage and motherhood than the ill-adjusted girl who has spent her time in special courses and doing special reading.

MILLICENT McINTOSH (1949)[3]

Women occupy a peculiar position in the history of twentieth-century schooling. Unlike immigrants, around whom school boundaries had been redrawn, or blacks, who forced an entirely new level of government participation, women easily assumed their places in public schools alongside men in a largely untroubled and uneventful manner. Women as a distinct group were not an important concern for educators in the elementary and secondary schools. On the contrary, women were usually better, more diligent students than men. They went to school more regularly, stayed to graduate more often, and provided fewer instances of problem behavior. They played no significant part in the discussions of IQ and group differences in intelligence in the 1920s. No educators seriously challenged the long-standing tradition of coeducation in American public schools, and apart from sexually linked vocational programs, most secondary-school academic curricula were entirely blind to the sexual make-up of the student body. Women made distinctions among academic subjects, more often choosing language rather than science and mathematics courses, but this did not fundamentally challenge their equal opportunity for education in the public schools.[4] Indeed, had women politely graduated from high schools and quietly retreated into the urban apartments and suburban bungalows and split-levels of America, they would have merited only occasional asides in most educational discussions. Instead of withdrawing after high school, however, women began in the middle of the nineteenth century vocally to demand and gradually to receive places in institutions of higher education—to which, by the twentieth century, they aspired in ever larger numbers. And it was in the realm of voluntary education, in the colleges and universities, that normally quiet women created a considerable ruckus.

The noise surrounding women's schooling was loud in the late nineteenth century when the pioneer women's colleges opened their doors, and state universities provided a variety of back-door entries into academia. By the first third of the twentieth century, most of it had subsided to the hum usually noticeable on football fields at halftime. Victory in the battle over women's education seemed certain by the early twentieth century, and although opponents were not completely subdued, American women were now attending colleges and universities in large numbers. By 1920, they represented almost one-half of all college enrollments, and in the forties, in the context of war-depleted male enrollments, the majority of all students in schools of higher learning were women.[5] It was at this point, however, that the noise that had subsided for half a century became once more audible, and in the 1950s, around the issue of women at college, it became clamorous.

The 1950s saw a particularly significant and heightened expression of what can best be described as the female paradox: the fact that women were receiving more education than they seemed to need. The female paradox was always latent in the nature of women's higher education in a society which continued to ascribe, though not entirely to confine, female roles to the family. After the war, as issues of effective manpower utilization and the use of educational resources for the most efficient national purposes surfaced, the paradox burst into the public media and became the subject of innumerable scholarly conferences, the matter of learned investigations, and the center of public debate.[6] By 1950, the proportion of women in institutions of higher learning plummeted to 30 percent, lower than at any point in the twentieth century, and the paradox of women's higher education penetrated to the very core of student culture as women students increasingly complained about the dissonance between their future lives and their studies. In this context, educators produced the first significant challenge to women's equal education in the twentieth century. For some, the remedy was to provide women with an education more appropriate to their specific needs in the family. Others sought remedies in a stricter attention to occupational preparation. But most educators urged that the liberal arts continued to provide the most judicious education for the many contingencies of women's lives. This view became the dominant ideology of the time, but it involved a crucial reinterpretation of the liberal arts as a form of utilitarian education with uncertain implications for women as well as for traditional education.

The debate about women's higher education was a particularly pointed instance of the pervasive dilemma of democratic schooling as educational institutions attempted to appraise the needs of their diverse constituents and to fashion a more relevant course of instruction. In the end, the most pertinent result of the debates had little to do with their intentions but lay in the fact that they helped to create the basis for a massive renewal of feminist aspirations.

I

Our memories of women's education in the 1950s have been indelibly marked by Lynn White's condemnation of traditional male-centered studies and his encomia to the culinary arts as the best possible womanly contribution to the future of western civilization.[7] Womanly gifts, White believed, ought fittingly to be honored in the college curriculum alongside, and perhaps even in preference to, the more traditional subjects. In fact, the discussions of women's education that began after the war were considerably more complex, and White's perspective was only one of many competing views. Nevertheless, White hit a sensitive nerve and helped to set the limits of the discussion by his extreme statement of a popular position. White's views could be seriously engaged in the 1950s because his search for a more suitable education for women appeared to grow naturally from accumulating and well-publicized observations about the desire of American women for self-fulfillment through matrimony and childbearing. From that perspective, White's proposals seemed both democratic and "progressive." They were also distinctly anti-feminist. At a time when most positions on women's higher education contained some open disavowal of feminism, White's perspective was firmly within the popular consensus.

The beginnings of higher education for women in the nineteenth century were not quite so completely an expression of feminist sentiment as we sometimes assume. As Helen Lefkowitz Horowitz has recently shown in an excellent study of the "Seven Sisters," the schools varied among themselves, and the sources for the development and growth of the schools most often associated with women's education were not exclusively feminist. Founders, faculty, and students had different and often complex visions of what the setting of higher education for women ought to be and the purposes it would serve. Nevertheless, by the early twentieth century women's higher education was closely associated with ideals of equality of intellect, if not necessarily of social opportunity, and women's schools especially stood their ground on providing an education equal in quality and in content to that of men.[8] This insistence was transformed during the twentieth century from a previous emphasis on a Latin-based classical curriculum augmented by the sciences to what we normally refer to as the liberal arts—a broadly nonutilitarian education whose stated purpose is a lively familiarity with the full range of civilized thought and activity in the sciences, the arts, the humanities, and most recently the social sciences. It was this liberal arts education that came to be the touchstone of female equal education.

College curricula for women had not remained exclusively liberal, but neither the substitution of teachers colleges for normal schools with their self-conscious utilitarian thrust nor the strong infusion of home economics courses and curricula, at state universities especially, seriously displaced the liberal arts college and curriculum as the standard bearer of women's equal education. The most serious challenge to the liberal arts came from an entirely different direction than the professional training which underlay the schooling of teachers and the increasingly professional orientation of home-economics departments. During the late 1920s and '30s, experiments in women's education began to suggest that women needed and colleges ought to provide specific instruction in the more exclusively "female" interests of their students. These programs were most often defined as "progressive" and frequently identified with the new progressive colleges like Sarah Lawrence and Bennington.[9] Designed to educate their students more adequately according to their needs and requirements, the progressive programs in higher education, like "progressive" secondary education, were meant to individualize instruction to make it more fully relevant to students' lives and aptitudes. And like the more individualized instruction in public schools, progressive colleges emphasized the fullest education of each student rather than the most proficient exposure to a predefined set of disciplines. Progressive colleges, like progressive education in general, usually also emphasized hands-on learning and stressed active, creative expression in the arts, an area in which women were believed to be especially interested or adept.

Constance Warren, president of Sarah Lawrence, defined the guiding criteria of this kind of education as flexibility and attention to students' development rather than concern with subject matter and subject content. "The liberal arts college," she advised, "would do well to heed the handwriting on the wall and make its program conform to the changing needs of youth in the modern world. . . . Young people gladly study about things which have meaning for them—which lie within their own experience and cast light upon familiar problems." This adjustment of the school curriculum to the needs of students had very specific implications for women. At Sarah Lawrence, for example, "while the emphasis is on individualized education, we find a predominantly feminine overtone." This, Warren added, had nothing to do with "the aggressive feminism" of the early pioneers of women's education. Instead, students could experience a "natural preoccupation" with marriage and preparation for their future roles in life. "The constant desire is expressed in every arena of the curriculum for more guidance, preparation and emphasis on family relationships and adjustments, on marriage, human biology, housing, woman's part in the economic world, on understanding oneself, children, and other people."[10]

Sarah Lawrence was an exclusive school, founded as a progressive alternative to traditional women's schools, and its attention to the individual came at a high cost to students and their parents. But its expensive education was costly in another way as well, as the progressive alternatives provided pressures for other schools to become more attentive to the needs of their students. A traditional "feminist" school like Vassar, for example, introduced a major in euthenics that followed on its earlier establishment of an Institute of Euthenics in 1926. Euthenics, or race development, heavily emphasized the responsibilities of women for human survival and continuity with instruction on child development, human relations, and family concerns. "The original purpose," according to the Vassar College catalogue, was "to educate women for their responsibilities as homemakers." Many schools in the late twenties, and especially in the thirties, moderately adapted to these pressures by introducing courses on the family, community relations, and child psychology, usually in more traditional departments like sociology, psychology, or education or, since these were themselves new fields and not everywhere available, in older disciplines like economics or philosophy. By the 1950s, the leaven of progressivism had influenced even the most traditional schools. The Vassar College catalogue, for example, declared that "certain courses are organized around the needs of the individual student. They help her to uncover what she is, and to say it honestly, and effectively, in whatever way she is best able to."[11]

The ferment of progressive education in the 1930s was, not surprisingly, often directed at women's education, since the question underlying women's higher learning had been latent for some time—what, after all, were women being educated for? The pioneers of women's education had always assumed that to be equally educated with men was its own reward, a manifest declaration of equality of ability. But by the 1930s, as Warren noted, students took "this for granted." As the small but innovative pioneer generations of the turn of the century gave way to a flood of female college students in the 1920s and '30s, the answer to the question was no longer clear. The earlier "spinster" generations were increasingly replaced by women who married after college, and after World War II this marrying propensity would be exaggerated as the interruptions of social life—of a protracted war following on the heels of a devastating depression—sent American women of all levels of education flocking to the altar. In the late forties, American marriage rates were at their highest point in the twentieth century, and college women as well as college men were leading the trend.[12] By the early fifties, the baby boom was in full swing. While college women continued to have babies later and less frequently than other women, they too were having them more often and earlier than at any time in the twentieth century. American college women had joined a trend that was making twenty a median marriage age for women for the first time in their recent memory.

The marital aspirations of American college women were already clear during the war, and that fact together with new progressive views provided the context for a reexamination of the value of higher education for women. Anticipating the crest of the debate by more than a decade, Robert Foster and Pauline Park Wilson published *Women After College*, in 1942. The book would subsequently be interpreted as an unrelieved lament on women's miseducation. In an intensive study of one hundred women, Foster and Wilson found that "the majority of these young women pursue a career for only a brief period, and then marry." By emphasizing not knowledge gained but the progressive objective of being a "well-adjusted member of society," the authors could only report that the traditional college curriculum had been entirely inadequate. "When . . . the average girl from Main Street or a metropolis goes to the average or above-average university, she is confronted with the task of selecting a course of study which has been made, tried, and tested by its adequacy in contributing to the successful technical and professional training of men. . . . One would suppose that an institution designed primarily to offer education for women in American life might take as its starting point an analysis of the cultural framework within which woman's life is lived; that such an institution would base its curriculum upon an analysis of the needs and problems confronting women within our contemporary world." By educating women like men, the liberal-arts colleges and universities had not

produced well-adjusted women but wives and mothers who were unprepared to cope realistically with the many twists and turns of their married
lives. "The findings of this study leave no doubt that education did little if
anything to prepare . . . the women of the group to meet their actual life
problems."[13]

By focusing on the average girl and urging that a realistic curriculum be
based on the actual cultural framework of women's lives, Foster and Wilson potently set the terms of the debate of the 1950s. Democratically alert
schooling has always to bear in mind some normative view of its population. As women's higher education became a more normal experience, no
longer the preserve of the exceptionally strong-minded or eccentric student, it too became concerned to define the requirements of the norm.
Even at Sarah Lawrence, where women could afford to be singular and
unique, that norm was increasingly defined in group terms by the most
salient characteristic of the group involved—their femininity. It was predictable that the norm would be defined to some degree not only by what
women were, but by what Foster and Wilson called the cultural context,
that is, by what they could be expected to do once they graduated. "The
higher education of these women failed to prepare them for household
responsibilities. It could, by orientation of viewpoint for certain courses,
by having an adequate euthenics program, and by basing its curriculum
upon a study of women rather than upon sacred tradition, make a contribution to woman, as a person and, particularly in her family household
experience." Foster and Wilson used the increasingly "progressive" definition of the liberal arts as preparation for "living" against the substance
of that curriculum. "The arts and science courses which are supposed to
contribute to personality development and to improve human relationships
did not perform their function in the lives of these women." The liberal
arts had not prepared women "for certain inevitabilities of their lives."[14]
Women After College had raised a powerful combination of issues which
would explode in the 1950s to define the discussion of women's higher
education: the educational requirements of the average woman; the problem of adjustment to the cultural context; and the irrelevance of a traditional curriculum as a fossil of an earlier, male-centered and feminist education.

By becoming more available and democratic, higher education in general
became subject to this kind of critique. Like other segments of the American educational enterprise, colleges and universities in the twentieth century had become part of an aggressive growth industry which often calculated successes in terms of expanding enrollment statistics. College
enrollments and graduations increased in each decade of the twentieth century, except for a brief decline during World War II. That growth was

often extraordinary, and a good part of it was due to the rapid normalization of higher education for women. Between 1920 and 1940, there was a four-fold increase in the number of bachelor and first professional degrees awarded to American youth. During these twenty years, the proportion of men receiving degrees to all men twenty-two years of age in the population rose from 3.5 to 9.7 percent. For women, the rise was even more dramatic—from 1.7 to 6.6 percent. The number attending colleges at any one time was far greater and usually more than twice as large as the number who graduated, and women especially often went to college without obtaining a degree.[15] Thus, college and university officials and policymakers became ever more alert to popular perceptions, and the issue of educating the norm or the average, heretofore largely a problem for the lower schools, became increasingly also a matter of concern for college and university officials.

So too, higher education became vulnerable to the fixations of pollsters. *Fortune* magazine, for example, had begun to poll public opinion on issues of higher education in the 1930s. In 1949 *Fortune* provided educators with some worries as the man-on-the-street not only expressed his strong preference for an education that was highly utilitarian but also fundamentally distinguished between the usefulness of college for men and women. Parents were much more eager to send sons to college than daughters, and of those who did want their offspring to attend, 57 percent wanted their sons above all to receive professional preparation, while the largest proportion of respondents (46 percent) believed training for marriage and family life should be the college's primary objective for daughters. When asked if "a college education is just as important for a girl who wants to get married right after graduation as for a girl who wants a career," 41 percent indicated that this statement was more false than true. As Karl A. Bigelow, one of the educators asked to comment on the results, noted, "The American people have a faith in higher education that should be sobering to those of us who are engaged in purveying that commodity—especially when we consider the highly practical results that are desired: training for a job first of all for boys, and that plus training for wife and motherhood for girls. Can we produce? Are we willing to produce?"[16] Not everyone found the poll results equally compelling, but it certainly provided food for thought and much fuel for the emerging discussions on women's education.

This vote for a utilitarian education by the paying parents and state taxpayers came in the wake of an even more alarming poll taken by *Time* magazine in 1947. *Time* was able to elicit some damning indictments of traditional liberal-arts education from former coeds. One college woman noted, "Many college women, like myself, make the mistake of not training for the most important career, marriage." Another echoed the sentiment: "As to my career, that of housewife and mother, college trained me ill. When I married, I had no training for coping with either a house or children. These things I have had to learn the hard way—and believe me, it has been hard." The criticism these women made went considerably beyond Sarah Lawrence's adjustment of the curriculum to student lives. That, after all, emphasized new areas for investigation and an adequate consideration of emotional expression and active participation in various creative activities, mostly the legitimate arts. What former coeds seemed to be saying was that college had some responsibility for training in the nitty-gritty components of household skills. "College could have helped me more in preparation for marriage and home management," one former coed insisted. "Some of my interests and attitudes were so very academic that I didn't take to the routine of a homemaker for quite a while." Even one woman who expressed general satisfaction with her academic training admitted, "I'd trade History of Civilization for a practical cooking and nutrition course."[17] By 1950, a total picture of the American college woman to replace the older image of the bluestocking was clearly in sight. Eager for husband and children, disgusted with angry feminists, and knee deep in diapers and dirty dishes, American women seemed eager to trade their much vaunted academic heritage of the enlightened best in Western civilization for a mess of pottage.

It was this undefined stew that Lynn White hoped to transform into a beautifully prepared and exquisitely presented dinner party, for which students at Mills College and elsewhere should be properly trained. As a historian, White had paid special attention to the contribution of small, seemingly insignificant and usually unhonored techniques and inventions. To these he attributed great consequences as fundamental instruments of evolving civilizations.[18] It was this perspective that White brought to women's potentials. Turning traditional values upside-down and investing academic studies with his own progressive education values, White hoped to endow women's ordinary endeavors with significance and save women's contributions from the realm of the trivial and despised. His book was, in fact, a sharp rebuke to the direction and values of male-centered and success-driven American culture, whose beligerent and desperate future he hoped to forestall through an appreciation and reevaluation of the womanly arts.

Genteel and learned, *Educating Our Daughters* (1950) placed women's education in a long and broad context of the Western and non-Western past, and its wide erudition and romantic visionary quality provided an ironic, but not altogether unsuitable, beginning to what would be a knock-down-drag-out fight over the direction and goals of women's education in the 1950s. White believed women could provide a balance wheel to the increasingly insensitive, technical, and hurried society of the late twentieth century. By cultivating the finer, small arts of pleasure, women's education, White believed, could stop an onslaught of incivility.[19] That basically conservative vision, in the guise of a more progressive and democratically responsive education, made *Educating Our Daughters,* with its explicitly paternal title, an appropriate opening to a decade in which conservative visions often became radical and progressive ideals took an unexpected turn. Above all, White had been alert to the developing waves of dissatisfaction and especially attentive to the democratic and populist sources that seemed to lie in popular representations of the voice of the new woman.

II

The direction of women's voices is not always the same as that of their feet. The expressions of interest in more adequate household preparation that *Time* magazine had discovered immediately after the war and to which White hoped schools would respond seemed to challenge the traditional liberal arts and especially the concept of equal offerings to men and women. But the war had also sent women in unprecedented numbers into the workshops and laboratories of the nation. Women also continued to go to college, where during the war they outnumbered men for the first and only time in the twentieth century, representing 55 to 57 percent of all those earning a college degree between 1944 and 1946. This was obviously a statistical fluke and the result of war-depleted male enrollments. But the conclusion of the war did not fundamentally change the long-term rise in attendance of women at colleges and universities. After the war, despite additional competition for enrollments from men benefitting from the GI Bill, women pursued college degrees in ever larger numbers. And while the proportion of women's enrollments declined relative to men between 1947 and the end of the 1950s, the increase in the number of women degree recipients kept pace with the increase among men.[20]

As significantly, while men returned to American industries after the war and more women than ever married, women did not retreat from the workplace. On the contrary, women's participation in paid employment increased steadily after the war. By 1957, 22 million women were in the paid work force, and they comprised one-third of the total number of American workers. Over one-half of these were married women, a higher proportion than at any time in the past. Single women were represented in the working population to the same degree as single men. Although only a small proportion of women with children under six worked for pay, nearly one-quarter of all women in the work force were widowed, divorced, or separated from their husbands. The configuration of women in paid employment was beginning to suggest a pattern of working for women and a life cycle of work that would exercise a powerful influence on the debates over women's education.[21]

The meaning of the statistics on working women was strengthened by the specific facts on the college educated. In 1950, college women worked more often and stayed in the labor force longer than noncollege educated women, and this was true for married women as well as those who were single. Of women between twenty and sixty-four, fully 47 percent of those with college degrees were employed outside the home as opposed to only 30 percent of those who had finished only the elementary grades. Part of this difference was the result of the fact that one-fifth of all college women were still single by their early thirties; and another, the consequence of longer delays in inception of childbearing among the college-educated. By the 1950s, however, younger college-educated women were less frequently delaying childbearing and were more often married than in earlier generations. More significant to the statistics on the higher labor-force participation of the college-educated were the facts that college women were more likely to reenter the job market in their middle thirties than those with less education and that they were much more likely to remain at work into their sixties. While paid employment figures for women increased in each year of the twentieth century, the increase after 1940 was much stronger and steadier than before. As the volume *Womanpower* (1957) issued by the American Manpower Council, observed, "The new significance of paid work in the lives of American women is largely a development of the last fifteen years." Increasingly, observers of the labor force were seeing women who had seemingly fled work for the security of husband, home, and children in their early twenties return to work in their thirties. "The most spectacular development of recent years," the authors of *Womanpower* noted, "has been in the employment of women over 30, most of whom are wives and mothers."[22]

In the 1950s, this pattern of employment strongly informed the discussions of women's higher education and began to provide a bridge between those who proposed that colleges should provide women with a distinctly female curriculum and those who stood firm on an education altogether the same as that for men. That bridge did not diminish the scope of the debate or blend the theoretical differences in the positions. It did, however, create a new and increasingly hegemonic position in the discussions, one that focused on the life pattern of women rather than on the old dilemma of career versus marriage. This new model of the educated woman as mother and worker at different points in her life distinguished most of the discussions of women's education among professional educators from the perspective that dominated the popular press. While magazines still purveyed a view of the "fifties" woman as stable, firmly rooted in a suburban home, patiently tending her three or four children, educators were discovering women in motion and a woman's life as a moving curve.

In his initiation address as chancellor of the Women's College of the University of North Carolina, Gordon C. Blackwell caught the sense of this discussion well: "Determination of the objectives of education for women is more complex than for men. The educational needs of women relate to her potential roles as homemaker, mother, citizen, worker and an attractively intelligent person. If a career becomes a reality for a woman, it may be of the split type occupying only a brief period before marriage, then a longer period after the children are of school age. Also many college women experience the two way stretch of home and job or home and college in tandem. These are the complexities of the educational needs of women." Robert L. Sutherland emphasized another feature of this discovery of women in motion—their often interrupted studies: "Educational needs of women do differ from those of men, but they don't differ in the sense that they need a different curriculum. They differ in the sense that you have to educate them when you can catch them." The Commission on the Education of Women of the American Council on Education became the most significant and influential educational body to adopt and develop this view in its many conferences and publications. As Althea K. Hottel, a director of the commission, noted in one such publication, "Preparation for different responsibilities and for the dramatic changes which take place in women's lives requires a long-range view of life and an awareness of the opportunities for personal growth at different stages in the life cycle. . . . The future tasks of education are to bring to each woman that understanding which will help her as she progressively makes crucial choices. . . . One can scarcely deny the need for a broad background of knowledge coupled with special skills." [23]

At the University of Minnesota, the model of women in motion provided the basis for the Minnesota Plan in which women were encouraged to return to school after they had formed their families, and younger women were provided with guidance to think of their lives as a whole. The president of the University of Minnesota observed, "The place, then, for modifying programs for the education of women is neither in the nature of the materials nor in their content; it is in recognizing that there is a tentativeness about women's commitments to intellectual life during the time they are twenty to twenty-five years old. . . . The need is for counseling and guidance in the period prior to this time of tentativeness and detachment that will make more certain a later return to the world of inquiry and academic life." The Minnesota Plan was probably the most explicit effort to adopt college education to the new model of womanhood as a life lived in stages. But other programs, including the Center for Continuing Education at Sarah Lawrence, mathematics retraining at Rutgers, programs in adult education at Brooklyn College, and the Radcliffe Institute for Independent Study, were also geared to the needs of women who returned to school after marriage and childbearing.[24]

Other schools also made various adjustments to the emerging consensus that women's needs were unique, depending on what aspect of this multi-role woman they chose to emphasize. In a wide-ranging survey of what American schools were doing to adjust to the new understanding of women's requirements, the American Council on Education found a variety of experimental programs and courses. Barnard College, one of the traditional "Seven Sisters," had introduced a course on women's roles in the modern world. The University of Michigan and Syracuse University had devised similar courses oriented to locating the roles and contributions of women in Western culture and history. Though largely traditional content courses, with a slant toward what we would today call women's studies and were then part of the general education movement, these additions often had a strong interdisciplinary dimension. Other schools were more ambitious. Students at the University of Illinois could take a minor in family and community affairs, while the University of Florida set up a whole department on family life. At Earlham College a "New Program of Home Arts and Family Life Studies," was announced for 1958–59, designed primarily for "the more than 80 percent of the women students who will become wives and mothers."[25]

While Earlham College was clearly following White's lead and designing a specific set of courses to educate women for homemaking, most of the program innovations of the fifties were, in fact, largely a continuation of progressive additions of the thirties, as students found it possible to make their studies more relevant with a potpourri of courses in psychology or sociology on the family and childhood. By 1962, about one-third of all colleges and universities made such studies available to women.[26] Most of these were a far cry from the cooking, home decorating, and flower arrangement courses White had advocated, but they did permit schools to see themselves as adapting to the multi-purpose woman.

The most radical innovations in course offerings were not in the liberal-arts colleges and universities and did not lead in the direction of greater preparation for home and family, though they implicitly addressed life-course concerns. At technical colleges and junior colleges across the country, schools with little resistance by discipline-bound faculties committed to older traditions of liberal education, a variety of offerings were being designed to meet women's specific needs in the labor market. In the 1950s, these schools were expanding and providing classes of students who had hitherto not gone beyond high school with new opportunities for life preparation. Places like Broome County Technical Institute, Vermont Junior College, and Georgetown Visitation Junior College were offering women courses in semi-professional pursuits that we have come to associate with the dual labor market—medical records, stenotypy, dental hygiene, business arts, and a host of similar subjects.[27] While making an obvious play for students oriented to acquiring specific skills, these efforts did not, in fact, challenge traditional institutions of higher education and their clientele. Nevertheless, the junior colleges were in their own way effectively appraising a situation in which women would require skills to enter and subsequently reenter the workplace. Directed in a way that the liberal arts could never be, the new range of college-linked courses spoke directly to the new model of women's fragmented roles. In so doing, of course, they not only prepared women for the realities of the dual labor market but also exposed them to its intrinsic hazards and limitations.

In a more limited way, liberal-arts institutions also made selective provisions of this kind. Small Agnes Scott College in Georgia, for example, with its limited faculty and resources but long-standing liberal-arts tradition, permitted its students to enroll in business economics at Emory University. Connecticut College for Women provided its students with opportunities to take courses in marketing and management as well as secretarial skills.[28] Most schools, of course, provided students with some courses if not whole programs geared to obtaining teaching credentials. Large universities were best able to diversify and provide these kinds of offerings to both men and women. Predictably, women usually chose different kinds of practical courses than men as they realistically assessed which would be most immediately useful, just as they did in high school. Some school officials were acutely aware of the danger of this differentiation of curriculum. The assistant dean of the faculty at Indiana University observed that there had been a trend, more marked for women than for men, to establish "dead-end programs, offering no basis for job advancement." At Indiana, "there has been a decrease of English majors . . . and an increase of majors in such fields as medical records, librarianship, medical technology, and dental hygiene. This has been going on long enough so that now some of these girls are coming back and wanting to do graduate study in such fields as biology and chemistry, but often they find they are not admissible to graduate schools without taking more undergraduate courses. Let's be sure we don't go too far in this."[29]

In fact, the overwhelming consequence of the model of women's lives as a moving pattern was to shore up arguments for the utility of the liberal arts. Despite some experiments like those at Earlham College and the strictly utilitarian offerings of junior colleges and technical schools, the debates on women's education and the growing hegemony of the life-cycle model confirmed rather than undermined the liberal arts as the preferred form of women's higher education. Most modifications in the 1950s were just that, small additions in courses on family or in secretarial skills, or special seminars attached to regular courses of instruction which addressed women's lives in the modern world, their prospects and dilemmas. In fragmenting women's lives into a panoply of roles and a long-term interrupted cycle, those involved in higher education made the broad, netlike construct of the liberal arts the almost necessary preparation for women whose lives could at any point take an unexpected turn; from work to rearing children; from family to community service; from marriage to divorce or widowhood. The Vassar catalogue, as befitting a school with a strong feminist past, noted that "most professional schools advise a student to obtain a sound foundation in the liberal arts as the best preparation for admission. This holds true of architecture, law, medicine, social service and teaching." But during the same decade, Vassar increasingly made provision for married students to continue their attendance at the college. Thus, while Vassar's catalogue became increasingly rich in descriptions of preprofessional preparation in expanded sections on graduate education, the same catalogue explicitly addressed the concerns of its students for matrimony: "The college is prepared to assist students who marry during their undergraduate careers in making appropriate plans for continuing their education. Permission to complete work for the degree at Vassar is given only if plans are reasonable in terms of a sound marriage and a sound educational program." Barnard College always expected to graduate more women than it admitted as freshmen, since its junior and senior ranks were augmented by women who took their final year in absentia or transferred from other "Seven Sister" schools as they followed their young, career-minded husbands to New York. Liberal-arts colleges were thus addressing the new "stretched" view of women's lives, eager to please the career-oriented as well as the married coed.[30]

The Commission on the Education of Women of the American Council on Education, which was heavily invested in the life-cycle model of womanhood, set the pace for the majority of opinion on the question of what kind of education was best suited for women's needs. "After a careful review of all evidence," the council concluded that "programs of education should provide broad and differential opportunities, they should not be limited by any stereotyped ideas of women and their roles." As the decade of the fifties progressed, the growing hegemony of the life-cycle view and the education appropriate to that complex model became manifest as conference after conference identified the liberal arts as the strategic preparation for women's multiple aims. In 1955, the American Association of University Women (AAUW) held a major conference geared to defining the "Goals of Women in Higher Education." In line with the spirit of the time, the goals were divided into three parts. Dorothy Woodward defined these in her keynote address (as summarized in the journal of the association): "A career being part (and she underlined the word *part*) of this triple goal, together with a happy home and service to her fellow men." The more than fifty college presidents, deans, and other interested educators in attendance unanimously agreed that only a broad education suited these purposes: "Liberal education is the best means to help women attain the three goals." Like most educators in the fifties, participants quickly zeroed in on the particular uniqueness of women's lives and concluded that "Women must regard the so-called interruption of the homemaking period, not as lamentable, but as a dignified and desirable phase of their lives. . . . Homemaking on these terms is as much a career as any other. . . . The home-versus-career dichotomy is false. The home makes a contribution as great as any other, greater than most." Finally the conference elaborated the consequences of the life-cycle view by noting that "The three goals—career, homemaking, community service—are sequential, not simultaneous. . . . We must see that in the pursuit of her aims, she sees that all must be done decently and in order." The AAUW conference had effectively highlighted the consensus of opinion on women's education in the fifties. Having adopted the complex, interrupted life-cycle analysis, it both defined liberal education as the most effective preparation and linked that education to the traditional female concern for family life and community service.[31]

The liberal arts increasingly became the solution to all women's concerns, meeting the needs of women whose "educational interests may be," in the words of psychiatrist Karem J. Monsour, "more of a hobby than a life and death matter" or, since few who spoke to the matter of women's education were willing to take quite so severe a position, not "as comprehensive as a man's." In the words of C. Easton Rothwell, who succeeded White at Mills, "let us think not only of her repreparation for a place in the labor force or in the realm of volunteer activities, but also of continuing education to expand the range of knowledge possessed by both the wife and her husband."[32]

The liberal arts were presented as a kind of investment in family intelligence, providing women with preparation for contingencies as well as resources for her multiple roles in the family and in the community. Put this way, the liberal arts, with perhaps a few courses on the family and child development and some education courses geared to primary school teaching, provided not only the breadth to fulfill all of a woman's requirements but also the very foundation for family togetherness. Esther Peterson, director of the Woman's Bureau, put it this way: "The significant influence of women in our society is felt first in the home. Women constitute the nucleus of family life. They are the first teachers of our children. For this reason alone, potential mothers should be educated to their highest capacities. It is staggering to think about the resources of knowledge, wisdom and patience needed to answer truthfully and effectively, hour after hour, and day after day, the steady flow of questions which pours forth from little children." Staggering indeed, but it is more staggering perhaps to contemplate women patiently marshalling their knowledge of Greek philosophy, music history, and organic chemistry to that task. By the late 1950s, however, Peterson's view was heard more and more often as the defense rather than the basis of an attack on the liberal arts. Mary H. Donlon, for example, declared, "We need to educate women, so that they may educate the young. . . . This is to say that we must cease shielding girls from hard courses. . . . We must recognize the folly of denying girls serious cultural education because they will marry, while at the same time entrusting children to their rearing. We must wipe out the crazy notion, that, in the adult world of ideas and of reason, women do not really count." The defense of the liberal arts in the late fifties reminds us of nothing so much as the arguments used in the early republic to provide women with the rudiments of literacy.[33] What had been introduced in the early nineteenth century to promote republican motherhood was replayed in the middle of the twentieth century in higher-education circles as a form of broad-based, responsible female citizenship.

It was ironic that the liberal arts curriculum, which was a banner of woman's equal intellectual attainment in the first half of the twentieth century, became by the 1950s the broad banner to wrap around the wide expanse of her life course and another form of the matronly girdle. It had not been predictable when Lynn White launched his attacks on the traditional liberal arts for women that those same liberal arts would become the strong arm of the hand rocking the cradle. In part, of course, by being put on the defensive, the proponents of traditional education found it effective to use the very weapons used against them. But more than this was involved. Once the life-course analysis became dominant, it was difficult to resist rationalizing the liberal arts as a fitting form of utilitarian education. Since the life-course model emphasized woman's uniqueness and took as its central phase woman's involvement in motherhood, the appropriateness of the broad education provided by the liberal arts became, in good part, an education for motherhood.

The stretched and sliced view of women's lives had by the late fifties and early sixties become nearly universal and increasingly defined not only the relevance of the liberal arts but also the very substance of women's lives in America. By 1963, it was embodied at the highest levels of government policy in the report on education by President Kennedy's Commission on the Status of Women. The commission report summarized the perspective developed in the course of the educational discussions of the fifties and noted that the "special responsibilities and life pattern of women impose special educational requirements." The report began not by looking at the situation of the young girl but at that of the mature woman, and the life-pattern analysis guided the volume as a whole. Among its conclusions, the commission noted that "the development of individual capabilities for effective participation at all levels of society is increasingly and continuously dependent on education. . . . It is especially true of women who bear such a large responsibility for the next generation in the Nation's homes, schools and communities."[34]

The decade of the fifties had begun with Lynn White's "progressive" assault on the liberal arts as insufficient for women's needs. By the early 1960s, women's needs had been so broadly defined that nothing but a new version of the liberal arts would do. As a result, the liberal arts, not in content but in purpose, had been so crucially redefined that instead of being the beginning of learning, they had, in effect, become the substance of learning. Above all, they had become the badge of an all-purpose preparation for life adjustment for women. Trained in the liberal arts, women would be fortified to fulfill the many roles and portions of their lives. It was altogether fitting, therefore, that in his 1964 inaugural address as president of Vassar College, Alan Simpson announced to the latterday descendents of feminists that "It is not a choice of marriage or a career, or of marriage and a career, but of a life lived in phases, with its opportunities for fulfillment."[35]

III

The discussions about the higher education of women in the 1950s generated much fruitful research, often illuminating panel discussions and a new vision of the life cycle of women. What they did not do was eliminate women's social role as a significant issue in the kind of education they should receive. Instead, the dominant view splintered woman's roles into worker, citizen, housewife, and mother and sliced up her life over time. This allowed various schools, administrators, and investigators to emphasize one role or a set of women's roles as paramount, and it allowed equal education in the form of the liberal arts to survive. But, instead of eliminating the confusion about women's education, it often magnified the problem for college women, and it sometimes led investigators to blame their subjects for not caring about their own future. In an essay entitled "The Uncommitted Majority," Paul Heist summarized this position: " 'Young women, you stand accused—held accountable for the days and years that are yet to come. While plagued with indecision, confusion and frustrations, many of you will not be meeting the essential social responsibilities nor satisfactorily resolving the demands of home and career; and perhaps more important, you will not attain acceptable levels of self-realization.' Words something to that effect may be inferred, if not directly borrowed, from the comments and apprehensive statements coming from numerous sources."[36]

Some of the more perceptive observers were keenly aware of the degree to which the educational discussions had not resolved but merely replayed the confusions women experienced in college. M. Eunice Hilton, dean of the College of Home Economics at Syracuse University, noted that the endless discussions suggested how little had really been solved: "This unsolved conflict causes continuing unrest for both men and women, and so there are endless attempts to solve it most of which seemed to deal with solutions to problems caused by the basic conflict rather than the resolution of the basic conflict itself. Resolving the basic conflict calls for changes in the real world, or changes in ideologies and mores." Hilton believed that the colleges needed to reconceive of their role, to challenge rather than to adapt to cultural cues: "It is the job of the educators to interpret the value of education for women, to bring realistic thinking into the educational planning of parents and students. . . . To change long habits of thought and action is no easy task; it is, in fact, an attempt to change a culture itself."[37]

Kate Hevner Mueller, a professor of education at Indiana University, was more specific. The colleges, she maintained, had not directed women to the liberating resources at their command: "Women have not capitalized in their curricula on the value of professional training as a liberating force in their education. . . . to hold that any kind of education for women is good in itself is a supercilious, debutante point of view, misleading to students and to society. The woman of the future will work, and only an education which fits her for this life work can free her from the crippling pressures of her surrounding culture."[38] Mueller urged the schools to provide women with a clear direction, a way out of the confines of a confusing culture with its mixed signals, a means to resolve and overcome dilemmas, and ultimately with a force for personal integration through professional self-definition. Mueller was beginning to voice what we might term the "new feminism" that sometimes struggled out of the confines of the debates of the 1950s, and her anticipation of the values of the seventies and beyond is a significant reminder of the peculiarly muddled context from which a renewed feminist view of women's equal education would emerge.

But Mueller was willing to do what few other participants in the discussions of the fifties were willing to do, to dismiss the issue of woman's family role as irrelevant to the direction and content of women's schooling. She did not dismiss a woman's need for family, for she hoped women would "demonstrate to their male colleagues that women can be happy and useful in both roles, just as the best men," but she dismissed the relevance of this role for issues of higher education. Moreover, Mueller was willing to do what others almost never did, to turn her gaze from both the average girl and the culture into which she graduated. In that sense, she, like Hilton, proposed that college education become a force for social change, not a reinforcement of the status quo. "Togetherness, that conspiracy of publishers, advertisers, and manufacturers, has had its day; but there will be other conspirators and women will fall for them because they, with their husbands, are spending their mental lives in the cute little prefabricated house which society holds out to them."[39] The force of Mueller's argument was of another order entirely than the one that confined the life-cycle analysis to viewing schooling as a preparation for women's inevitabilities.

Unlike Mueller, or even White who had challenged the dominant values of the existent culture, most educators who participated in discussions and investigations on women's higher education in the 1950s were mesmerized by the society as it was. And the most fundamental feature of that society and its culture as it concerned women was matrimony and motherhood. Most commentators believed that they were taking their cues entirely from female students themselves. Thus in 1950, Bancroft Beatley, president of Simmons College, which had a highly developed prevocational program for women, noted that it was "obvious that the typical college woman looks forward to marriage, a home of her own, and children." He added that despite the "pernicious" influence of the feminist movement, "fortunately for the character of American home life, the great majority of the current generation of college women readily acknowledge that their goal is marriage and children and see no conflict between accomplishing that objective and living a rich life." Beatley's appraisal, with its rebuke of feminism and its encomium to the satisfactions of family life, echoed through the decade. So did the repeated assertions that educators were merely taking women at their word. At mid-decade, Mervin B. Freedman reported that at Vassar, "Marriage at graduation or within a few years thereafter is anticipated by almost all students: the percentage who state that they are not likely to marry, or who are quite uncertain about it, is negligible." By the early 1960s, the refrain was still strong. Carl Binger, a psychiatric consultant at Harvard, reported that college women were "interested, first and foremost, in finding a mate," although he added significantly that "they do not shout this from the housetop. They often spend a good deal of their time and energy in trying to conceal it from themselves and others."[40]

Investigations of women students in the fifties and sixties went to great pains to document how deeply women were invested in marriage goals and organized their researches around this premise. In one study, Jane Berry surveyed 677 college women from fifteen colleges and universities across the country. She found that most women ideally expected to have their first child before twenty-five and that two-thirds of all women interviewed wanted at least three children. The degree to which the governing beliefs of the period guided the investigation itself is suggested by Berry's opening statement: "As might be anticipated, marriage occupies the most prominent place in the life plans of the young women cooperating in the study." The governing beliefs were also evident in Berry's summary of the reasons why women went to college at all: "To be a more well-rounded person and a better wife and mother; to occupy my time until marriage; to be prepared in case I have to earn my living; to meet men that have the same interests, and would be good husbands and fathers."[41]

Other investigators made the same assumptions about women's goals but were not quite so sanguine as Berry about the results. At a time when the discovery of talent was a national security issue, a variety of investigators were keen to discover which college women were able to swim against the tide. At the University of California at Berkeley, the Center for the Study of Higher Education was concerned to locate the sources for independence of thought and sponsored a large-scale assessment of the motivations of college students. On the basis of these investigations, Paul Heist concluded that "It is apparently difficult for most college women to emphasize, let alone substitute for matrimony, . . . maximum education and preparation for meaningful careers and contributing citizenship. . . . The opprobrium of not being engaged as a college senior is almost stigmatic on many campuses." "Most girls," of superior ability, Heist found, "were giving serious thought to their educational and professional futures at the freshman and sophomore level, but they expressed less intensity of commitment for further education by the time they were interviewed as seniors." In another article, Heist made this point even more emphatically: "Among these students of the highest calibre, and often the most impressive potentiality, it was surprisingly infrequent to find a young woman genuinely committed to a discipline, a professional future, or a career. . . . For those senior women interviewed, not already married, all saw marriage as a culminating goal of great if not first importance."[42]

Heist brought considerable insight to his discussion of the major sources of this lack of professional motivation among college women: the lack of role models, fears about demonstrating "masculine" characteristics, feelings of female inadequacy, and the general cultural assumptions that confined women to positions of passivity. Despite Heist's clear and often acute discussion of the large cultural inhibitions on women's success in college and after, he, like so many others, noted that "curriculum and faculty perspectives must allow examination of and preparation for the realities of living. . . . Higher education should accept greater responsibility for providing a more adequate foundation for combining intelligent motherhood, homemaking, citizenship, and the achievement of a meaningful life." Heist made this point again in another place when he observed that "an education outside of the context of the culture and life of the particular person is not an education."[43] Thus, despite his own regret at the outcome, Heist felt compelled to accept the democratic position that higher education should address students' needs and, therefore, reinforce the life-cycle inevitabilities of women.

At Vassar, a large-scale investigation of college women was also under way that documented the specific intra-campus sources for the goals, objectives, and self-definitions adopted by most women students. Under the direction of Nevitt Sanford, the Vassar study, like that at Berkeley, reinforced the image of marriage-hungry undergraduate women. The peer culture at Vassar set "concrete limitations as to permissable life-styles and life-work ambitions. Vassar girls, by and large, do not expect to achieve fame, make an enduring contribution to society, pioneer any frontiers, or otherwise create ripples in the placid order of things. Future husbands should mark out and work directly toward a niche in the business or professional world." The Vassar peer culture strongly enforced larger cultural norms. "Not to marry is almost inconceivable and even the strongly career-oriented girl fully expects that someday she will be a wife and mother. Not only is spinsterhood viewed as a personal tragedy but offspring are considered essential to the full life. . . . In short her future identity is largely encompassed by the projected role of wife-mother." John H. Bushnell concluded his discussion by noting that "it is not surprising to discover that Vassar students view the suffragette movement and the whole issue of Women's Rights with indifference. . . . For these young women, the 'togetherness' vogue is definitely an integral theme of future family life, with any opportunities for independent action attaching to an Ivy League degree being willingly passed over in favor of the anticipated rewards of close-knit companionship within the home-that-is-to-be."[44]

Many investigators traced college women's absorption in marriage goals back to their precollege experiences where women preferred some studies to others, and college-going was less urgently desired by them than by men of the same background.[45] Some investigators moved in the other direction, however, away from the cultural context to locate the sources of women's choices in their daydreams, fantasies, and inner selves. In one especially influential study, Elizabeth Douvan and Carol Kaye did not look at the culture at large or the subculture of the schools, but at the prevailing differences in male and female "identities" for the sources of college women's marriage orientations. "The identity issued for the boy is primarily an occupational-vocational question, while self-definition for the girl depends more directly on marriage." As a result, "Boys often phrase college aspirations as vocational aspirations. . . . For many girls, college obviously is an end in itself, only dimly conceived in an instrumental light." Douvan and Kaye concluded that "Girls' phantasies about college are not simple in content. The dominant theme is a social-sexual one, but other themes— travel and geographic mobility, transformation of the self, social mobility, and a general sensuous longing for experience and the exotic—figure in their thoughts as well. . . . The dream of college apparently serves as a substitute for more direct preoccupation with marriage; girls who do *not* plan to go to college are more explicit in their desire to marry, and have a more developed sense of their own sex role. They are more aware and more frankly concerned with sexuality. . . . Since we find nothing to indicate that girls who plan to go to college are late developers socially or sexually, we infer that their sexual interests take some alternative form— and our guess is that they inform the college dream."[46] Douvan and Kaye thus attributed college women's marital preoccupations to their natural identity formation. While noncollege women could be more direct and blatant, college women had to sublimate their needs. In this analysis, college-going itself became a marriage substitute, a socially induced delay which quite understandably became suffused with marital fantasies.

There was a way in which the discussions of the 1950s and early 1960s became a case of the dog chasing its own tail. The more educators believed women undergraduates to be directed toward marriage, the more women undergraduates seemed to want nothing more than marriage. In fact, there was no reason not to expect the results they found. Women, like men, wished to marry and rear families. That was obvious in the fifties and was disturbing only if investigators either questioned the utility of higher education for women or sought to rationalize the courses the colleges were offering. A few saw clearly past the tempest to assert that the colleges' concern was not with what women wanted but with what schools had an obligation to provide. Margaret Clapp, president of Wellesley College, came quickly to this point at one of the organized conferences on women's higher education: "I think it is important to study the motives, interests and opportunities, and problems of women, and of individuals, but when we talk about institutions of higher learning, it seems to me we have to consider not only the students who are there . . . but also the responsibility of these institutions, as the custodians of a very painfully acquired body of knowledge over the centuries, to transmit that knowledge and to advance the knowledge and truths over time. These are the central responsibilities of institutions of higher learning."[47] Clapp was standing on the very high— and unprogressive—ground of traditional education in which students were taught what others believed they needed to know for the sake of the continuity of learning, not the development of the student. In the fifties, with issues of life adjustment and personality development all around, it was difficult to appreciate the degree to which Clapp's old-fashioned position provided the very vantage for the future progress and advancement of women.

Mirra Komarovsky was almost alone among researchers to grasp the crucial contribution that this traditional perspective would make for women's future liberation. But Komarovsky refused to accept the traditional perspective for its own sake. She brought greater flexibility and a more reformist spirit to her inquiries. In *Women in the Modern World* (1953), Komarovsky's answer to White's *Educating Our Daughters,* she was eager to provide students with an updated education that spoke directly to women's concerns and their roles in the modern world. But Komarovsky was interested in enlightenment, not practice, and she knew a red herring when she saw one. She disavowed a strictly functionalist education as misguided and weak. "Every educator worth his salt wants an education which will function in the future life of the student, but function on what plane? Is deepening of intellectual awareness to be dismissed as nonfunctional?" "Whether or not it is an intrinsic feature of the 'practical' courses, the fact remains that they sometimes tend to stress adjustment to the existing institutions rather than their critical appraisal."[48]

In a study that touched all the bases and went out of its way not to fall into any pre-existing pattern, Komarovsky used her extraordinary insight and materials drawn from teaching at Barnard College to expose the futility of a debate over an education specifically appropriate to women, whether it was narrowly or liberally defined. "The very education which is to make the college housewife a cultural leaven of her family and her community may develop in her interests which are frustrated by other phases of housewifery. . . . We want our daughters to be able, if the need arises, to earn a living at some worthwhile occupation. In doing so, we run the risk of awakening interests and abilities which, again, run counter to the present definition of femininity." *Women in the Modern World* was a brilliant and tolerant but often angry defense of the liberal arts, not as the protector of past tradition, not as the only solution to the fragmented life course of women, but as the crucial basis for the development of human potentials. Above all, *Women in the Modern World* was a defense of education— equal education—as an independent cultural force. Komarovsky rejected a strictly old-fashioned curriculum, as she rejected one whose only relevance was for the status quo. She saw intellectual substance in courses on the family and family relationships. She fully aired the many complaints of former alumnae bored by housework and frustrated by childrearing, who felt untrained for their tasks, and she explored in depth the confusions of undergraduates whose goals for the time being seemed limited to marriage. But Komarovsky, unlike most of those who discussed the education of women in the fifties, saw that the solutions to these problems did not lie in fundamental changes in the college curriculum or in the rationalization of the liberal arts as a package deal for fragmented womanhood; the solution, if it existed, lay in the society and not in the schools. The schools could stimulate thought, they might even promote dissatisfaction and confusion, but they could not solve the dilemmas of women's lives. "It is the job of the college to provide a soil for the fullest fruition of intellectual gifts." It was the responsibility of the society to provide women with the opportunity to permit these gifts to flower in any direction women might choose to use them.[49]

For Komarovsky, women's dilemmas were not the result of too many choices but of too few, of too many inevitabilities and too many traumatic transitions. Komarovsky observed the "sharp discontinuities" that "characterized the whole pattern of their lives. . . . The incentives and the psychological adaptations at one stage of her life cycle did not make for a smooth transition to the next one."[50] Komarovsky refused to accept any easy solution to women's plight. While she urgently sought social and cultural reforms, her study of women remained largely unresolved and very unlike the tidy package deal that was offered to young undergraduate women in the form of the liberal-arts curriculum of the 1950s and '60s.

IV

Komarovsky's insight and sensitivity to women's dilemmas were not widely shared in the fifties and early sixties. And many researchers found themselves blaming students for the confusions that surged around them and their education. Since they believed themselves to be merely observing what women themselves wanted, investigators and educators alike seemed unaware of the degree to which they contributed to the choices women made. "The perceptual horizons," of senior women, Paul Heist asserted, "did not extend much beyond the romance of matrimony," and they were largely unconcerned about what lay beyond. "A great majority of college girls, both freshmen and seniors . . . either profess or seem to be generally *unaware* of the problems portrayed by so many. . . . the great majority live in an anticipatory haze, with romantic notions about matrimony and home, soon to follow." Jane Berry found that very few undergraduates had done any reading "on the subject of marriage-motherhood-career-relationships."[51] Having disdained planning for the future beyond marriage, women were themselves to blame when they felt unhappy, frustrated, unprepared, or unfulfilled later in life.

In fact, educators provided women with little firm ground from which they could move beyond the college haze. At a time when Erik Erikson was developing a strong male model of the life cycle that emphasized internal integration and linear progress culminating in psychic health, educators concerned with women were proposing that women held various roles at different times in a fragmented life course which was somehow to provide self-fulfillment at each separate and unique juncture. I do not want to diminish the contribution to a more realistic appraisal of women's lives that emerged from the discussions of the fifties, nor to underestimate the potential confusions for male students who were required to resolve work commitments during the vulnerable period of adolescence. Nevertheless, it is useful to compare the models of manhood and womanhood that coexisted in the fifties. For men, schooling was perceived as fundamental in defining future and lifelong directions in the work sphere with the accompanying ego strengths that came as work choices helped to define selfhood. Erikson also saw the college period as full of confusion for men, but the successful emergence from adolescence into adulthood promised firm resolutions through work choices that issued in lifelong self-definitions. For women, schooling was a partial preparation for work before childbirth and after children matured. But early adulthood would be filled with family-centered concerns to which schooling might or might not contribute, depending on how wise a woman was in using her broad, all-purpose education. In this schema, college provided women with neither a context for firm lifelong commitments nor the occasion for working through personal conflicts.[52] Instead, the college period generated conflicts in roles and self-conceptions.

Women, in fact, seemed to suffer from just such deep conflicts in self-conception in college, and neither the life-cycle model nor college life generally provided much guidance. At Vassar, for example, "many seniors . . . experience a sense of conflict between what they have been educated for and what awaits them. They seldom can define this conflict for themselves or elaborate its details, but it is present nevertheless, and it often contributes to the perturbations and doubts of the senior year." And "many seniors rush into marriage . . . as a way of resolving the dilemmas thrust upon them by graduation." Komarovsky provided the most detailed documentation of this conflict from memoirs of and interviews with students and alumnae. One senior wrote: "I get a letter from my mother at least three times a week. One week her letter will say 'Remember that this is your last year at college. Subordinate everything to your studies. You must have a good record to secure a job.' The next week her letters are full of wedding news. This friend of mine got married; that one is engaged; my young cousin's wedding is only a week off. When, my mother wonders, will I make up my mind? I wouldn't want to be the only unmarried one in my group. It is high time, she feels, that I give some thought to it."[53]

The conflict between studies at college and the marriage goals of students was latent throughout the discussions of the fifties, despite the often glib denials by educators who believed that, unlike earlier belligerent feminists, women no longer saw a conflict between career and marriage. Indeed, Komarovsky documented the degree to which this perceived conflict often led undergraduate women to mask their achievements and discount their abilities. A transfer from a coeducational institution noted that "everyone knew that on that campus a reputation of a 'brain' killed a girl socially. I was always fearful lest I say too much in class or answer a question which the boys I dated couldn't answer." Another noted, "Quite frankly, I am afraid to go into some kind of business career because I have a feeling that I would cheat myself out of marriage. . . . This fear has led me to revolt mentally against the sort of life toward which my chosen subjects here at school are leading me." Komarovsky reported that "65 percent of the coeds at a large Western campus thought it was damaging to the girls' chances for dates to be outstanding in academic work." At Berkeley, Paul Heist found that "even the young female intellectuals, scholars and creative students alike, will forego the satisfactions of their academic pursuits in order not to jeopardize their opportunity for marriage." And David Riesman observed that "In the better institutions, they scorn dilettantism while yet understandably regarding deep intellectual involvement as a potential threat; looking perhaps at some of the unmarried women on the faculty, they may fear that such involvement would cut them off from the life of a normal, average woman, and they are persuaded that it is more important to be a woman than to become some kind of specialist." At Vassar, the "emphasis on combining good marks with a reasonably full social life is so strong that some students who, in reality, have to work hard to maintain an impressive grade-point ratio will devote considerable effort to presenting an appearance of competency and freedom from academic harassment." The grind was no more popular in the fifties than in the twenties, but for women it was not merely the overachievers but achievement itself that seemed to be the threat.[54]

It is striking how much hiding and dissembling emerges from the discussions of women's conflicts over academic success versus popularity, career versus marriage. This need to distance themselves from the implications of their studies suggests the degree to which the "academic line" that the liberal arts provided an excellent overall preparation for marriage and family was not assimilated to college women's sense of themselves or into their own subcultures. Indeed, perhaps the real theme of women's education of the period was not the fragmented life course but the divided self which it enforced. Women found the signals emanating from most schools confusing, and they were apathetic to the larger debates around them, not because they were uninterested, but because they provided little help. To advise women that their lives would be repeatedly interrupted could have had little meaning to women in their late teens and early twenties whose own visions of the future were culture-bound and rosy. Women undergraduates were little different in that sense than men. In examining the personality characteristics of male and female students, one group of investigators found that "few students have the kind of personal autonomy, or independence, or even, perhaps, social alienation which permits them to defer for long their vocational or marital aims, in the interests of following other pursuits. The majority of students soon forego experimentation with roles, and any questioning of basic values, in order to secure as soon as possible a relatively definite plan for the work of the future."[55]

It was not, therefore, surprising that the University of Minnesota could only interest 100 young women in a plan aimed at assisting women to a more realistic preparation for their long-term life course. "Our biggest problem is to get young women who are undergraduates to concern themselves with planning for continuing education." Recent Vassar graduates were also uninterested in "the matter of what life will be like in 15 or 20 years." Like their male peers, women accepted the values of the culture and directed their undergraduate lives accordingly. Unlike male undergraduates, however, for whom a college education made professional sense, women's college careers offered no momentum toward the future. David Riesman, in his usual perceptive way, grasped this fact clearly, noting that women in good liberal-arts schools "have a hard time thinking of themselves as pursuing a career (rather than a job) after the children are grown. . . . Instead they pursue the liberal arts with the thought—and they do think about it—that on graduation they will get the sort of job open to any reasonably intelligent and attractive A.B. . . . even very gifted and creative young women are satisfied to assume that on graduation they will get underpaid ancillary positions."[56]

Women undergraduates were absorbed in the here and now where dates and future marriage formed the most urgent part of their self-image as women, an image defined by their culture and strongly supported by student subculture. In that context the skidding age of marriage for women proved a real threat and caused seniors who were not yet engaged to be married to experience what was popularly known as "the panic." College women did not need sociologists to tell them that in a culture where "the individual man's equipment for achieving success in most semi-professional and professional occupations increases in effectiveness over the post-college years toward a high point during middle age or later. The individual woman's success, however, is culturally declared to be at a maximum very early in life—during the years covered by college careers, to be exact." In thus clearly pinpointing the differences in the goals of men and women in college, Walter Wallace had described the source of women's fear, but none of the available theories could effectively allay either that fear or the conflict implicit in women's attendance at college. The emphasis on the liberal arts shielded the colleges, not the women, from that fact. To have expected women to find a means psychologically to integrate that fear with their involvement in day-to-day academic activities was to demand of the students what their teachers could not achieve in their theories. As anthropologist Florence Kluckholn reported, "Time and again when young women of college age or younger are asked what they know about household management, child rearing or cooking, the answer immediately given is, 'I'll do that when I have to.' And with the matter brushed off in this indifferent phrase they turn back to books on economics, social organization, Italian literature, or nuclear physics."[57] The conflicting signals of which so many undergraduate women complained led to a split attention rather than a fundamental integration of purpose or to the devising of long-term plans.

Certainly, some women studied home economics or dental hygiene, and very many took at least some education courses in the expectation of holding the jobs that Riesman recognized to be part of their acceptance of the limits of opportunity.[58] But for most women in liberal-arts programs, the split response, scurrying for a date and reading Dante, was the better part of wisdom, since they needed to succeed on the terms the society had set for them, and very few, even college presidents, seemed able to provide alternative visions. Even the woman with a Radcliffe Ph.D. expressed a deep conflict between professional and marital goals and "the majority of married Ph.D.s [gave] up full-time work to pursue their professional interest on a part-time basis, intermittently, or not at all." For most the conflict had "invariably been resolved in favor of family and especially children."[59] It was small wonder then that the average undergraduate woman saw little point in thinking too hard about the problems of her own fragmented future. College women preferred not to anticipate the degree to which their family chores would remain isolated from the academic pursuits of their youth. It was not until later, isolated in suburban tracts, that so many well-educated women would question the wasted resources of their college days. It was that sense of unfulfilled possibilities, of studies once seriously and equally pursued with men but now largely a matter of nostalgia, that Betty Friedan would resurrect into the basis of female anger and rebellion.

History offers us the delicious dessert of hindsight, and looking back from a period when women's education, in the seventies and eighties, seems to have settled into an aggressive pattern of equal rights for professional preparation, it would be easy to dismiss the concerns about higher education for women in the fifties as a tempest in a teapot. But the issues raised and the failure to resolve them by educators and students alike are in fact of considerable consequence. In the early fifties, women's higher education seemed genuinely threatened initially by competition for place that defined the postwar educational world and in which the allocation of precious educational resources seemed to be best spent on men. This was no mere illusion since women were often passed over in favor of men for admission to coeducational schools.[60] The hard-fought battle for equal education for women also seemed threatened by attempts at fundamental curriculum restructuring like those proposed by White. Above all, cultural perceptions about women's lives fundamentally challenged the utility of higher education for women, and in so doing they challenged the meaning of higher education itself.

Schools exist within a specific cultural context that defines the problems they are asked to address and usually sets the limits of their effective action. The higher education of women was an almost pure instance of this relationship. In a society that had liberated women sufficiently to pursue higher studies but had not actively encouraged them to use those studies, the latent problem of what their studies were for could not long be ignored; certainly not after college-going became a popular, even a normal, experience for middle-class women. If that society, moreover, continued to define womanhood by distinctly family-linked attributes and rewarded women who demonstrated those attributes with early marriage, the issue became clearer yet. Why, after all, should women go to college when all they really wanted was motherhood? If the colleges saw themselves as mere functional appendages of the existing culture—as became increasingly the case in the twentieth century—the question almost necessarily led to Lynn White's answer, or it resulted in a rationalization of the liberal arts as an all-purpose education, good for mothers, citizens, and workers.

What prevented White's answer from guiding the future development of

American college education for women had less to do with the leadership of the colleges or their foresight in preparing women for a contingent future than it did with the conservative orientation of the colleges and the vested interests of their faculties. The liberal-arts college is as much an expression of the research university as it is the well-conceived and articulate bastion of liberal learning it pretends to be. Drawing its faculty from research-oriented university departments defined by disciplines, the college could not easily acquiesce in a heavy reinvestment and reorientation of purpose. This is not entirely to disparage the sincere commitment that many (especially the older women's schools) had developed to the liberating traditions associated with their broad, sexually neutral offering. It is rather to suggest that the history of how the liberal arts became central to the curriculum, rather than either function or utility, explains their persistence and dominance over time. After all, men as well as women in the fifties were told that the liberal arts provided the best preparation for their futures (which in their case meant the professions or business), as indeed they did. But this was not so because they were the only or necessarily the most effective preparation but because by the 1950s and '60s graduate school requirements were built around the assumption that only liberal-arts students qualified for admission. The liberal arts were the best preparation, not because they were intrinsically useful or wise, but because they fulfilled professional-school requirements that had grown up in the context of the specific history of American higher education and because by the mid-twentieth century the liberal arts had become the curriculum of choice among the leading institutions.

The liberal arts had become such a fact of life by the 1950s that they could be used to cover all the bases, as they were when they were made to fit the multi-purposes of women's fractured life cycle. But, in adopting a moderately progressive, student-oriented aura to their historically derived curriculum, by emphasizing that the liberal arts provided an education for living, the colleges were largely deceiving themselves, not their students. As one young woman brilliantly assessed the situation: "It seems to me college prepared me to be another female college professor."[61]

In that sense, the fifties proved to be a holding period for women's education. Despite some modifications in curriculum and some broad experimentation in semi-professional as well as in explicitly family-centered curricula, the preservation of the liberal arts as suitable for women as well as for men proved to be a godsend to a female minority. It permitted some women, at least, to use the broad base of the liberal arts to enter the male-dominated professions. From these, they began to stir the caldron for women in general with notable consequences for future efforts at equality of opportunity in the society. And it allowed some women to become college professors who would continue to defend the liberal arts and women's equal right to the social access—as well as the enlightenment—they provide.

In the 1950s, however, the focus of educational discussion was on the majority, on the average girl and her inevitably average culture. That concern cannot be altogether dismissed in a society where education, even higher education, is aimed at the mass. Here the liberal arts were less successful, providing most women with little guidance or direction, encouraging conflict and even a false sense of passivity. It is doubtful that the majority of women ever saw in their liberal-arts education the all-purpose relevance with which educators invested it. Instead, the majority accepted training in the liberal arts as the price of attending college. The liberal-arts program provided them with a sense of immediate purposefulness while they pursued the goal of finding a marriage partner. As Walter Wallace concluded in his study of student socialization at a midwestern liberal-arts college, "the strictly academic side of college life . . . is likely to be viewed by women as a set of bothersome regulations which has to be put up with while one gets on with the real purpose of college." Douvan and Kaye were undoubtedly correct when they described college life as infused with glamour that was the deflected haze of romantic ambitions for marriage. American colleges in the 1950s and early '60s provided women with the glamour of independence without the immediate urgency of decision. As a result, the utility of attendance in a liberal-arts program for women proved to be a lie since it closed rather than opened options. By providing the illusion of training without the substance of direction, it allowed women to invest all their eager expectation in marriage because their studies, good in themselves, bore no clear fruit in long-term preparation. In the absence of goals other than marriage, women found their education not useless but unusable. It was no small wonder that women panicked in their senior year. Without the immediate prospect of marriage, the liberal-arts degree in itself directed women after college only to their traditional outlets— teaching or the typewriter.[62]

The liberal-arts programs of most undergraduates also strongly underwrote the existing culture, not because they prepared women for the many-faceted roles they would soon play (as educators wished to believe), but because they gave women time and room to devote themselves to the task of finding a mate as the final diploma of graduation. Women in college were surrounded by a romantic haze which educators did little to dispel in insisting that the liberal arts were, after all, a fine preparation for motherhood. Millicent McIntosh's assertion that Greek archaeology was good for mothers only reassured women that they should be good mothers not good archaeologists. In so thoroughly shoring up the view that motherhood was, after all, the best game in town, most educational defenses of the liberal arts in the fifties and early sixties succeeded in reinforcing the status quo as thoroughly as a more specifically family-oriented education would have, all the while encouraging conflicts not in orientation but in self-perception based on self-deception. Unresolved, those conflicts in college provided the fertile source for angry feelings of betrayal that ultimately issued in revolt.

A few women struggled out of the illusions to grasp at the essence of their genuinely equal education, but that minority did so at a cost not unlike that of the original pioneers of women's education who were viewed as personally eccentric and socially maladroit. Unlike those early pioneers, however, the minority in the fifties and sixties could expect little social support from their peers at school and no reinforcement from the school culture. And like them, they were in the long run forced to question the cultural assumptions and social categories with the genuinely liberating vantage that a liberal-arts education had, in fact, provided them. In the end, the liberal arts had provided the few with the wherewithal to question the condition of the many, and the tradition of equal education resulted in demands for a wider equality in which that education would make sense.

In a society which had come to value education for the solutions it offered to the problems of the individual and for its specific utility, for the manner in which it helps to define and support the culture, as well as for promoting excellence and critical thought, it is perhaps an inevitable consequence of an open-ended education that it should encourage confusion as well as enlightenment. In the 1950s it allowed women to bide their time as well as to strive for self-realization. By the twentieth century, the American faith in education had become so broad—and so demanding—that it could no longer (if it ever could) satisfy the many elements whose needs it sought to meet. In that sense the paradox of women's higher education in the 1950s is an instance of the much wider paradox of democratic schooling in the United States, where real changes may issue from the peculiar dialectics of educational institutions oriented to protecting cultural arrangements and social hierarchies while they educate the mass.

6

Imitation and Autonomy:
Catholic Education
in the Twentieth Century

The fundamental theory of liberty upon which all governments in this Union repose excludes any general power of the State to standardize its children by forcing them to accept instruction from public teachers only. The child is not the mere creature of the State; those who nurture him and direct his destiny have the right, coupled with the high duty, to recognize and to prepare him for additional obligations.

Pierce v. Society of Sisters (1925)[1]

The first step for successful Catholic teaching is to convince every Catholic teacher and have her treasure this conviction in the very core of her heart: that what we have done, what we are doing, the way we have done it, and the way we are doing it has not been and is not inferior to what the public schools have done and are doing, the way they have done it, and the way they are doing it. We have great room for improvement, but we have by no means been failures.

MONSIGNOR JOHN J. FALLON (1936)[2]

Whatever else they have achieved, schools in the twentieth century have succeeded in interposing between parents and their children, for longer and longer periods, the authority of a public and state-controlled agency. In the nineteenth century, when public schooling first became a prominent form of democratic reform, it was clothed in the language of republicanism and citizenship, since these ideas exemplified the governing perceptions of the public interest in childhood nurture.[3] By the early twentieth century, when what was called the crisis of childhood began to haunt progressive reformers and other observers of the general social crisis created by immigrant-fed cities and industries, it was logical, indeed predictable, that the public authority available through the schools would be expanded. Health, vocationalism, sexual behavior, social adjustment, Americanization, all fell under the widening umbrella of the school. While historians have recently been attracted to the belief that public life has been on the decline since the early nineteenth century, a close look at the schools would suggest an equally potent and parallel trend as more and more aspects of private life—and among them some of the most intimate—have increasingly become a matter of public and state interest. One of the most significant of these has been childhood, over which parental authority has been systematically eroded during the past century.

The Catholic church was early and anxiously aware of these developments. Eager to protect its flock from the allurements of a secular and alien culture and the spiritual hazards of exposure to schools controlled by a Protestant (and usually hostile) majority, the Catholic church took decisive steps to shore up its own authority. In so doing, the church believed it also strengthened the authority of parents over what in the twentieth century became the millions of young Catholics who in the absence of alternatives would necessarily have gone to the public schools. The alternative schools created by the church certainly expressed a public interest, though the interested public was the more discrete and identifiable minority community. As such, the Catholic schools, while under ecclesiastical jurisdiction and control, provided a school alternative more closely related to the specific concerns of parents and the Catholic community. Thus, Catholic schools provide a useful and necessary perspective for understanding school development in the twentieth century. As significant alternatives for ethnic Catholics, they help to describe both the range of possibilities available to outsiders in the twentieth century and the real boundaries of those choices.

Catholic schools were established in the nineteenth century, shortly after initial moves to provide a common schooling for all Americans. The early struggles to provide a separate education for a self-conscious minority figure prominently in the public debates of the nineteenth century. By the early twentieth century, the church's right to retain control over Catholic children and to maintain schools for them was an established fact and part of the American self-definition of how minority rights, at least in religion, were constitutionally provided for. That fact (despite repeated challenges and close calls), also suggests the degree to which the exercise of public authority over childhood in modern America has been neither monolithic nor completely successful. At the same time, the growth of Catholic schools in the twentieth century has not been entirely self-generating. While the Catholic schools flourished on the basis of the huge, early twentieth-century migrations, the enormous scope and extent of the school system eventually created under church auspices was far more than a simple response to the numbers of Eastern and Southern European, French-Canadian, Irish, German, and Mexican immigrants. On the contrary, the church's schools grew because the public schools grew. Every expansion of public authority; every extension of the compulsory school age; every addition of programs, services, and activities to school offerings expanded the efforts of Catholic educators on behalf of their own and the complexity of Catholic schools.

Had the public schools not expanded aggressively in the twentieth century, it is safe to say that the Catholic schools would have remained what they were in the nineteenth century—an important but limited expression of community autonomy. Instead, they became an enormous and complex alternative school system which, by the middle of the twentieth century, enrolled around 14 percent of all children of school age and included in 1962 almost five and one-half million children.[4]

Throughout the twentieth century, parochial schools paralleled and imitated changes in the public schools, but with an important difference: the Catholic schools have never succeeded in providing enough places for all Catholic children or even all Catholic children who wished to attend. Despite the often repeated slogan, "Every Catholic Child in a Catholic School," the universalism of Catholic doctrine could not fully compete with the powerful American tendency toward universal education, and of all Catholic children, the number of those who attended Catholic schools has rarely exceeded 50 percent throughout the twentieth century. As a result, despite the explicit drive to autonomy and the less openly avowed competition with the public schools, the Catholic schools have always been dependent on the existence of the public schools to achieve their goals and objectives. Ironically, the success of the Catholic schools—their ability to maintain academic standards and to open lines of social mobility to graduates— would have been impossible without the safety net provided by the public schools. The history of Catholic schools in the twentieth century is therefore very much part of the story of modern public schooling. Not only did Catholic schools serve an enormous public and respond to the same social pressures as public schools, but they also were manifestly dependent on public school authority in order to maintain their own autonomy.

I

The Catholic viewpoint on education, with its strong Christocentric emphasis, roots in Thomist rationalist psychology, and dependence on papal pronouncements and conciliar decisions, provided a strongly conservative center of gravity for educational goals and practice. The church schools— parish-based parochial schools, central diocesan schools, and private schools owned by religious orders—all shared self-conscious and well-articulated traditional objectives. These served as the firm basis of all teaching, not only in matters of religion. Catholic schools have heavily emphasized throughout the twentieth century issues of character formation, self-discipline, the unity of learning, and high levels of academic proficiency. And Catholic educational values continued to emphasize subject-matter mastery, especially on the secondary-school level, long after public schools redirected their vision from the subject to the child. As William McGucken observed, "For the Catholic secondary school, development of the Christian virtues is obviously of greater worth than learning or anything else. . . . But if the school does not attend to intellectual training at all, is not concerned with the fact that its students are not mastering grammar or reading or whatever may constitute the high-school curriculum, then it is not merely a poor school; it forfeits the right to be called a school at all, even though it may be successful in developing the virtues of a Christian character. The Catholic secondary school has the specific function of training for intellectual virtues." [5]

While public schools increasingly emphasized personal adjustment and group socialization, the Catholic schools have throughout most of the twentieth century resisted the implications, though not always the forms, of "progressive" educational practice. In one area, however, the Catholic schools found that the public schools were increasingly in agreement with their own long-standing commitments—schooling was to provide training for living. Education for living had become for educators in the twentieth century a highly ambiguous, not to say slippery, term. Citizenship, personal adjustment, vocational training had all been subsumed in the idea, and Catholic educators, like their secular peers, were not immune to the enticements of that ambiguity. Nevertheless, the Catholic schools could and did always claim that what distinguished them first, last, and always from the public schools was the Catholic emphasis on Christianity as the central core of living. Its lessons about this life as well as the next stood at the center of the school curriculum and provided the integrative force for everything else the schools might offer. Catholic educators most often defined the distinction between their schools and the public schools as the difference between idealism and materialism. This crucial distinction was the governing rationale for Catholic-school separatism and was made in every comparison, latent or manifest, between Catholic and public schools. One of the best and least polemical of such descriptions was contained in a statement by the Notre Dame Study of Catholic Elementary and Secondary Schools in the United States issued in 1966: "Every Catholic school teaches religion. . . . In all, symbols associated with the liturgy and prayer are intimately associated with the school day. The very presence of the religious is in itself a dominant, unforgettable symbol. Here are persons set apart from the world reminding that world not merely of sin, of justice, and of judgment, but also of the unavoidable choice between the holy and the unholy, between the things that are of time and the things that transcend time." [6] Education for living for Catholic educators inhered in the lessons of that choice and the meaningfulness of the supernatural as part of life.

Yet, throughout the twentieth century, the issues and dilemmas with which the church schools had to contend came from within the realm of time and from the very material forces against which the church schools provided, in their view, the greatest resistance. Those conditions included issues of ethnicity and language, enrollment and curriculum, academic evaluation and professional success, school finance, and above all the competition and imitation forced upon the religious schools by the public authority exercised through the publicly supported schools. That is to say, whatever the strength of the church's conservative and traditional commitments, the church schools had to operate within a universe of problems common to Catholic and secular schools and one in which more often than not these problems had first been addressed by the public schools.

These social constraints operating on Catholic education were clearest in the area of secondary education, on which so much educational energy was expended in the first half of the twentieth century. While American high schools had become independent institutions by the late nineteenth century and were by the early twentieth century educating large numbers of boys and girls whose schooling would end before college, the liberation of the Catholic high school from its elite connection to the colleges and seminaries did not really come until the 1930s. Until that time, Catholic secondary education was limited to the few students eligible for the specific classical subjects deemed essential preparation for college-level work. The reason was partly organizational. Most Catholic elementary schools were based on the parish. Small, homogeneous, and locally financed, the elementary school and its parish could provide only limited resources and limited demand (especially in the new immigrant communities) for high-school enrollment. It was the colleges and universities run by the religious orders, not the parishes, that provided the academic instruction, often on a boarding plan, which absorbed the ambitious and talented students whose sights were set on further education.[7] Parish schools could and often did attach some additional instruction beyond the sixth grade, but this was usually limited in the late nineteenth and early twentieth centuries and rarely provided more than a few years of training beyond the primary grades. Most parishes simply could not afford the full-fledged program of an accredited high school with its complex curriculum and costly variety of specialized subjects.

The initial impulse for high-school development came in 1884 at the Third Plenary Council of the Catholic Church of America in Baltimore. The council issued strong and obligatory injunctions that required all

Catholic children to be educated in Catholic schools. That requirement in the context of the impending upward-age revision of the compulsory school-attendance laws meant that Catholics would have to provide schools for those who had passed beyond the elementary grades. Indeed, the Third Plenary Council expressed its hopes and approval for the creation of central high schools under diocesan control. By authorizing diocesan schools, the council facilitated the eventual appearance of publicly as opposed to privately run Catholic high schools.* Such schools would not depend on the local parish and its limited enrollment and financial possibilities, but would cover larger areas under the supervision of the bishop and benefit from his access to much wider resources. The central high school would be connected to the parish elementary school, not the college, and would connect high schools to parish schooling just as had happened to American secular schools when high schools were cut loose from college and university auspices. The new diocesan form of organization also made the high school, as translocal institution, a context for vastly expanded inter-ethnic contacts which would eventually facilitate the assimilation forestalled by parish schools with single nationalities. In 1904, a committee of the National Catholic Education Association presented a set of resolutions specifically addressing this issue. "The time seems opportune for a more general effort on the part of Catholics for the establishment of Catholic high schools." But the resolutions also noted that "while the high school is intended mainly for pupils who do not go to college, it would fail of an essential purpose did it not also provide a suitable preparatory curriculum for those of its students who either desire to prepare for college, or would be led to do so, were such preparatory curriculum offered."[8] That last provision would exert a powerful and continuing influence over the nature of the diocesan schools, but it took the church a long time to implement effectively the resolution in general.

By the early 1920s, there were still only thirty-five diocesan high schools, and most of the 130,000 Catholic secondary-school students in 1,500 institutions attended private schools. By that fairly late date in the history of the American public high school, "In Catholic circles . . . the status of the parochial and even of the central high schools was [still] dubious."[9] Not until the period between the wars were the possibilities of Catholic secondary education first realized in a real expansion of enrollments. The impulse for this expansion came from sources outside the church, from the political and social events in the wider society. These can be summarized as the push provided by the enforcement of child-labor and compulsory

---

* As used throughout this chapter, "private school" refers to schools run by religious orders rather than by the Catholic church through the parish or diocese.

education laws and the pull increasingly exerted by the Catholic public as prosperity and permanence made the Catholic immigrants of the nineteenth century (Irish and German), and increasingly also those of the twentieth century, eager for the social and professional advantages of advanced schooling.[10] The maturation of the huge Catholic immigrant community in the context of enforced compulsory education laws required the church to provide institutions for the adolescent children of the native-born Catholics and the young immigrants who had come at the turn of the century. In this way the church could fulfill its own objectives by providing every Catholic child with Catholic schooling and prevent those children from seeking places in the secular, materialistic schools provided at public expense.

By the middle of the twentieth century, the church had succeeded in those goals to a large extent. In 1949, nearly one-half million students attended Catholic high schools, and that number grew to over a million by the early 1960s. The Catholic response to the social and cultural demands for secondary education was vigorous and especially notable in the expansion of the new "democratic" diocesan high schools, most of them in large cities. These grew from 35 in 1922 to 150 in 1947 and then to 344 in 1962. The central diocesan high school, in line with its original objectives, was able to offer more varied programs to a more complex and heterogeneous mix of students than had been possible either in the parish-based school or in the privately run academies. As the Reverend John P. Breheny, assistant superintendent of the Archdiocese of New York, observed in the 1940s, "The central school can better provide the type of secondary education needed today, because it is in a position to offer a more varied program of study. Where academic high schools conducted by religious communities exist, the central school should be more concerned with meeting the needs of the student who lacks the finances or the scholastic intelligence to attend such a school." [11]

Nevertheless, the church's success in opening its high schools to the multitude of its parishioners was limited. Catholic secondary education, much more than American secular schools, remained heavily dependent on private schools. In 1947, more than one-third of the Catholic secondary schools were still owned and operated by private religious orders. By 1962, diocesan schools still accounted for only 16.6 percent of all high schools. At that point, parish high schools accounted for 37 percent and private schools for another 37 percent. Moreover, almost one-half of the diocesan schools in 1962 were less than ten years old. The diocesan schools tended to be much larger than the others and therefore enrolled nearly one-third of all secondary-school students, but 38 percent of all Catholic secondary-school students continued to attend private high schools.[12]

The heavy dependence on private schools throughout the twentieth century had very significant implications for Catholic secondary education, and continued to distinguish Catholic high-school experience from that of the public schools. While the overwhelming majority of American high-school students attended district public schools in the great period of expanded enrollments (1920 to 1960), Catholic secondary-school students attended at least four different kinds of schools—Catholic central high schools (diocesan), local parish schools with high-school departments, private academies run by religious orders, and public secular schools. In fact, despite the best efforts of Catholic officials, most Catholic secondary students have throughout the twentieth century attended public, non-Catholic schools. In 1962–63, only one-third of all eligible Catholic students went to Catholic high schools. By contrast, over one-half of all eligible Catholic elementary-school students went to Catholic schools.[13] Twice as many high-school-age Catholics went to public schools as went to Catholic schools, and of those who did attend sectarian institutions, at least one-third attended privately owned schools which were largely untouched by the more democratic orientation of central diocesan schools or the public high schools.

The failure of large numbers of Catholic secondary-school students to attend Catholic schools was only marginally the result of choice. Andrew Greeley and Peter Rossi found in their investigation of former Catholic school students that many of those who did not attend Catholic schools, even after World War II, had no school which they could attend. Greeley and Rossi concluded that "with the single exception of respondents from the West, approximately 70 percent of those for whom schools were available did in fact attend them." Despite some continuing variation among ethnic groups, the majority of Catholics by the middle of the twentieth century wanted to fulfill church directives and canon law requirements and send their children to Catholic schools. That many could not was not usually the result of financial considerations. Catholic schools have never been technically free of cost. In fact, however, until recently costs have been low, and the church often provided schooling free of charge to those who could not afford what in the case of most diocesan and parochial schools were very modest tuition charges.[14] Instead, the inability of many Catholics to attend Catholic high schools resulted from two related factors— lack of sufficient facilities and selective admissions and promotion policies.[15]

It was paradoxically these same conditions that provided Catholic schools and their students with an advantage not available to those in public schools. Lack of sufficient places in secondary schools allowed Catholic high schools to remain firmly, though not entirely, academic in orientation, and provided greater access to social mobility for Catholic school students. The Catholic high schools were, in fact, able to provide some curriculum differentiation without investing heavily in vocational education or fundamentally altering their commitment to a far more uniformly academic program than was possible in public schools, because Catholic schools could select their students and expect the public schools to take care of the rest.

As Catholic secondary education expanded, it was confronted by the same problems that beset public schools, problems usually associated with divergent student aptitudes and objectives. Catholic schools, especially the large, heterogeneous, and usually urban schools, could not altogether resist the pressure to modify the once sacrosanct, Latin-based, classical curriculum. And beginning in the 1930s, but especially after World War II, Catholic secondary schools began to offer a wider choice and more opportunities for students not oriented to college. The scientific course and the commercial course, as well as the general course took their place beside the Latin, college-preparatory curriculum in the comprehensive high school. By the early 1950s, one-half of all large diocesan high schools had such a four-course arrangement. But a vocational curriculum established by public schools for their least able students was an extremely unusual offering in Catholic schools.

Moreover, while the majority of central schools called themselves comprehensive, a careful examination of the structure of the available curricula reveals that most of the alternatives were largely variations on the academic curriculum, the most significant alteration being a reduction or elimination of Latin requirements. In 1940, a survey of Catholic secondary schools showed that three-fourths of all subject offerings were in the traditional "five academic fields: English, social studies, foreign languages, math and science," and that "The subjects required of all pupils are, as a rule, those demanded by colleges for entrance." As Brother William Mang, the author of the survey, also noted, "Industrial arts and graphic arts are generally neglected." In 1951, a comprehensive survey of central high schools by the Reverend Edward F. Spiers demonstrated the rather narrow range of alternatives provided by the parallel curricula in most schools. The major difference was the modification in the Latin requirement. In fact, only students in the watered-down general curriculum had no Latin requirement and considerably reduced requirements in math and science (only one year of each). Even students in the commercial curriculum were required to take two years of Latin, and students in the scientific course as well as the classical course (both college-oriented) continued to pursue a full four-year Latin sequence. Comprehensive schools had broadened their offerings, but these were overwhelmingly in the addition of modern languages, the fine arts, and commercial subjects like typewriting, shorthand, and bookkeeping. Of the vocational subjects, only mechanical drawing was available in even one-third of the schools. In 1980, when James Coleman and his associates compared public- and parochial-school students, they found that seniors in Catholic schools had taken more semesters of math, science, English, and foreign language than public-school students.[16]

In part, the failure to provide vocational and technical courses was the result of financial restrictions and the heavy dependence on religious orders for teachers. But Catholic educators also offered philosophical and ideological resistance to real watering down of academic subjects. In emphasizing "solid instruction and vigorous mental discipline," Catholic schools often took their stand on older psychological grounds than the prevailing child-centered and utilitarian values dominant in the public schools, and they insisted on the unity of learning rather than pragmatic responsiveness to the goals of learners. "In striving to make the school meet present needs," George Johnson noted in 1919, "there is a danger of becoming too practical and utilitarian. Secular education is prone to despise cultural values. . . . The doctrine of formal discipline is being generally scouted and the cry is for specific education. . . . Though the effects of formal discipline have been exaggerated in the past, the fact has yet to be conclusively disproven. Culture, or the building up of individual character, is best accomplished by means of general and not specific training." In 1936, the Reverend John F. Dwyer asserted, "No better way of imparting these skills [mental discipline] has yet been found than the old classical course. Certainly it is not in those schools and systems where a false theory of democracy dictates the curriculum, and makes the slowest boy of the slowest class the norm of the group's achievement; and where the curriculum is solicitously fitted to take in the lazy dullard who belongs in school only by the fiat of American education law."[17] It was, of course, the fiat of American law that forced the Catholic secondary schools to expand in the first place and to adopt, if only moderately, new academic subjects and objectives. But for many Catholic educators, that law was always an external constraint, not something voluntarily chosen or preferable to older, more elite perspectives.[18]

Dwyer's views were not unchallenged during the period of greatest Catholic secondary-school expansion. But he was also not alone in his firm and reasoned stand. Catholic educators like Dwyer provided an articulate opposition to the pressures for extreme curriculum differentiation and dilution of academic standards that overwhelmed public education. And even when Catholic high schools bent to necessity, they continued to require a heavy dose of hard academic subjects of their students. As Monsignor John J. Fallon explained in 1937: "Within the past few years . . . a feverish desire to give pupils something to do and to express physical activity in the same credit terms as mental activity has led to our present state of chaos." But, "Our high schools stand as an example of that unit offering a solid intellectual training. . . . If high schools offering a general multiplicity of courses with complex curricula, differentiated for numerous types of students are required, we cannot hope to compete."[19]

Similarly, the heavy dependence of Catholic secondary education on private schools meant that the academic curriculum was often the only curriculum available to a large proportion of students. Most private schools continued in the older, college-preparatory tradition and provided students with no alternatives to the classical sequence. In addition to intellectual resistance and organizational realities, the lack of adequate high-school facilities always meant that Catholic schools, even those organized along parish and diocesan lines, could select and exclude students who did not fit the preferred school norms and academic values. Even though the selection process was never based exclusively on academic criteria, the potential for success in old-fashioned academic courses was always at least part of the evaluation process. In a broad survey of high-school admissions policies, Sister Mary Janet concluded that "It is generally known and regretted that today there are not enough Catholic schools to accommodate all the boys and girls seeking admission to them. . . . Some schools refuse boys and girls of low scholastic standing in the elementary school, or of low mental ability as determined by intelligence tests administered in the high school or in the elementary school." In fact, only one-half of all the high schools she surveyed admitted students of below normal IQ. Other investigators found similar results. Sister Mary Degan, after conducting her own survey of admissions policies, observed that "Children with a 'low IQ' or 'low ability students' appear to present an almost insurmountable problem for many administrators. . . . 'Low ability' certainly puts its possessor in the unenviable position of being a 'problem child' in the eyes of many." When the Reverend Thomas Frain conducted his survey of a very large number of Catholic high schools, he discovered that the IQ of Catholic-school students did not at all approximate an expected normal curve of abilities. Instead, on the basis of the records of 141,618 students (22.1 percent of all Catholic high-school enrollments for 1954–55) "the percentage of rapid learners exceeds greatly the percentage of slow learners." Not unexpectedly, "The percentage of rapid learners was greatest at private Catholic high schools."[20]

A decade later and in the wake of enormous school expansion and a doubling of Catholic high-school enrollments, selective admissions were still firmly in place. The study conducted by Notre Dame University in 1962–63 revealed that all secondary schools were forced to reject applicants because of space limitations and that the number rejected was equal to 30 percent of all ninth-grade enrollments. Fifty-three percent of all secondary schools imposed academic criteria when filling their limited places;

68 percent used admissions tests. Private schools were the most academically selective, with 70 percent imposing academic standards for admission
and 80 percent requiring entrance tests. Diocesan and parish schools were
less academically demanding but almost one-half of these schools also had
academic standards for admission, and more than half required entrance
examinations. The median IQ for twelfth graders in this huge sample was
in the seventy-fourth percentile.[21]

In addition to initial selection procedures, Catholic schools also dropped
students who failed to demonstrate the necessary ability, competence, and
social conformity. One study found that in large diocesan high schools
over one-third of the schools investigated dismissed students for failure.
Those who did not, admitted that "they usually leave themselves. We advise them to do so but do not require it." Certainly all Catholic high schools
did not require their students to demonstrate extraordinary academic potential, and many had no restrictive admissions, but taken all together, the
Catholic high schools, unlike public schools, could maintain academic
standards by developing a more uniform student body than public schools.
Again and again, Catholic-school administrators admitted that students of
low ability (usually judged by IQ and/or performance) were urged to go
to public vocational schools. "The policy," according to one of Sister Mary
Degan's respondents, "has been to encourage those with failing grades to
attend the public high school to take up manual arts courses." Another
testified that "students showing indifference in elementary grades are not
urged to attend. Usually they are advised to follow vocational training by
eighth grade teachers." The prevailing hierarchy among Catholic secondary-
school educators was well defined by one student of the central high schools,
"Students who do not show aptitude for college preparatory work are
advised to try the commercial course. If unable to follow either course,
they are advised to transfer to a public school which offers Industrial Arts."[22]

Catholic high schools were selective in other ways as well. Although the
data are spotty, they indicate that ethnic background played some role in
the ability of students to enter and to graduate from parochial high schools,
at least in the first half of the twentieth century. For male students, attendance at Catholic high schools before World War II was overwhelmingly
a phenomenon of boys of Irish and German descent whose parents were
native-born. Thus, in a study of twenty-one Catholic high schools for boys
in 1936–37, Brother William Mang found that more than one-half of the
students were from one of these two groups—three in ten were Irish, two
in ten were German. Of the rest, another two in ten were "American,"
while only 10 percent were Polish with other new immigrant groups falling
behind that percentage. Moreover, at a time when the large majority of
Catholic parents were immigrants, almost three-quarters of the fathers and

more than three-quarters of the mothers of all students in the sample were American-born. Only 17.5 percent of the parents of all boys were born in foreign countries, and only 10 percent were born in countries where English was not the mother tongue. This pattern of attendance could not be explained by economic factors alone, since almost one-half of the fathers were in occupations on the low end of the scale. Fewer Catholic-school boys came from homes where fathers were professionals and proprietors than public-school boys, and considerably more came from families in which fathers were in the building, machine, and printing trades or in transportation service. While graduation rates were slanted toward the more economically advantaged, admissions were more ethnically selective.[23]

Language, economics, and ambition undoubtedly played a significant part in the ability of Catholics to attend Catholic high schools. And all these factors worked against the attendance of those from the newer Catholic immigrant groups. As Bernard J. Weiss has observed about newer immigrants at Catholic colleges, "feeling even more alienated from their larger society than their predecessors because of language difference and exposure to a radically different environment, Southern and Eastern European immigrants made a conscious effort to preserve their traditional rural values centering on the family and the promotion of ethnic group solidarity." To this end, ethnic parochial schools centered on the parish were an asset, while high schools and colleges were not. "This tended to create inhibitions among their youth regarding higher education, since it could produce an eventual break" from their ethnic roots and milieu. There is also some evidence that youth from poor foreign homes may have been discouraged from developing ambitions for high-school attendance by sisters and priests, some of whom were eager to preserve ethnic boundaries. Thus in Chicago, few Polish parochial-school girls were to be found in the normal school. While 65 percent of the graduates of English-speaking parishes attended high schools, less than 35 percent of those from Slavic parishes (Polish, Lithuanian, Slovak) went to high school. The former superintendent of the Chicago parochial schools observed about this situation in 1925 that "The pastors of these places do not encourage it."[24]

The reluctance of newer immigrant groups to attend Catholic high schools coincided with the strong tendency for parochial high schools to be selective, excluding all but those who were highly motivated, ambitious, talented, and had the language and social advantages of native-born parents to propel them further along the Catholic-school path. Ethnic group self-selection and school admissions criteria allowed the Catholic high school both to function as a meritocracy, by recruiting the most talented (and most Americanized) from among the newer immigrant parishes, and to respond to the ambitions of older ethnic groups, especially the Irish, for social and economic mobility through Catholic higher education. It also made the Catholic high schools into extremely effective agents of a specifically Catholic form of Americanization, since the dominance of the Irish (and to a lesser extent the Germans) provided a particular social context in the secondary schools. If the diverse paths of assimilation described in Chapter 3 may be generalized to the parochial high-school case, the dominance of Irish Catholics in many eastern schools may have created an environment similar to that of New Utrecht High School in Brooklyn, where Jews provided a model of assimilation for other ethnic groups because of their number and power in strategic campus affairs. After World War II, the greatly expanded facilities and increased attendance of newer immigrant groups in Catholic high schools probably diminished this pattern and eclipsed Irish control, although it never entirely overcame the patterns that the Irish had established during a critical period of immigrant experience.[25]

Whether the result of choice or necessity, the lack of available places for all who might wish to attend permitted Catholic secondary schools to benefit certain ethnic groups especially and to maintain a much more uniformly academic program and higher standards of performance than the public school. But this greater uniformity did not mean that all Catholic secondary-school children had similar experiences. On the contrary, Catholic-school differentiation was as powerful as that in the public school, and even more sharply hierarchical. While comprehensive public high schools channeled students into separate curricula, Catholic students often wound up in different schools. In a somewhat exaggerated schema, we could describe the social structure of Catholic secondary-school children as an elongated pyramid. At the top were those who attended elite, and almost exclusively college-oriented, private schools. Below them were the students in diocesan and parish high schools who were themselves divided between those in the prestigious classical run and others in the general or commercial curricula. Finally, at the bottom, pushed out of the Catholic track entirely, were the majority of Catholic students who for one reason or another chose to attend public schools or could not get into or remain at the Catholic schools. These latter may have been excluded for academic or disciplinary reasons, because their IQs were low or grades inadequate, or because they were considered uncooperative or undisciplined. In the first half of the twentieth century, many of these were from newer immigrant groups who may have chosen not to attend or could not make the grade because of social or language difficulties. Many of these last may have dropped out of school entirely or, if they did attend high school, remained in vocational or general tracks in the public schools until they graduated or left. In either case, they were failures by Catholic-school criteria and, if they judged themselves in those terms, must certainly have considered themselves inferior to the Catholic secondary-school students.[26] This is an obviously exaggerated diagram which pays no attention to the many different factors involved, but it does suggest the ambiguous costs and benefits of an educational philosophy that clung tenaciously to an older view of high-school education. Those costs were individual and social, because those left out of Catholic high schools were throughout the first half of the twentieth century mostly the poorer and newer ethnic groups. This tended to support and to confirm what is usually not talked about in Catholic literature, the ethnic hierarchy within Catholic education.

Catholic high schools were forced to become more democratic by public policies which required that Americans stay at school longer, but it was always only a partial democratization. As a result, Catholics in Catholic high schools had a more uniform education than non-Catholics received, and Catholic schools used the safety net of the public schools to catch the less desirable fallout. Ironically, Catholic-school autonomy and the ability of Catholic schools to retain their more traditional emphasis came at the expense of the public schools which, by providing vocational and watered-down curricula and lower standards, were forced to educate those students the Catholic schools could not or chose not to teach.

In the long term, the particular nature of Catholic differentiated education paid off in social success. As Andrew Greeley and Peter Rossi (and others) have found Catholic high-school students of all social levels have gone on to college at higher rates than comparable non-Catholics and Catholics who did not attend Catholic high schools. Greeley and Rossi also found that "Catholic school Catholics had increased their social class margin over other Catholics." Although Greeley and Rossi do not explain this phenomenon of Catholic school "overachievers" as the result of selective admissions policies, it is difficult to conclude otherwise.[27] Throughout the twentieth century, the church has been committed to the schools as levers of Catholic achievement and especially as creators of Catholic leadership. The schools sought out those with potential. Even the committee of the National Catholic Education Association, which in 1904 had advocated diocesan schools for the multitude, was careful to retain the classical curriculum for those who desired it or "would be led to do so." As one educator observed in at once defending Catholic school academic standards and observing that this neglected the needs of many, "We need Catholic leadership and the only place to recruit it is among the best minds. . . . It would simplify matters tremendously if we could confine ourselves safely to the academic high school. However, it does not seem that we can do so in conscience." In fact, Catholic high-school students have been assisted up the social ladder through a concerted, though not entirely predesigned, program in which academic success is encouraged, potential achievers sought out, and academics in general emphasized in a far more traditional school environment than that available to the public-school student.[28]

II

The traditionalism of Catholic schools did not go unchallenged. Catholic educators also participated in a culture that emphasized the right of all children and adolescents to extended education. The democratization of the public high schools and the attendant philosophy that emphasized the unique potential of each child exerted a considerable influence on Catholic educational discussions and practice starting in the 1930s, becoming especially strong after World War II. That influence is clear in the organizational meetings of the National Catholic Education Association and in the large number of dissertations in education sponsored by The Catholic University of America. Many doctoral students absorbed the "progressive" spirit of all education schools and condemned what they saw as the retrograde emphasis on the college-preparatory curriculum. Moreover, although most Catholic thinkers and educators rarely spoke John Dewey's name without venom, progressive education left its marks on Catholic-school practice.

The greatest kinship between Catholic educational thought and progressive ideals lay in the diffuse area which Catholic educators sometimes referred to as the education of the whole child—or more simply, character education. This view of education came close to the heart of Catholic pedagogy. "If the process of education . . . is to fulfill its function of developing the whole person, a principle which has universal approval," Sister Mary Janet Miller insisted, "the Catholic educator considers that task incompletely performed unless knowledge of God and our duties to Him are included in the educational program." In addition, "Schools consider it their duty to aid parents in every other phase of the development of children—in matters of health, of home life, of leisure pursuits, as well as in intellectual skills and habits." "Character," George Johnson observed, "must reveal itself in the midst of tangible circumstances. . . . The function of the Catholic school should be understood in the full light of the Church's mission. It is not merely a preparation for higher education, but a preparation for Christian living."[29] Indeed, the education of the whole child, long articulated as an ideal of progressive education, was the fundamental purpose behind Catholic-school separatism. Had Catholics believed that schools merely taught academic skills, the threat of public education would have been greatly reduced. It was precisely because Catholic educators hoped to maintain control over this much broader and more basic educational socialization that Catholic schools were necessary. By controlling the education of the whole person, Catholic schools hoped that Catholic children would remain part of the minority subcommunity.

Catholic education insisted on the fundamental unity of the human personality, values, purposes, and their profound relationship to God. It was, therefore, God who had to be placed at the center of all instruction. In this, of course, they differed from the secular schools, which, bound down by what Catholic educators called naturalism and materialism, had placed the child and his growth at the center of modern instruction. Still, whatever the intellectual distinctions, Catholic educators could and did acknowledge the potential kinship between their expansive definition of education and the progressives' equally broad objectives. This agreement was implicit in the official "Objectives of the Catholic High School," issued by the National Catholic Education Association in 1944. Catholic secondary schools had seven objectives: "To develop intelligent Catholics. To develop spiritually vigorous Catholics. To develop cultured Catholics. To develop healthy Catholics. To develop vocationally prepared Catholics. To develop social-minded Catholics. To develop American Catholics."[30] While the objectives gave priority of place to intellectual and spiritual factors, the inclusion of citizenship, vocational, social, and health concerns was very much in line with the views of the larger education community of the mid-twentieth century.

Many Catholic educators also acknowledged in theory the desirability of greater democracy in education. Catholic universalism did not only insist on the spiritual equality of Christians and the church's responsibility for their instruction, American Catholics also participated in the democratic culture and ideology of their society. "Now the day is past," Mother M. Juliana declared in 1931, "when the elementary education is sufficient equipment with which the boy or girl can face the world. . . . A Catholic high-school education is essential for the Catholic adolescent. . . . Let us all, priests, Religious, and loyal laity be up and doing to provide free high schools wherever possible to give our boys and girls their rightful due in this age of the high school and college." This perspective became more insistent by the forties and fifties. In the postwar period, the failure to provide adequately for Catholic students of all intellectual levels increasingly became a matter of discontent for Catholic educators. As Sister Mary Janet put it in 1949, "There is no principle of Christian social justice which can justify a passive acceptance" of the failure to supply a full education to all Catholics including the academically deficient. "There is, however, fundamental Christian truth in the ideal of respecting all types of human abilities, talents, and interests and in helping to educate youth for Chris-

tian family life and Christian occupations of all kinds in addition to educating the potential scholars."[31] The Catholic secondary schools never succeeded in developing to the point where all Catholics could find a place.
Nevertheless, the Catholic schools did diversify their offerings, and exclusionary policies were described apologetically or condemned by the second
half of the century.

As their schools began to provide places to a wider variety of students,
Catholic educators participated in the larger educational discussions about
individual differences and individualized instruction, and they were quite
as eager as public school officials to embrace the greater instructional specificity and efficiency promised through mental tests. In 1948, Brother Louis
J. Faerber dedicated his doctoral dissertation on the provision for low-
ability pupils in Catholic high schools to "Mary, Heavenly Shephardess,
God's Provision for Individual Differences." Basing his strong advocacy of
Catholic secondary education for the remedial student on the American
concept of democracy as well as the Catholic belief in the equality of souls,
Faerber noted that "Basic equality of educational opportunity . . . means
that in secondary school each pupil be given *equal chance to gain that
kind of education from which he can best profit*. It does not mean giving
all pupils an identical education." Faerber went on to urge just the differentiation in curriculum—and in the same terms—as was common practice
in the public schools. He abhorred the "humiliation" to which low-ability
students were subjected when "trying to cope with a program of abstract
academic subjects far beyond the reach of their minds," and "the practice
. . . of shunting off those pupils who fail to the public schools." Faerber's
thesis, and many others which appeared thereafter, suggests both the degree to which the Catholic high schools had so far avoided the worst dilemmas of the public schools and the penetration of secular educational
beliefs in all their details into the ideas of many Catholic educators. The
extent of that penetration is indicated in the results of a workshop at Catholic University in 1948. The conferees concluded that "It is the responsibility of Catholic secondary schools to take all the children of secondary
school age who apply for admission," and that the Catholic schools had
the obligation to "help them to grow and develop to their maximum capacity, spiritually, mentally, physically, emotionally, and socially." Perhaps even more pointedly, the Reverend Michael J. McKeough noted, "One
of the gratifying developments in recent educational thinking is the evidence that teachers realize that they are teaching children rather than subjects." In responding to student differences, Catholic high schools had taken
a long step in the progressive direction. "Today," he continued, "there is
more general agreement that the real test of our educational efforts is the

effect we have on the whole child, and not merely on his mastery of certain segments of knowledge."[32]

The studies of individual differences also suggest the extent to which Catholic schools had already been influenced by democratic conditions. As a result of the "fantastic" increase in high-school enrollments, Faerber observed, "more and more student bodies of Catholic high schools have been found to be composed of a widely heterogeneous group, in many cases representing the greater part of the range of individual differences among pupils, going all the way from 70 to 75 IQ to 150 IQ or more." Sister Mary Degan made a similar observation: "In spite of limited and overcrowded facilities, we are now receiving into our regional Catholic high schools pupils of varying abilities, needs, backgrounds, and prospects. Each school has the responsibility of giving to its pupils the education which will be most useful to them."[33]

By the late forties and fifties, the IQ had become as much a part of the administration of Catholic as public schools. Indeed, it was an essential ingredient in the conceptualization of differing aptitudes as well as proposals for program differentiation. Like the public schools, Catholic schools used mental tests as diagnostic tools. Even more than in public schools, Catholic administrators used them to separate students, channeling some toward further advancement in the Catholic school hierarchy and depriving others of further religious education. In 1949, Sister Mary Janet found that less than one-half (48 percent) of all secondary schools denied using the IQ as a measure of admissability. When making assignments to different curricula, in the larger schools especially, the test was used almost mechanically. Faerber also noted that "Sometimes the I.Q. carries such an exaggerated importance for teachers that they tend to accept it as the single index for discovering the child's complete status in his process of growth and development." Most investigators of admissions and tracking similarly discovered this heavy dependence on the results of IQ testing. Sister Mary Degan found that only 22 percent of the large diocesan high schools she surveyed did not give IQ tests to their students for placement purposes. Indeed, the IQ test had become so ubiquitous and essential that many eighth graders were "drilled in the various types of the test."[34]

As the Catholic high school became more complex, IQ testing became the most commonly employed means for providing more homogeneous groups and better calibrated instruction, but it was not the only technique adopted from secular thought and progressive practice.[35] Vocational guidance and extracurricular activities also became standard in Catholic as well as public high schools. As early as the 1930s, the National Catholic Education Association devoted considerable attention to issues of school guidance. This was in line with the increasing pressure for diversified courses and better tailored instruction. "Educational guidance of the future," the Reverend Paul E. Campbell observed at a meeting of the association in 1935, "must concern itself with fitting subjects to persons rather than with fitting persons to subjects." Although Catholic educators resisted pressures for extreme program diversification, more modest efforts to provide students with a degree of personalized direction did not provoke similar opposition. "Vocational guidance can well be imported without vocational education," the Reverend Kilian J. Heinrich observed. "The former does not need special buildings, shops, and establishments." Catholic secondary schools could oppose fads in new courses and still adopt progressive techniques that seemed to make instruction more useful and effective. By the late forties, one study noted that "well-organized guidance programs are considered essential to the proper functioning of the modern high school."[36]

Despite a commitment to a continuous tradition and belief in unvarying truth, Catholic educators readily appropriated "scientific" techniques useful to pedagogical efficiency. "We frequently emphasize the fact that our educational position is conservative," the Reverend George Johnson of Catholic University noted. "We must recognize that there are certain eternal truths and first principles that never change, certain values that are ageless, certain elements in our social heritage to which children in every generation have an inalienable right." At the same time, "it is our sacred obligation to do all in our power to promote the scientific study of education and to utilize the findings of scientific pedagogical experiment, for the purpose of increasing the effectiveness of our work."[37] That flexible traditionalism, as it were, was a functional and practical necessity for Catholic educators. Throughout much of the twentieth century, Catholic educators were defensive about Catholic schools. This was not only because the schools were an expression of minority status and therefore conspicuous or because they were often overcrowded and underfinanced, but also because their manifestly conservative positions on education looked more and more old-fashioned in the context of the ever-changing educational scene. Catholic educators and schools tried hard to select from among the multitude of new trends those that seemed to fit their needs without compromising their own specific objectives. Homogeneous grouping, IQ testing, guidance, and extracurricular activities were easily made to fit these requirements.

Extracurricular activities were an especially good example of the fit between certain progressive practices and the Catholic search for updated forms of schooling, and they were eagerly embraced by the schools on both the elementary and secondary levels. They fit the Catholic view of education of the whole person and were an effective instrument for registering the differing interests of students without forcing a heavy investment in alternative curricula. One student of parochial schooling noted that "Various extracurricular activities are now considered almost an integral part of the whole process of elementary schooling," even though Catholic schools often had to "go to the people" for the extra funds that were required to run these programs.[38] Another student of the activities put the issue bluntly: "Catholic schools are directed by Americans and teach children for life in contemporary society, and so are bound to be influenced by the same sociological trends as public schools. Accordingly, extracurricular activities are part of the school life of nearly every Catholic high school in America. Catholic education, however, is developed, philosophically from an unchanging set of principles which, while not preventing it from being sensitive to changes in the society in which the school functions, should prevent Catholic educators from succumbing to fads or even to extremes that tend to eviscerate the basic purposes of the schools." Sister Mary Margarita Geartts proceeded to demonstrate how extracurricular activities were justified by the principals of the 300 high schools in her survey. While most of the justifications were not at all specific to Catholic education, Geartts managed to rationalize their relationship to eternal Catholic truths by referring to "the whole Child" as "the subject of all Christian education, within and beyond the school." No outright progressive proponent of student interest and child psychology could have done better. Indeed, Geartts's dissertation is a study in how Catholic educators could bend almost any instrument to their purposes and find ample Catholic theory to cover the stretch marks. Most of the principals of the Catholic schools in Geartts's study agreed with her, and the overwhelming majority (227 of 262) said that extracurricular activities were "essential" to secondary education. Geartts concluded that "The major objectives of the activities are in harmony with both the general Catholic philosophy of education and the explicit objectives of Catholic schools."[39]

There is no reason to question Geartts's conclusion, but it might be useful to indicate that other factors besides Catholic philosophy often underlay the absorption of techniques developed in public education by Catholic schools. The most important of these was the latent competition that Catholic schools experienced from public schools. As the public schools grew into modern, youth-centered institutions, they developed, more or less successfully, means for maintaining student interest and involvement. Although Catholic schools could depend on their unique program to maintain parental allegiance, the same was not always true for adolescents or children. Geartts was well aware of this and noted in passing, "Today, the entire American educational scene is dominated by the policy of making school life as interesting and attractive as possible, in order to keep young people in high school until they graduate." Catholic secondary schools had to provide at least some of the things available in public schools, even though "the public school has more and greater variety," if they hoped to keep their population.[40]

Pride, too, was a potent factor in Catholic-school education. Catholic educators did not want their schools to be just good enough, but to be at least as good as those attended by non-Catholic youth. For most of the twentieth century, educators had good reason to be defensive about the nature of the education provided by church schools. As professionalization and centralization proceeded in public schools with their reliance on degrees and certifications, Catholic schools lagged behind. The dependence on religious orders whose sisters often had briefer and more informal periods of preparation and the fragmentation of control by parishes, ethnic groups, and teaching orders in the early twentieth century provided a basis for criticism by public-school advocates as well as by Catholics concerned about the schooling of their children. In Chicago (which had the largest parochial school system in the world), for example, historian James Sanders notes that "the evidence more than suggests that each religious order thought of its schools as a system in itself." Another problem was the often overcrowded and underfinanced facilities of many schools where up-to-date equipment was missing and classrooms sometimes burst at the seams and overflowed into basements and cafeterias. So too, low pay scales undermined the Catholic schools' ability to compete for competent lay teachers. As early as 1884, the Third Plenary Council had noted "If hitherto, in some places, our people have acted on the principle that it is better to have an imperfect Catholic school than to have none, let them now push their praiseworthy ambition still further, and not relax their efforts till their schools be elevated to the highest educational excellence." And in 1922, Joseph Hamill observed that "If our schools are to survive, Catholic children must be given, in addition to their religious training, as good a preparation for their lives here below as they can obtain in the state schools." That refrain would echo through the first half of the century. By 1959, Neil McCluskey still betrayed an acute sensitivity in defending parochial schools. "Local conditions many times justify temporary compromises that may not be in perfect accord with either the Catholic ideal of education or standard academic practices."[41]

Catholic educators chafed at more than financial stringency and low levels of teacher preparation. Catholic educators lived within a larger social universe and an educational world in which revision, updating, and a dependence on "scientific principles" were considered necessary and beneficial. Since most educational research was conducted on and about the public schools, it is not surprising that the direction of influence for most of the twentieth century was from the secular to the Catholic schools. This trend was observed by the Notre Dame study of Catholic education as late as 1966: "At present the study of administration, curricular, and guidance problems by persons serving the public school so far outstrips that under Catholic auspices that the influence of the first is dominant. . . . The inevitable result is that in several areas there is no recognizable difference between the two systems."[42] Catholic schools revised their programs, added various services, and made their schools more attractive as well as effective for all the reasons the public schools did—because they were confronted by the need to adapt to larger and more varied student populations and because educational development in this century has been rapid as research, professional development, and revised educational concepts have made change a byword of schooling. Catholic schools faced an additional reality, however: the public schools with which Catholic schools compared themselves and competed.

The enforced balancing act of Catholic education—between conservative and unvarying philosophical commitments and technical innovation—was captured in an unusually effective way in a small book by the Reverend Laurence J. O'Connell entitled *Are Catholic Schools Progressive?* O'Connell set the tone and direction of his discussion in the preface: "Progressive education, a product of the twentieth century, poses a problem for Catholic educators. Undoubtedly, many of its methods are superior to the methods of the traditional Catholic schools, and yet Catholic education must not, even in the name of technical progress, compromise the philosophical and theological principles on which it is established." Having said that, O'Connell proceeded to harmonize most progressive techniques with Catholic principles, as Geartts had done with extracurricular activities. At the same time, he tore unmercifully into John Dewey, the bête noire of Catholic education. Dewey's naturalism bore the brunt of the attack. But O'Connell attacked along a wide front—the denial of sin and depravity, the denial of the duality of man's nature, the denial to the teacher of an effective role in shaping the child's mind and will, the denial of real ends and purposes to education. O'Connell concluded that "From a philosophy which is as naturalistic as that of John Dewey and of the other philosophers of progressive education, it is difficult to see how anything could come which is good and acceptable to the orthodox Catholic teacher. And yet the philosophy which has prompted and stimulated growth in progressive education can at times be divorced from the practices to which it has given rise. Thus it is possible to accept improvement in techniques and methods while continuing to reject uncompromisingly the philosophies which have given them birth." It was more than possible, since O'Connell showed how much those changes had already been firmly incorporated into Catholic pedagogy. He probably did not appreciate the degree to which the use of progressive techniques had long ago been divorced from genuine progressive philosophy in the public as well as the Catholic schools. It is significant, nevertheless, that Catholic educators still felt impelled to go to considerable length by the middle of the twentieth century to disavow progressive philosophers and philosophy. No mater how pragmatic Catholic schools were in using the techniques of the public schools, they felt deeply uncomfortable in the presence of progressivism's original beliefs.[43]

The "Americanization" of Catholic schooling was manifest by the 1950s, and it was perhaps best enunciated in a public-relations instrument issued by the Department of Superintendents of the National Catholic Education Association in 1950. In a large pictoral album, *These Young Lives,* student experiences at all levels of Catholic education were examined and illustrated. Catholics were shown in their own elementary schools and colleges, in programs for adults and the handicapped, in kindergartens and seminaries. The Catholic schools had indeed come a long way from their scrappy beginnings. Firmly in the center of the volume was a section on "Life Adjustment Education." Life adjustment, the volume announced, was a necessary part of Catholic education because "Despite compulsory education laws and the fact that the United States is committed to the principle of secondary education for all, less than 80% of the elementary school children entered high school. More than 40% of those who did enter quit before graduation. . . . A program of Life Adjustment Education may supply what these students felt was lacking and may help to keep students in school much longer." Despite their initial hesitation, the Catholic schools had firmly entered the world of twentieth-century American education.[44]

III

The vulnerability of Catholic educators to the broad currents of American educational development often had paradoxical effects, and the flurry of studies in the 1950s that advocated greater attention to women's household roles in Catholic colleges for women are interesting for what they suggest about the boundaries between Catholic-school autonomy and American educational ideals. Clearly derived from secular ideas that had penetrated schools of education, these studies seem not so much incongruous with Catholic practice as redundant and inappropriate. Catholic education in the United States has always been especially sensitive to gender differences, and Catholic schools have, as a result, historically attracted more girls than boys. It is not clear whether parents chose the more traditional moral and religious instruction of the parochial schools for their daughters because women are usually assigned a strategic role in family and cultural maintenance. It is likely, however, that traditional Catholic families were drawn to the church's commitment to separate instruction for boys and girls, especially during the vulnerable adolescent period. As we have seen in Chapter 3, a secular women's high school like Bay Ridge in New York City drew heavily on the female Catholic population of Brooklyn for its students. The Catholic church tried wherever possible, and especially at the secondary-school level, to provide men and women with separate schools. One Catholic educator observed, "The ideal . . . is separate schools; a lesser ideal is separate classes in the same school; and full coeducation is tolerated only when circumstances leave no alternative." In 1949, women made up 56 percent of the Catholic high-school population, and the majority of Catholic high-school students went to single-sex schools. That proportion changed only slightly, declining to 55 percent in 1962–63. Of these women, almost one-half of all those heading for college were bound for Catholic colleges.[45]

Separate instruction for young Catholic men and women meant that most Catholic women would be fully introduced to the guiding Catholic perspective on women's matrimonial obligations. Indeed, the fundamental role of religion and religious instruction gave Catholic educators a central focus for socialization to separate sex roles largely missing from public high schools where academic courses were both coeducational and not explicitly geared to instruction in values and morals. For Catholic women not oriented to religious vocations, marriage and family served as the sure basis for moral instruction in Catholic doctrines and viewpoints. No doubt, this partly explains why college was less attractive to Catholic women than men. In 1963, one survey found that 64 percent of high-school men in Catholic schools but only 47 percent of the women planned to go to college. The disparity also existed between the goals of Catholic and Protestant women. Far more non-Catholic than Catholic women persisted in college for four years, and a higher proportion of Protestant than of Catholic graduates were women. One Catholic woman observed, "A man really needs a college education for a job. A woman does too, but if she gets married, that is usually the end of it. She usually doesn't use her education, and if she doesn't I think it was a waste of money."[46] While such views may well have reflected ethnic or even prevailing American distinctions in comparative educational utility, they were strongly enforced by Catholic beliefs about the separate social destinies of men and women.

At the college level, Catholic institutions did not so much devalue women's learning as provide it with a firm and unambiguous context. In 1961, Sister M. Madelva tried to answer those who questioned the utility of educating Catholic women by insisting that Catholic women were educated for "intelligence, courage, charity in a militant Christian minority." "Our daughters," she noted, "through their Christian education, know that the perfection of love is service: in the normal vocation of woman as a wife, a mother; in the state of greater perfection as a religious; but always, always as a teacher, a compassionate, merciful, normal woman finding the fulfillment of life and of education in selfless understanding and love and care of others." Mother Grace Dammann was more direct: "The most far-reaching result of a really Catholic education must be the training of the wives and mothers of the next generation to understand exactly the part that home training and home life play in the upbuilding of a Catholic conscience and a 'new order.' "[47]

Catholic women's colleges are the largest single constituent of American Catholic higher education and, in the words of Robert Hassenger, "their growth has been the most conspicuous feature of Catholic college development in this century." Rooted in older convent traditions with their ideals of piety and chastity, Catholic higher education for women in the United States also participated in the dynamic flowering of women's higher education in general and was deeply influenced by those developments in the late nineteenth and early twentieth centuries. The majority of Catholic women's colleges were founded in the wake of the pioneer secular institutions. Nineteen were established between 1899 and 1915, and fifty-six more by 1930. By then the liberal arts prevailed in most women's schools, and Catholic educators, fully aware that young Catholic women from ambitious families were being drawn to the education available at secular schools, and responding to the growing demand for lay teachers at Catholic schools, modeled the Catholic colleges on these. As a result, Catholic colleges for women both inherited the secular ideals of comparable women's schools and rested securely on the Catholic emphasis on the differences between the sexes. For both sexes, the classical curriculum and the liberal arts represented "the unity of truth and the hierarchy of truths."[48] Thus, the Catholic colleges for women were like secular schools and different—strongly planted in the liberal arts, yet not rooted in a tradition of equality but in the Catholic view of distinct goals and responsibilities for each sex.

Although Catholic colleges for women differed among themselves, they imitated secular institutions throughout the twentieth century. By the forties, in their "endeavor to keep pace with the state universities," many colleges had introduced a wide variety of courses not clearly part of the traditional liberal arts—secretarial science, social service, library science, nutrition, finance, and the by then traditional offerings for women in home economics, nursing, and teaching. These, Mother Grace Dammann noted, were in response to the "acute job-consciousness in the minds of students and parents." In most cases, Catholic educators hoped to keep these in balance with a broad, liberal-arts education, but college brochures and bulletins also clearly appealed to the practical, vocational goals of their students. One announced in bold capitals, "LEARNING FOR LEADER-SHIP, in science, in education, in liberal arts. Unusual leadership opportunities will be yours." In another, potential students were told, "Your day is here, Catholic women and girls. Public life needs you." In so advertising themselves, the Catholic schools had moved with their young women quite consciously into the modern world of work. In one survey of Catholic-college personnel services, eighty-four of ninety responding institutions reported making some forms of vocational guidance available to students. These included career days, facilities for interviews, letters of recommendation, investigations of jobs, and other placement services.[49]

Catholic colleges for women were responding to the changes in women's lives after the Second World War and were providing Catholic women with some of the career opportunities available to their secular counterparts. It was in this context that in the 1950s and early '60s discussions of the higher education of Catholic women, very like those that beset women's higher education in general, took place. The flutter of studies in the 1950s that focused on better and more explicit preparation for women's household duties suggests how closely bound Catholic education was to the larger currents of educational development in the twentieth century. The criticism of Catholic colleges for women often echoed that of secular schools. One study noted, "Colleges, particularly for women, have often been reproached for placing undue emphasis on a career rather than on the art of becoming a well-instructed and successful parent." Another student of women's colleges observed, "Catholic colleges for women, very much like their non-sectarian counterparts, have followed colleges for men too closely especially in their intellectual phases. Sufficient importance has not been attached to the principle that the higher education of men and women must differ in certain fundamental respects."[50]

By the 1950s and early '60s, Catholic colleges for women were attacked for being both too professionally oriented and too much like men's schools. As Sister Mary Audrey Bourgeois noted, "For the majority of the girls 'the lives they lead' means a life spent in a family group. Professional life is usually only the interlude between the scholastic years of the students and the years centered in the family group." In fact, Catholic women's schools often had extensive offerings related to issues of marriage and family life, but just as in secular institutions, these were presented in the usual academic manner. Such instruction was offered under the rubric of Christian marriage, family, or introductory sociology courses. Very few concerned household instruction or practice, and one critic, Sister Mary McGrath, complained that some topics were scarcely treated at all, "such as Sewing,

Meal Planning, Home Nursing." Indeed, most of the instruction came from within the traditional framework of Catholic courses in theology, philosophy, or religious practice and focused on conventional Catholic concerns—abortion, birth control, the church and the family, divorce, mixed marriage. Students did not seem especially interested in the courses that were available. "In one woman's college, with an enrollment of 770, the total number of students taking the three marriage courses offered, Child Care, Home Management, and the Family is fifty-two."[51] In estimating general interest, McGrath concluded that not more than one student in ten enrolled in at least one of these marriage-related courses.

Catholic colleges provided their students with ample opportunity to consider the importance of their future family lives. Conferences, extracurricular activities, religious instruction, and the peer culture bound graduates of Catholic colleges into the mainstream of Catholic values and American culture. Far more than was true at secular colleges, the objectives and goals of Catholic institutions were full of the concern to provide "adequate and proper preparation for marriage and family life." Indeed, the aims were usually "so fundamental in the entire purposes of the governing religious community's philosophy that it was not considered essential to make it a particular point of issue."[52] Well before the liberal arts were rationalized at secular schools as an excellent preparation for motherhood, Catholic colleges had assimilated them to their purposes. A fundamental liberal-arts education, Grace Dammann observed in 1942, was meant "to arm her with an intelligent idealism regarding the details of her life and a sane and balanced outlook will enable her to avoid decisions reached by emotional paths rather than by reason and by faith. Women so educated will be able to create in their homes an atmosphere permeated with good taste, sound evaluations, and fundamental faith. Nothing short of such liberal training will enable those with God-given influence and opportunity to have that impact upon the general social life of the country which will raise its moral standard and help to save it from the materialism which engulfs it."[53]

The Catholic colleges had thus evolved a precariously balanced environment in which strong academic goals meant to provide Catholic women with an education equal in quality to that of secular schools were surrounded by a traditional Catholic culture. In that context, it seemed peculiar that students of education, like McGrath, should have found the colleges for women insufficiently oriented to family values. McGrath, for example, would be satisfied with nothing less than a fundamental reorganization of the curriculum and a commitment by all subject-matter instructors "whether it be literature, history, economics, or any other subject," to rendering their courses relevant to the future wifely roles of their students. "It means setting up a curriculum that will impress upon young women the dignity and worth of homemaking, and, at the same time, prepare them for both the joys and the difficulties of their vocation to family life."[54]

Yet, in many ways, McGrath's complaints were not peculiar at all. Despite the broad orientation to marriage and family, many Catholic colleges for women, like their secular counterparts, had provided women with a strong and sexually neutral liberal-arts education. As McGrath was well aware, "administrators, in their anxiety to keep the curriculum strictly conformed to the liberal arts pattern," chose not to provide practical courses as alternatives. Often in order to protect "the academic dignity" of their programs, even home economics departments were "hidden away in an obscure section of the building."[55] The dignity and academic respectability of Catholic colleges, like other parts of the parochial school system, had been hard fought and dearly bought, and the Catholic colleges for women were not eager for a Lynn White-like invasion of their programs. Moreover, the Catholic opposition to the intellectual dilution implied in a vocational program for women, like that of vocational education in secondary schools, made a successful incorporation of McGrath's proposal highly unlikely. Many of the best Catholic colleges were carefully balanced, providing on the one hand a distinct atmosphere in which women's separate destiny and religious purposes were amply promoted and on the other an academic rigor based on a classical foundation which resisted intrusions from more practical curricula. In addition, Catholic schools always had to be aware of their secular counterparts. At a time when most secular schools for women were becoming more like the Catholic colleges in their promotion of a generally female atmosphere around a hard core of solid academic courses, the Catholic colleges maintained their precarious balance.

IV

In one area, Catholic schools followed the American pattern only too well and in direct opposition to basic doctrinal commitments to the equality of Christian souls. Throughout most of this century, Catholic schools have been racially homogeneous. Indeed, before the 1960s Catholic schools usually followed a clearly segregationist policy that served the aims and prejudices of white parishioners but not those of black students and their families. In 1949, Sister Mary Janet found that 40 percent of all high-school principals did not answer when asked whether they admitted black students to their schools, and 22 percent admitted that they did not. Only one-third of all respondents claimed to admit blacks. At that time there were over 300,000 black Catholics in the United States, but only 5,620 blacks in Catholic high schools. Of these, fully 80 percent were in "42 schools in which Negroes formed the major part of the student body."[56]

The Catholic church's official doctrine was, of course, highly inclusive and a papal encyclical in 1939 had expressed the church's "special paternal affection, which is certainly inspired by heaven, for the Negro people . . . for in the field of religion and education . . . they need special care and comfort." But the church had not succeeded in effectively providing for the educational needs of black Catholics or in breaking down the racial barriers that underlay American social practice. At a time when little more than a handful of blacks were attending Catholic high schools, 13–14 percent of the students in those same schools were non-Catholics. In one study of admissions, Sister Mary Degan found that of the fifty-five high schools that openly barred blacks, forty-six admitted non-Catholic students. One Catholic administrator, without being specifically asked about this issue, volunteered in answer to the question "Do you admit to your high school *all* Catholic students who apply if they hold an eighth-grade diploma?" "Yes, if white and we have room." In another case, a school administrator noted that "local prejudice would exclude Negroes." In 1971, according to the National Catholic Educational Association, 5 percent of all students in Catholic elementary schools were black and 85.5 percent of all elementary-school students were in all-white schools. Minority teachers never taught in all-white schools.[57]

In fact, the Catholic record in matters of racial justice in the schools was weak until well into the 1960s when federal policy enforced compliance to equal-rights rules. Despite attempts to proselytize and some special efforts to reach out to disadvantaged blacks in both the South and northern cities, the Catholic schools remained overwhelmingly white, and the American Catholic church was securely anchored in the views of the surrounding culture. As Father Emerson Moore, a black New York priest, noted in 1974, "To a great extent, the church today has remained a silent spectator in the cause of social change for black America. . . . As a first step, it must look to the needs and abilities of its own black members." Catholic schools were generally hostile to industrial education, but, except for dependent children in institutional schools, its few vocational programs were often aimed at schools attended by blacks. One of the only industrial schools established by the church was for black girls in Chicago. In that same city well into the 1930s, the children of the growing black community could attend only one specially designated school, a policy that provoked considerable anger among Chicago's black Catholics. At St. Elizabeth's, established in 1922, where "the pupils were all Negroes," a high-school curriculum was attached to the parish school, and it gave special attention to "shorthand, commercial drawing, radio, music, arts and crafts, public speaking." Students at St. Elizabeth's were also "taught to be race conscious—conscious of the contribution of the Negro to culture and of their obligations in meeting the ever present prejudice toward him without hatred." Similarly, at St. Malachy's, which opened a two-year high-school department in 1941 and was overwhelmingly black, students gave "special attention" to "the study of Negro history and Negro culture."[58] Unquestionably, this emphasis on black pride was a positive step by the Catholic church, but it existed in the context of a school system that was overwhelmingly segregated and where black children were specifically barred from attending local territorial schools and white high schools.

In Raleigh, North Carolina, a very different archdiocese, most of the students at sixteen black Catholic schools were not Catholic, and their schooling by the church was part of its effort to convert blacks—an effort begun after the Civil War. In North Carolina, schools built for blacks even after 1954 continued to follow a segregationist policy. During the late fifties, the church began openly to oppose segregation, and the schools were officially "open to integration," but as late as 1957, a school specifically designated for blacks opened its doors.[59]

It was in Louisiana, in the dioceses of Lafayette and New Orleans where about one-third of all black Catholics lived, that the greatest effort by the church on behalf of blacks might have been undertaken. But the failure there was representative of the record of the Catholic church in general. In 1867, immediately after the Civil War, the church moved quickly to open the first school for blacks in Lafayette. But by 1893, the Catholic hierarchy had firmly established a policy of separate churches and prohibited mixed congregations. A committee of black citizens issued a strong denunciation of this action as a "positive violation of Catholic principles," but the protest proved futile. Throughout the first one-half of the twentieth century, the Catholic church in Louisiana continued to maintain segregated facilities. In 1941, the diocese of Lafayette had the largest number of Catholic grade schools for blacks in the United States. All thirty-two of the schools were segregated, and all were elementary schools only. Segregation made it difficult to provide an education to blacks beyond the eighth grade because blacks could not effectively commute to schools at a distance. When Holy Rosary School became (and long remained) the central black high school in Lafayette, it not surprisingly offered industrial arts as well as the more standard college-preparatory, business, and general curricula.[60]

As was the case elsewhere, black Catholic schooling in Louisiana was hampered by financial problems. Since so much Catholic education was dependent on voluntary contributions, the lack of support for blacks among the more affluent white Catholic community imposed severe limits on black education. As Loretta Butler observed in her study of Lafayette schools, "officials recognized that in operating separate schools the presence of deep-seated prejudices made securing funds for their maintenance difficult. Not only was it difficult to secure funds, but it was also difficult to recruit the needed personnel." The order of black nuns in Louisiana, the Holy Family Congregation, could not find colleges willing to train them, and as a result, Catholic schools for blacks had persistent staffing problems. In most ways, Catholic schools for blacks paralleled public schools for blacks in Louisiana—underfinanced, overenrolled, and inadequately staffed. By 1960, the situation in Louisiana had changed very little, and segregation still prevailed in the Catholic schools. In that year, Butler concluded, "there have been limited efforts throughout the history of the area to educate the Negro Catholics. The efforts have been too little and too late. The schools are inadequate and unable to provide for the needs of the community."[61]

In Louisiana, as in North Carolina, the school policies of the universal Catholic church had been largely an American product. As was true for public schools, local conditions of racial prejudice inhibited any impulse to move beyond the inherent restrictions of a racially separatist society. Significantly, in the Catholic schools of Chicago where segregation was not legally mandated, separate education also prevailed for a good part of the twentieth century. By the forties, in the words of James W. Sanders, who has written the best and most complete study of that city's parochial schools, "The Archdiocese made considerable progress in opening the Catholic churches and schools to Negroes, though no truly integrated parishes developed, and resistance remained intense." As black migration to Chicago increased after World War II, "changes in racial composition of a neighborhood almost invariably left the Catholic schools with empty classrooms." One black mother, educated at St. Elizabeth's High School (a black school), tried to enroll her daughter in five different high schools in 1946, but complained that "they all turn me down saying they don't take Negroes." Even in 1960, when a high school for boys was opened and was advertised as "without regard to race, creed or color," it was clearly exclusively a black school.[62]

In most big cities, the Catholic church's policies during the early twentieth century were based on the ethnic parish, as churches and schools were reserved for distinct subcommunities. In most cases, these separate schools for white ethnics were defined as expedient and conformed to the demands of various national groups to maintain language and cultural ties through the generations. They were a positive force for group cohesion and intergenerational continuity. In the late nineteenth and early twentieth centuries, children were often taught exclusively in the language of their homes. And well into the twenties and thirties, Catholic children were being taught religion in Polish, French, Lithuanian, Italian, Magyar, and other languages by sisters familiar with those languages.[63] The schools for blacks followed the logic of that development. In Chicago, for example, black Catholics were taught black history and pride in black culture. At the same time, there were certainly no obvious linguistic reasons for maintaining national parishes for blacks, and the separate schools were not established in response to black nationalist demands. In contrast to white ethnics, most black Catholics resented and abhored the policy of separatism imposed by the church. Moreover, blacks, unlike other national groups, were not given the option of choosing a territorial parish until much later than any other group. And the inability of blacks to attend the more ethnically heterogeneous high schools was a clear indication of the fact that blacks were not treated like other ethnic groups. Separate schools were provided to keep blacks from white schools and not to provide them with a more efficient pathway to Americanization.

In the United States, the Catholic church always moved along a thin line between universalism and the complex cultural and historical backgrounds of its parishioners. Serving an extraordinarily heterogeneous community whose loyalties were often to the distinct Catholic churches of their past, the church hierarchy frequently compromised its larger visions to political expediency. While ideally, "every Catholic child belongs to Christ's Mystical Body," immigrant children also belonged to distinct communities, to which their parents hoped to bind their children's loyalty. It was through the forms of ethnicity that parental authority was expressed and enforced. The church and its schools pragmatically supported that loyalty and through it helped to bind future generations to itself.[64]

The ethnic parish and its associated school came into being in the late nineteenth century in response to the demands of various groups (first German and later Eastern European) for self-direction and language continuity. Ethnic conflict within the church and resentments about Irish control of the hierarchy forced the American church to a de facto fragmentation of parishes by ethnicity. In the church's efforts to maintain the loyalty of the new immigrant groups, the Catholic schools became critical allies and were especially attractive to parishioners because they offered instruction in diverse native languages. The Catholic hierarchy often opposed the introduction of foreign languages, like German and Polish, into the public schools in order to highlight the distinct advantage of the parochial schools. As James Sanders notes, "Catholic leaders understood full well that many children were sent to parochial schools not only for religious reasons," but in order to maintain ethnic solidarity and to permit parents to exercise some control over the future of their children. In Chicago, which by the late nineteenth century had become a great ethnic city, parochial schools were overwhelmingly ethnically isolated, so much so that the school system was irrational—some schools were largely empty while others burst with excess students. Many neighborhoods experienced a situation like that of the Back-of-the-Yards district: "Though the Irish school lost three-fifths of its pupils between 1900 and 1910, the nearby Eastern European institutions struggled to provide for 4500 more. Yet, for a Pole or Lithuanian or Slovak to attend St. Rose before the last Irish child departed was unthinkable. As one school stood almost empty, another stood a-building just across the street or down the block."[65]

This situation, the result of pride and first-generation conflict, also meant that the schooling of Catholic children could be very uneven, with some schools well-financed, furnished, and staffed, while others were poor in all those ways. The local autonomy that inhered in the de facto ethnic parish had both benefits and costs. It also led to internal conflict, when, for example, Polish sisters were engaged to staff a Lithuanian school and were accused by parents of instructing their children in Polish ways. "Many of the conflicts between Catholics of different nationalities," James Hennessey concluded, "revolved around control of parochial schools."[66] The founding and maintenance of schools in this context was never smooth or easy, but the parish schools succeeded in supporting ethnicity and language in ways that the public schools never could, however ethnically homogeneous their populations might be.

In many ways, parochial schools in the early twentieth century were transitional institutions, providing an effective passage for immigrant parents and their children as they adapted to the American environment. In allowing for language continuity especially, but also for specific forms of ethnic ritual and for nationalist sentiment, parochial schools permitted immigrant communities to educate their children without losing a sense of the past. At the same time, the church's success among different ethnic groups was not uniform, and its schools were likewise differentially supported. Among the newer immigrant groups, Poles, for example, expressed their religious piety and fervent nationalism by sending their children to parochial schools, at least on the elementary level, while Italians, at the other extreme, were highly reluctant to do so. Most Italian children went to public rather than to parochial schools. Different groups responded differently to parochial schools, as they did to the Catholic church in America, based on various views of the church's authority and its guidance, views lodged in the past and in the often tense contemporary relations among groups. For some, the parochial schools were an eagerly accepted compromise between the past and the future, while for others they represented a choice with no obvious benefits.

The intense ethnicity of the turn of the century that church schools had expressed and helped to define began to subside by the interwar years, and by the mid-twentieth century it had receded so that Catholic-school attendance was only marginally related to ethnic concerns. In the 1920s and '30s, a period coextensive with high-school expansion, Catholics moved out from poor city centers to more middle-class areas, both inside the cities and in the suburbs, where ethnicity became less important and the opportunities for economic and social success more so. And within older city districts, the constant movement of population eroded strongly etched ethnic parish boundaries and led to compromises and alliances. Although ethnicity did not by any means disappear, the parochial schools' role as ethnic reinforcers declined. Increasingly, the original ethnic appeal of parochial schools was replaced by other factors, including locality, religious sentiment, and mobility. By the middle of the century, ethnicity was no longer central to Catholic-school attendance. Some groups, like the Irish, continued to support Catholic schools in disproportionate numbers, and some, like the Italians and Hispanics, were still not sending their children to the church schools in proportion to their significant number. But longer residence, even for the Italians, appears to have enhanced the chances that children would attend Catholic schools, probably because these groups, like the Irish before them, began to appreciate the schools' other advantages. By the 1960s, parents of parochial-school students were overwhelmingly native-born and over one-half had native-born parents. Language maintenance as well as memories of the European churches were now largely a thing of the past. While ethnicity cannot be altogether discounted as a stimulant to continued parochial-school support, its central significance had been eclipsed by other issues.[67]

In 1955, sociologists Peter and Alice Rossi found a correlation between aspirations for mobility and Catholic-school attendance. Catholic high-school students were less likely to express values and ideals of family cohesion and loyalty than Catholic students in public high schools. Students in Catholic high schools adopted a whole range of attitudes associated with upward mobility more frequently than those in non-Catholic schools, and these differences superseded differences in social and economic background. Rossi and Rossi concluded that "If these data are to be taken as indicative of Catholic secondary schools in general, it would appear that the more mobility-conscious Catholics are to be found within such schools." In another study, Andrew Greeley and Peter Rossi found that overall, "those who went to Catholic high schools . . . were more likely to go to college than were those (Protestant or Catholic) who went to public schools, with the exception of Protestants attending public schools whose father had gone to college."[68]

Rossi and Rossi's suggestions have found support elsewhere. Although most surveys of parents' reasons for sending their children to Catholic schools confirm that the stated reasons are heavily religious, not academic, the educational backgrounds of parents with children in Catholic schools tend to be higher than Catholics whose children went to public schools, and more parents of Catholic-school children were committed to sending their children to college.[69] Although the differences were rarely very large, they fall into a pattern. By the mid-fifties, parochial schools were ethnically complex, and the parents of students were often themselves of mixed ethnic background. Instead of ethnicity, parental decisions to send children to parochial schools had become an expression of Catholic middle-class identification and a sign of aspirations for children.[70] By the fifties and sixties, as Will Herberg argued some time ago, ethnic subcommunities had been Americanized as they emphasized religious identification over ethnicity and merged into what he called the triple melting pot (Catholic, Protestant, Jewish). Catholic universalism thus overshadowed older loyalties. Herberg may have exaggerated the evanescence of ethnicity as more recent literature has suggested, but Catholic schools, at least, no longer played their earlier strategic role in maintaining ethnic loyalty. Where Catholic schools had once expediently permitted themselves to serve as an arm of the ethnic community and of the poor, they had become in the mid-twentieth century an expedient conduit for the expression of mobility aspirations.[71]

It is, no doubt, in this fashion that the Catholic schools have begun to function for blacks as well. By the early sixties, St. Clair Drake and Horace R. Cayton had concluded that for many blacks, "one of the primary attractions of the Catholic Church is its educational institutions. . . . Many parents felt that the parochial school offered a more thorough education in a quieter atmosphere with adequate discipline and personal attention." As sociologist James Coleman's recent comparison between public and parochial schools has suggested, black children in Catholic schools were closer in college aspirations to whites than those blacks in public schools. By the late 1970s, when Coleman conducted his study, the revolution in the schooling of blacks of the post-1965 period affected Catholic schools just as it affected public schools, and Catholic schools were integrated institutions, indeed, better integrated than public schools. As with the progeny of European immigrants, so too with blacks; the Catholic schools had become channels for upward mobility and an expression of aspirations for middle-class status. Andrew Greeley came to this conclusion in studying the achievements and career paths of minority students in Catholic and public schools: "Perhaps the principal reason for choosing a Catholic education for one's child if one is a minority family is that the child's chances of graduating from college are perceived as being enhanced by such an educational experience."[72]

It is perhaps ironic that a school system based on separatist principles that embraced even more discrete separatist policies early in the twentieth century should serve a half century later as a means for assimilation and social mobility. Yet the church had never proposed that its schools would be anything less than American. The Catholic parochial school had always been a blend of ideals and expediency as it maintained Catholic identity while acceding to the demands of its constituents for the type of education they desired. At the same time, and unlike the public schools, the Catholic schools had always assumed that Catholic education was something more than a complete adaptation to its constituency. The church had an obligation to teach along certain lines, and it had the responsibility to raise up the children of Catholics to be practicing Catholics and leaders in the American Catholic community. In this sense, the church schools had always anticipated their recent successes, both in subsuming ethnic identity to Catholic universalism and in the social success of their students.

The Catholic schools' ability to adapt to American circumstances gave them a peculiar advantage over secular schools. As I have suggested, that advantage, especially as it operated through the high schools, was in good part the result of admissions standards. But the advantage was as much in the eye of the beholder as in anything specific to the schools. The Catholic schools, as alternative schools, tended to attract more like-minded attenders than the public schools. As a result, their selectivity existed even before examinations or applicant screening procedures. The parents of Catholic-school students, including blacks, had to be previously motivated to choose the Catholic over the public schools. Early in the century those objectives were often ethnic; in the middle to late twentieth century, they are frequently academic and social. Although the stated objectives for choosing Catholic schools usually emphasized moral and religious aims, academic and vocational motivations figured in those choices by the mid-1960s, and may actually have subsumed religious aims by the 1970s. By that point, the "old-fashioned" more academic emphasis of Catholic schools had become a distinct advantage to mobility-conscious parents. In 1966, when parents of Catholic-school students were asked to evaluate the importance of various factors in schooling, 90 percent of almost 25,000 respondents believed that training children in self-discipline and hard work was either the most important or a very important reason for sending their children to Catholic schools.[73] By the second half of the twentieth century, the very traditionalism of the Catholic schools, which twenty years before had been a source of defensiveness, had become a source of strength. In the highly competitive school situation in which going to college has become the common denominator of middle-class life, the selectivity of the Catholic schools and their more uniform high-school curriculum had become an important advantage. And many of the descendents of immigrants and blacks for whom the church schools had no place and had made no provision were clamoring for the admission of their children.

V

Local control of American schools has historically aimed to provide parents with a measure of authority over the content and direction of the education provided to their children by the state. That ideal has never been entirely realistic or realized as Michael Katz and other historians have demonstrated. Long before the complex urban realities and the professional self-consciousness of the twentieth century rendered the concept of local control increasingly anachronistic and largely rhetorical, local communities were torn apart over issues of schooling. Class, religion, and culture have throughout the American past created fissures in communities which translated into differential access and varying influence over the schools.[74] American Catholics more conspicuously and effectively than other groups have withdrawn their children from this politically charged environment. In establishing alternative schools, Catholics acted on the implicit principle of community control, although that community was defined not only territorially but also ideologically and culturally and on the basis of the American constitutional protection of religious freedom.

The Catholic system of education that was eventually erected on those grounds has become an enormous enterprise, comparable in many ways to the public school system. And like that system, it has been buffeted by the ideas, practical concerns, and dilemmas of mass education. Ethnicity, racial divisions and hostilities, financial worries, issues of administration, and scientific pedagogy have affected the Catholic schools no less than the public schools. But Catholic schools have operated in this context with two essential differences. First, they could depend on the existence of the public school system to provide solutions to problems the Catholic schools chose to ignore or failed to resolve. Second, they could depend on the explicit consent of the families who sent their children to Catholic schools.

The first of these differences resulted in a large measure of defensiveness for Catholic educators throughout much of the twentieth century. Catholic schools were never as fully innovative as public schools; nor were they ever as fully democratic. Certainly, the Catholic schools became both more democratic and more innovative as the century progressed, but Catholic philosophy as well as limited resources always provided brakes on these forces. In an educational world where progress and expansion were by-words of success, Catholic educators were always looking over their shoulders at what the public schools were doing and how their own programs were perceived and evaluated by Catholics and the larger American community. Pride as well as minority self-consciousness forced Catholic schools to operate in ways that were never entirely autonomous and, indeed, often heavily derivative.

At the same time, public policy allowed Catholics a fundamental measure of minority autonomy. This was both because the democratic commitment to religious pluralism permitted the schools to exist and because the public schools' more expansive policies allowed the alternative schools to exist on their own terms. By providing an education to Catholics who elected not to attend Catholic schools or to those whom the Catholic schools rejected, the public schools allowed Catholic schools to remain far more homogeneous than their integrated philosophy alone would have made possible.

This strengthened the overall effect of the parental consent available to Catholic schools. Catholic-school selectivity and the explicit consent of parents protected Catholic schools from many of the most difficult issues of democratic education. This was further effected by the hierarchy of institutional command and authority, for which Catholic schools could call upon the long tradition of Catholic belief and practice. In other words, Catholic schools had all the advantages of local control without most of its disadvantages. The Catholic hierarchy and a uniform and continuous Catholic philosophy permitted Catholic education to develop systematically while still being connected to Catholic consensus and parental assent. By subscribing to Catholic schools, parents at once exercised their own authority over their children's education and voluntarily submitted that education to the hierarchy and structure increasingly necessary to the complex institutional situation of twentieth-century schools. The parents of children in Catholic schools supported those schools financially, morally, and willingly. And their attitude provided the schools with a much fuller measure of authority over the children under their jurisdiction. It also, and not inconsequentially, strengthened the students' own sense of satisfaction with the schools and their disciplinary expectations, since parental support enhanced students' confidence in the excellence and value of their education.[75]

This does not mean that all Catholic parents were completely satisfied with their children's education.[76] It does mean that extreme dissatisfaction would force parents to take their children to another, usually public, school, an option not easily available to the parents of children in public schools. That action would only minimally affect the Catholic school. Catholic au-

thorities always had the power of church doctrine to inhibit mass with-
drawals and usually forestalled widespread dissatisfaction by seeking ways
to imitate the public schools and to provide Catholic children with the
more obvious advantages of public schooling. Throughout the century,
Catholic schools diffused some of the positive threat of public-school competition through selective imitation.

In the long run, these differences provided a dual advantage for the
Catholic school system. Its ability to exercise the democratic option of
alternative schooling by providing a less fully democratized education resulted in considerable social success. As James Coleman's report comparing public and parochial schools has demonstrated, Catholic-school children performed better than public-school children on all measures, regardless
of class and ethnic background. Indeed, the Catholic schools have allowed
their students to transcend in their performance many of the social divisions of class, race, and ethnicity while the academic performance of public-
school children usually echo these. That is to say, Catholic education has
homogenized the backgrounds of its students, performing the function that
public schools were intended but have often failed to achieve. Ironically,
the less democratic Catholic schools have registered greater democratic
gains.[77]

It would be easy and tempting to ascribe these gains to the specific forms
of Catholic instruction, and the Coleman report has initiated a trend in
just this direction.[78] The recent emphasis on standards, basics, and academics has at least in part adopted the Catholic schools as a latent model.[79]
Certainly, the parochial schools' emphasis on a more uniformly academic
training, on standards, and on discipline have contributed to their instructional successes. But other factors have provided Catholic schools with
advantages largely absent in public schools—parental trust and consent,
mobility aspirations, selective admissions and promotions, and a willing-
ness to ignore the requirements of those less able or submissive. Coleman
has in fact used the result of his study to suggest that the public school
system and idea be fundamentally restructured to allow parents greater
options in selecting schools for their children, so as to provide the consent
and homogeneity of interest now largely absent from territorially defined
schools.[80]

As so often in the history of American education, the historical dialectic
between outsiders and the public schools has entered a new phase. As the
Catholic schools leave behind a long period of defensive imitation of the
public schools, those concerned with public education have discovered new
ways of learning from the minorities in their midst. The influence of outsiders on American education may have just begun.

Conclusion

Since the mid-nineteenth century, education has served the many household gods of American society—citizenship, morality, mobility, assimilation. In the twentieth century, those gods have sometimes changed their form and for many have become justice and equality, but the secular religion of education has remained both an arm of the state and a meaningful part of general culture. The demands that Americans have made of education and their faith in its efficacy render the religious analogy not at all far-fetched or trivial. In the absence of a common or state-supported religion, the schools have taken the place of the church as a means of control and as institutions through which the population has sought its variously defined roads to salvation. Indeed, the emotional and even evangelical tone adopted by critics of school developments often makes the language of school reform far more volatile than that attached to other institutions and this lends credence to the analogy. As institutions with ecclesiastical roles in the society, the schools have aimed to unify and integrate the nation and to direct the behavior and beliefs of its complex population in regular and socially acceptable ways. Since the late nineteenth century, few have questioned the logical consequence of the schools' ecclesiastical functions—that all should be exposed to its teaching; and by the early twentieth century, this view was enforced through laws requiring and enforcing attendance.

At the same time, however, the American school has become a quintessentially liberal institution, embodying the traditional liberal belief in freely available opportunity and the potential for individual self-realization within broad social bounds. And the school represents the liberal confidence in the possibility of individual improvement and social progress. Liberalism has always assumed as well that individuals and the groups to which they belong would have differing interests and needs that could be accommodated, or at least negotiated, through effectively operating institutions. The liberal tradition of American schooling has organized its religious functions in very specific ways. For, if the church attempts to prepare its worshippers for the glories of the next world and to contemplate something larger than themselves, the liberal tradition of American schooling has always maintained a steady gaze on this world and on its very human rewards. As a result, American schooling has always been at the crossroads of potential conflict—conflict among groups, conflict in aims, and the conflict between social reality and social ideals. Above all, American education has never been able to free itself of the conflicts inherent in a liberal institution serving an enormously diverse population.

The progressives stood at the juncture of the two traditions, and they shaped their modern form. Progressive reformers understood at once the power of the public school's enforcement potential, and they naturally and unhesitatingly drew on its promise of unity and the by then already strong faith in the promise of education. They expanded the public school's power by extending and intensifying its realm. As liberals, however, they expected schooling not only to define and maintain social goods but also to encourage individual achievement, and they expected the schools to serve the needs of their diverse constituencies. They were also caught in the real restraints of their own time as they attempted to re-create a coherent society through revitalized education, turning often back to a remembered past, but also forward to the threats and possibilities that diversity might bring. In their attempt to balance the liberal and ecclesiastical dimensions of schooling, they expressed and shaped the dilemma that has been part of schooling in America ever since.

Nowhere perhaps is the dilemma of schooling clearer than in the education of outsiders. In many ways, it is the presence of outsiders in the schools that has brought the conflict to the surface in school history, as it did for social progressives. But the problem of schooling outsiders was more than the occasion for the emergence of conflict between the aims and functions of schooling. Outsiders both defined the cultural landscape within which schools operated and forced the schools to develop and to define themselves in that context. In other words, cultural pluralism is inseparable from the history within which the dual traditions of American education took shape, and the problems associated with pluralism permit us to examine the evolving tension of those traditions.

In the late nineteenth and early twentieth centuries, the social problem of immigration became the preeminent issue for the schools. It was, in fact, a problem that the schools were in the best position to handle because they could offer both the experience of control which seemed necessary and because they based their appeal in the language of social opportunity to which immigrants and their children would respond. By the second and third decades of the twentieth century, the magnitude of immigration and its diversity had stimulated the schools to seek out ways to refine their mechanisms of incorporation. The IQ and its many associated consequences were the result. The IQ did two things that seemed to balance the public school's ecclesiastical and liberal functions. First, it allowed the schools to function pedagogically in the process of expanding—and to do so with minimal strain. It allowed schools to include masses of very different students and to regulate their programs simply and efficiently. Concurrently, it allowed the public school not only to define itself as a bulwark of liberal opportunity at a time when schooling became more strategic to success, but also to pride itself on its growing fulfillment of this ideal, especially in escalating high-school enrollments. Because the IQ allowed educators to reify their mastery of individual differences by clothing that concept in the language and technology of a science, it seemed for all intents and purposes the very perfection of the liberal ideal. Each student could realize his or her full potential, now no longer a matter of guesswork or approximation, through the various opportunities provided by the schools. I have suggested that the costs of this schema were great because individual potential was carelessly enumerated and because it was grounded in a limited and stingy conception of group abilities. But it would be a mistake to underestimate the genuine effort at meritocratic standards this involved. In a liberal society, and through the channels of its liberal institutions, the problem of locating and providing for diverse talents was and remains a challenge, one that Jefferson pinpointed long ago as a legitimate part of social policy.

In the social sphere of the high school, the tension between control and diversity had very different results since students continued to define themselves according to ethnic criteria which had no role to play in school policy. The complex and bustling world of high-school activities was in fact a liberal's dream, since choice and inclination defined associations and commitments. Whatever control was exercised was not lodged in the homogenizing intentions of educators, but in the demanding and differentiating criteria of peers. However educators may have predefined students in order to structure their classroom learning, and no doubt this had some effect even on social life—on how students saw themselves and each other—this provided only the very widest limits on social relationships. Schooling could go only so far in America in determining associations and preferences. As the streets spilled into the schools, so the schools would help to encourage certain kinds of stratifications along ethnic lines, and provided students with different experiences of assimilation and imitation. Thus, if the church analogy is correct, American high schools were far more Protestant than Catholic, since they provided enormously diverse social experiences that reflected a variety of factors including class and local conditions.

If American schools have enforced controls in exchange for opportunity, this was, until recently, hardly a bargain blacks were encouraged to make. While the 1920s saw some easing of the customary American belief in the possibility that all could be commonly educated, that belief had scarcely at all been applied to blacks. Blacks have been more than outsiders in America; they have all too often been defined as the archetypical other, embodying those qualities from which we dissociate ourselves. Since education has been essential to our self-identity as a people, both the denial of education to blacks and the common assumption that they were incapable of benefitting from schooling helped to confirm their position. In the South, of course, race controls were lodged in laws, customs, and perceptions that kept blacks away from the resources and fellowship of the white community, and the schools were symbolic of the emphatic separation of races. But blacks in the South, as elsewhere, had long registered their faith in the American religion of education and eagerly embraced its liberal promise. During Reconstruction and into the early twentieth century, blacks enthusiastically availed themselves of schools when these were provided. They did so again during the New Deal when the federal government provided a variety of means for black education.

Despite its unorthodox forms, the crisis education of the New Deal agencies and of the army tended to confirm the ecclesiastical and liberal nature of education in the United States, especially for blacks. In the first place, it encouraged blacks to renew their quest for admission into what Mary Bethune and many of her contemporaries called "the American program." Education was both a path to and a sign of that admission, and nowhere was the signal concern for incorporation more forcefully expressed than when hundreds of thousands of blacks sought literacy and other forms of schooling. For blacks, the possibility of schooling resonated with issues of citizenship, and the New Deal, by defining the government's obligation to its citizens in broadly social terms, encouraged those resonances. If Roosevelt and his advisers would do nothing about lynchings, segregation, or poll taxes, they held out to blacks the possibility of education and employment. This was, of course, liberalism at its best. The New Deal agencies in many instances proclaimed that they were offering opportunities to all—irrespective of color. And they used federal regulations to remove barriers which had blocked access to federal programs.

They rarely went beyond that liberal ideal, but even that was far more than blacks had been offered before. Because the New Deal proclaimed equal access in its regulations and because it included blacks in its accounting, it offered the hope of genuine change. This process of ecclesiastical inclusion was quite as significant as the liberal goods that were offered. The war and the army completed the process, not because they offered reform or encouraged opportunity, but because the experience made the societal costs of black exclusion from education clear. The army also made clear that blacks could be educated like whites, and, in so doing, it redefined the meaning of mental aptitude and its measurement. By the late 1940s and early 1950s, in the context of growing anxieties about national security, the exclusion of blacks from American education was no longer tenable, and in 1954 ecclesiastical inclusion was enshrined as a national philosophy. After that, the issue for blacks would become very like that of immigrants in the early twentieth century whose special problems became matters of concern for the schools as well as for the society. After that, too, issues of access gave way to questions about liberal outcomes, although the problem of inclusion has hardly been laid to rest.

The war also had an enormous impact on women, and it was not surprising that the schools would register women's confused status. The ecclesiastical function of American schooling was always clear for women. Women's inclusion in the schools was rarely questioned after the schools became firmly established in the social landscape. There was no reason why it should be. As fundamental constituents of the society and essential to its orderly functioning, especially to its moral life, women had been schooled beside men for 100 years by the time the war brought the dilemma of women's education to the surface in the 1950s. But while women operated easily in the schools, their position in the society was more complex and conflict-ridden. In many ways and for a long time they had hardly any role to play in the liberal ideology of American life because liberalism and its commitment to opportunity was a marketplace ideal. While women held a variety of jobs and had economic functions, opportunity for self-realization apart from the private sphere of the family had never been designed to include women. The common assumption was not that they were incapable of public success (although this always lingered on the fringes of popular as well as serious thought) but that they were uninterested in it and that it was unbecoming.

During and after World War II, however, women's marketplace participation grew enormously, and by the 1950s the liberal function of education for women became the subject of extensive discussion in colleges and universities. In the 1950s, institutions of higher education as interpreters and enforcers of culture articulated the usual conceptions of women's role.

But they introduced a new liberal-utilitarian issue into the discussions. Some educators, like Lynn White, hoped to use the colleges as a setting for specific women's concerns. For White, the very liberalism of the schools required that they adapt to the specific needs of women by redefining their educational offerings and their aims. Others, more aware of the increasing role women were playing in the economy and committed to improvement in women's social role, hoped to balance the older assignments of marriage and motherhood with a new appreciation of women's role in the workplace. In enforcing that balance, they sought to alert and prepare women for the inevitable problems this would bring. Neither group ever questioned women's legitimate place in higher education, although in the 1950s there were some faint voices willing to do even this and thus dismiss women altogether from the fellowship of educational salvation. On the contrary, most educators who considered the problem hoped to make education more useful and usable than it was. In fact, the discussions were soon to collapse of their own weight as some women capitalized on the real opportunities the schools offered to challenge the very concept of a defined women's interest. In that sense, the liberalism of the schools operated in a very different way than the discussions about it, since higher education allowed women to be schooled largely equally to men and underwrote eventual social access. If the schools tried to control women by continuing to define motherhood as women's dominant goal, the schools' offering apart from their values had encouraged some women to reject those ideals and had forced others to confront the very conflicts inherent in their education.

It is ironic, but appropriate, that American educational liberalism has permitted the development of an alternative to the public schools that is deliberately ecclesiastical and which in no small way opposed liberalism as a tenet of education. In fact, the legitimate existence of alternative schools has been an emblem of liberalism in education. The Catholic schools have operated apart from American public schools for almost 150 years, and during the twentieth century they have become, as a group, a flourishing and important institution which, in a final irony, have often allowed their students to garner the rewards of liberal society to a greater extent than their publicly educated peers. This is largely because parochial schools have been better able than public schools to monitor their students by providing rewards only to those whose conformity to the ecclesiastical demands and the homogeneous programs of the church schools has been greatest. In other words, parochial schools, although heirs of the Catholic tradition of universalism, have by no means allowed their charges the equal access available in the public schools. Indeed, it is the public schools' greater pluralism that has permitted parochial schools their greater exclusivity and homogeneity. In the parochial schools, the price of success has been a far more stringent and confining conformity. At the same time, even the Catholic schools have adopted some of the liberal techniques of public education, which they describe as progressive, despite the usual disdain for progressivism's philosophical underpinnings expressed by Catholic educators. Like public schools, parochial schools have accommodated as well as prescribed, adapting not only to the American pedagogical environment but also to an American population whose desires have been directed to this world's goods as much as to the rewards of the next.

Critics of public schools and the schools themselves have recently adopted various lessons from the success of the parochial schools, especially their emphasis on regularity, uniformity, discipline, and control. In so doing, they have begun to move toward a much more deliberately ecclesiastical model than has generally been the case in the twentieth century. That this may come at the expense of educational liberalism is in many ways too bad. Despite its many infelicities and clumsiness—sometimes even stupidity—the American liberal tradition of education, which in the twentieth century is inseparable from the progressive tradition, has allowed American schools to open their doors wide at all levels to the population as a whole and at its best to admit that talent comes in many shapes. It has also proclaimed that the individual's goals are quite as important as those of the society. That this democratic ideal has often been derailed (because it operates together with the ecclesiastical function of our schools and is culturally configured) ought not to deprive us or our children of the hope it holds out now and in the future. For Dewey, liberal goals required liberal means, and educational progressivism involved both. The history of education in the twentieth century has made us aware of how complex and contradictory both these ideals can be and that neither can guarantee the equality that has become a modern commitment. But an enforced equality which operates within the strong ecclesiastical tradition of the schools has its own costs. As American schools continue to operate within the shifting balance of their liberal and ecclesiastical traditions and as new groups and issues help to define their meaning, it is clear that despite our criticism and discontents the American faith in education will continue—possibly because we have no alternatives, probably (one hopes) because it still offers the best means toward a humane future. For all of us who care about the schools and the society they create, that faith will almost certainly require that we confront the many continuing dilemmas of American democratic schooling.

Appendix 1:

Data Collection, Sampling
and Analysis

The sample for Chapter 3 was created and composed in the following manner:

1. On the basis of information derived from "Nationalities of Students 1931–47," Student Nationality File, Bureau of Reference, Research, and Statistics, Teachers College Library, Columbia University, I composed a list of New York high schools and their nationality profiles for the years 1931–47. These statistics had been based on the nativity of students' parents. I then made a shorter list of schools with various ethnic configurations that would illustrate the range of ethnicity patterns in the three most significant boroughs, Manhattan, Brooklyn, and the Bronx. With the approval of the New York City Board of Education and the Office of Borough Superintendents, I contacted each school, usually the assistant principal's office, and ascertained the location and availability of yearbooks. In several cases, the yearbooks could not be found or were not made available for the years of inquiry. The seven schools chosen were, therefore, those which had the following characteristics: (a) demographically interesting, geographically dispersed; (b) at least two from each borough; (c) yearbooks extant; (d) I was able to obtain access to the yearbooks.

2. In each of the selected schools, I used only the yearbooks for the June graduating class. Most schools also had January graduates, pupils who were held back or more rapidly advanced toward graduation. I decided against these as less representative of the cross section of graduating seniors. I collected complete information for the June graduates of every other graduating class between 1931 and 1947 for each school. Where there were gaps in the yearbook holdings, as was sometimes the case, I chose the next available year.

In preparing the materials for transfer to computer tape, it became clear that the number of the students for whom I had information was larger than time or money would allow me to analyze, and I decided to divide the sample in half, choosing four years for each school instead of eight. The four years were chosen to capture the experience in each school over the longest possible time period. Whenever possible the years were the same for all schools. When this was not possible, the next closest year was chosen.

Following is a list of the schools (with the name of the principal whose kind permission made the research possible), yearbook titles, and years used for the sample:

George Washington High School, 549 Audubon Avenue, Manhattan (Samuel Kostman, principal), *The Hatchet,* 1934, 1939, 1941, 1947;

Seward Park High School, 350 Grand Street, Manhattan (Noel Kriftcher, principal), *The Almanac,* 1933, 1939, 1941, 1947;

New Utrecht High School, 1601 80th Street, Brooklyn (Michael Russo, principal), *The Comet,* 1933, 1939, 1941, 1947;

Evander Childs High School, 800 East Gunhill Road, Bronx (John McCann, principal) *The Oriole,* 1933, 1937, 1941, 1945;

Theodore Roosevelt High School, 500 East Fordham Road, Bronx (Phillip Lefton, principal), *Senior Saga,* 1933, 1939, 1941, 1947;

Bay Ridge High School, 350 67th Street, Brooklyn (Joan Leonard, principal), *Maroon and White,* 1933, 1935, 1941, 1947;

High School of Commerce (now Louis D. Brandeis High School), 145 West 84th Street, Manhattan (Murray A. Cohn, principal), *Commercial Caravel, Caravel, Commerce,* 1931, 1933, 1939, 1941, 1947.

In the case of the High School of Commerce, five years were chosen in order to get one coed class. It turned out not to be large enough to be usable as a female sample for the school.

3. After the schools and years were selected and the data collected, the sample was composed of all students whose names appeared in these issues of the yearbook. Every activity of each of these students was categorized according to one of twenty-five categories: no activity, football, basketball, track, other sports, president of student body or senior class, other political office, editor in chief of the student newspaper, other newspaper staff, other publications, physics, chemistry, other science, other academic, program committee, Arista, debate, celebrity, social activities, service, glee club, orchestra, drama, yearbook, religion, other activities. Of these activities, program committee and debate have not been used because they were not usable indicators—either the numbers were too small overall, or the activities were not sufficiently widespread among schools.

4. Activity information, the ethnicity of the student (as determined by name), the sex of the student, and the various details of the school affiliation were entered on an IBM computer at the University of California, Berkeley. All the computations were performed using the SPSS (Standard Program for Social Science) program.

5. The computations were made in aggregated form across all schools, aggregating the classes in each school, and by year in each school. Because the numbers were too small in this last set of calculations, no conclusions were possible about changes over time within each school.

Appendix 2

TABLE A. Extracurricular Participation by Men by Ethnic Group and Activity

ALL SCHOOLS

	Jewish	Italian	Black	Irish	German	Native	Other	Undecided	Total
Some activity	(2369)	(834)	(98)	(216)	(246)	(679)	(427)	(206)	(5075)
	46.7	16.4	2.0	4.3	4.9	13.4	8.4	4.1	100.2
Football	(68)	(38)	(2)	(10)	(12)	(31)	(14)	(5)	(180)
	37.8	21.1	1.1	5.6	6.7	17.2	7.8	2.8	100.1
Basketball	(100)	(10)	(13)	(12)	(5)	(26)	(19)	(9)	(194)
	51.5	5.1	6.7	6.2	2.6	13.4	9.8	4.6	99.9
Track	(144)	(56)	(30)	(24)	(23)	(30)	(30)	(12)	(349)
	41.3	16.0	8.6	6.9	6.6	8.6	8.6	3.4	100.0
Other sport	(323)	(91)	(8)	(40)	(39)	(111)	(71)	(27)	(710)
	45.5	12.8	1.1	5.6	5.5	15.6	10.0	3.8	99.9
President	(16)	(3)	(0)	(4)	(4)	(9)	(3)	(0)	(39)
	41.0	7.7	0.0	10.3	10.3	23.1	7.7	0.0	100.1
Other political	(310)	(97)	(10)	(43)	(37)	(94)	(54)	(27)	(672)
	46.1	14.4	1.5	6.4	5.5	14.0	8.0	4.0	99.9
Yearbook	(93)	(23)	(0)	(10)	(8)	(39)	(8)	(7)	(188)
	49.5	12.2	0.0	5.3	4.3	20.7	4.3	3.7	100.0
Editor in chief	(13)	(3)	(0)	(0)	(1)	(4)	(1)	(0)	(22)
	59.1	13.6	0.0	0.0	4.5	18.2	4.5	0.0	99.9
Other news	(127)	(15)	(2)	(8)	(13)	(35)	(9)	(11)	(220)
	57.7	6.8	0.9	3.6	5.9	15.9	4.1	5.0	99.9
Other publications	(85)	(20)	(1)	(6)	(8)	(27)	(7)	(5)	(159)
	53.4	12.6	0.6	3.8	5.0	17.0	4.4	3.1	99.9
Physics	(30)	(3)	(0)	(0)	(2)	(8)	(4)	(3)	(50)
	60.0	6.0	0.0	0.0	4.0	16.0	8.0	6.0	100.0
Chemistry	(46)	(6)	(0)	(2)	(4)	(14)	(5)	(7)	(84)
	54.8	7.1	0.0	2.4	4.8	16.7	5.9	8.3	100.0
Other science	(58)	(11)	(1)	(0)	(5)	(17)	(7)	(11)	(110)
	52.7	10.0	0.9	0.0	4.5	15.4	6.4	10.0	99.9
Other academic	(288)	(81)	(4)	(10)	(22)	(72)	(47)	(22)	(546)
	52.7	14.8	0.7	1.8	4.0	13.2	8.6	4.0	99.8
Arista	(273)	(59)	(3)	(11)	(16)	(62)	(42)	(18)	(484)
	56.4	12.2	0.6	2.3	3.3	12.8	8.7	3.7	100.0
Glee club	(69)	(35)	(5)	(10)	(9)	(38)	(15)	(2)	(183)
	37.7	19.1	2.7	5.5	4.9	20.8	8.2	1.1	100.0
Orchestra	(76)	(24)	(3)	(1)	(5)	(16)	(10)	(6)	(141)
	53.9	17.0	2.1	0.7	3.5	11.3	7.1	4.2	99.8
Drama	(64)	(13)	(0)	(2)	(7)	(16)	(2)	(5)	(109)
	59.2	12.0	0.0	1.8	6.5	14.8	1.8	4.6	99.9
Social	(150)	(34)	(4)	(10)	(14)	(46)	(23)	(10)	(291)
	51.5	11.7	1.4	3.4	4.8	15.8	7.9	3.4	99.9
Service	(1792)	(651)	(78)	(172)	(183)	(404)	(332)	(152)	(3764)
	47.6	17.3	2.1	4.6	4.9	10.7	8.8	4.0	100.0
Religious	(9)	(19)	(0)	(15)	(6)	(26)	(12)	(1)	(88)
	10.2	21.6	0.0	17.0	6.8	29.5	13.6	1.1	99.8
Other activity	(454)	(173)	(20)	(50)	(69)	(174)	(98)	(45)	(1083)
	41.9	16.0	1.8	4.6	6.4	16.1	9.0	4.2	100.0
Celebrity	(44)	(14)	(2)	(2)	(7)	(32)	(6)	(2)	(109)
	40.4	12.8	1.8	1.8	6.4	29.3	5.5	1.8	99.8
# in each group	(2947)	(1141)	(132)	(289)	(349)	(878)	(552)	(280)	(6568)
% male population	44.9	17.4	2.0	4.4	5.3	13.4	8.4	4.2	100.0

Note: The figure in parentheses gives the number of men of each ethnic group involved in the designated activity. The percentage represents the proportion of all male students involved in the designated activity who were members of each ethnic group.

TABLE B. Extracurricular Participation by Women by Ethnic Group and Activity
ALL SCHOOLS

	Jewish	Italian	Black	Irish	German	Native	Other	Undecided	Total
Some activity	(3305)	(1119)	(125)	(199)	(318)	(1227)	(657)	(349)	(7299)
	45.3	15.3	1.8	2.7	4.4	16.8	9.0	4.8	100.1
Football	—	—	—	—	—	—	—	—	—
Basketball	(130)	(68)	(9)	(21)	(18)	(73)	(44)	(23)	(386)
	33.7	17.6	2.3	5.4	4.7	18.9	11.4	5.9	99.9
Track	(2)	(2)	(0)	(0)	(1)	(0)	(0)	(0)	(5)
	40.0	40.0	0.0	0.0	20.0	0.0	0.0	0.0	100.0
Other sport	(332)	(107)	(17)	(25)	(20)	(106)	(58)	(33)	(698)
	47.6	15.3	2.4	3.6	2.9	15.2	8.3	4.7	100.0
President	(9)	(2)	(0)	(0)	(0)	(3)	(0)	(1)	(15)
	60.0	13.3	0.0	0.0	0.0	20.0	0.0	6.7	100.0
Other political	(332)	(107)	(17)	(25)	(20)	(106)	(58)	(33)	(698)
	47.6	15.3	2.4	3.6	2.9	15.2	8.3	4.7	100.0
Yearbook	(171)	(52)	(3)	(18)	(22)	(99)	(47)	(21)	(433)
	39.5	12.0	0.7	4.2	5.1	22.9	10.8	4.8	100.0
Editor in chief	(9)	(2)	(0)	(1)	(1)	(2)	(0)	(0)	(15)
	60.0	13.3	0.0	6.7	6.7	13.3	0.0	0.0	100.0
Other news	(173)	(23)	(1)	(4)	(18)	(57)	(22)	(11)	(309)
	56.0	7.4	0.3	1.3	5.8	18.4	7.1	3.6	99.9
Other publications	(86)	(10)	(1)	(2)	(4)	(24)	(4)	(5)	(136)
	63.2	7.4	0.7	1.5	2.9	17.6	2.9	3.7	99.9
Physics	(2)	(3)	(0)	(0)	(0)	(0)	(2)	(0)	(7)
	28.6	42.9	0.0	0.0	0.0	0.0	28.6	0.0	100.1
Chemistry	(16)	(7)	(1)	(6)	(3)	(16)	(9)	(6)	(64)
	25.0	10.9	1.6	9.4	4.7	25.0	14.1	9.4	100.1
Other science	(76)	(37)	(2)	(9)	(7)	(40)	(26)	(21)	(218)
	34.9	17.0	0.9	4.1	3.2	18.3	11.9	9.6	99.9
Other academic	(573)	(246)	(15)	(29)	(55)	(189)	(129)	(50)	(1286)
	44.5	19.1	1.2	2.2	4.3	14.7	10.0	3.9	99.9
Arista	(519)	(89)	(5)	(23)	(32)	(157)	(81)	(37)	(943)
	55.0	9.4	0.5	2.4	3.4	16.6	8.6	3.9	99.8
Glee club	(176)	(55)	(23)	(12)	(12)	(94)	(50)	(21)	(443)
	39.7	12.4	5.2	2.7	2.7	21.2	11.3	4.7	99.9
Orchestra	(44)	(25)	(5)	(2)	(3)	(28)	(14)	(9)	(130)
	33.8	19.2	3.8	1.5	2.3	21.5	10.8	6.9	99.8
Drama	(133)	(20)	(1)	(7)	(14)	(64)	(29)	(16)	(284)
	46.8	7.0	0.4	2.5	4.9	22.5	10.2	5.6	99.9
Social	(430)	(211)	(12)	(45)	(46)	(199)	(109)	(57)	(1109)
	38.8	19.0	1.1	4.1	4.1	17.9	9.8	5.1	99.9
Service	(2494)	(789)	(90)	(126)	(222)	(866)	(432)	(244)	(5263)
	47.4	15.0	1.7	2.4	4.2	16.5	8.2	4.7	100.1
Religious	(17)	(67)	(1)	(18)	(4)	(31)	(11)	(4)	(153)
	11.1	43.8	0.6	11.8	2.6	20.3	7.2	2.6	100.0
Other activity	(979)	(585)	(31)	(129)	(137)	(554)	(341)	(155)	(2911)
	33.6	20.1	1.1	4.4	4.7	19.0	11.7	5.3	99.9
Celebrity	(20)	(5)	(0)	(7)	(5)	(20)	(5)	(6)	(68)
	29.4	7.3	0.0	10.3	7.3	29.4	7.3	8.8	99.8
# in each group	(3842)	(1363)	(163)	(228)	(387)	(1407)	(771)	(429)	(8598)
% female population	44.8	15.9	1.9	2.7	4.5	16.4	8.9	4.9	100.0

Note: The figure in parentheses gives the number of women of each ethnic group involved in the designated activity. The percentage represents the proportion of all female students involved in the designated activity who were members of each ethnic group.

TABLE C. Extracurricular Participation by Men by Ethnic Group and Activity
GEORGE WASHINGTON HIGH SCHOOL

	Jewish	Italian	Black	Irish	German	Native	Other	Undecided	Total
Some activity	(289)	(27)	(27)	(28)	(58)	(148)	(68)	(34)	(679)
	42.6	4.0	4.0	4.1	8.5	21.8	10.0	5.0	100.0
Football	(8)	(0)	(1)	(4)	(2)	(7)	(3)	(1)	(26)
	30.8	0.0	3.8	15.4	7.7	26.9	11.5	3.8	99.9
Basketball	(17)	(1)	(8)	(3)	(1)	(6)	(4)	(1)	(41)
	41.5	2.4	19.5	7.3	2.4	14.6	9.8	2.4	99.9
Track	(13)	(5)	(11)	(3)	(4)	(7)	(1)	(2)	(46)
	28.3	10.9	23.9	6.5	8.7	15.2	2.2	4.3	100.0
Other sport	(43)	(7)	(5)	(8)	(6)	(26)	(11)	(6)	(112)
	38.4	6.3	4.5	7.1	5.4	23.2	9.8	5.4	100.1
President	(1)	(0)	(0)	(0)	(0)	(4)	(1)	(0)	(6)
	16.7	0.0	0.0	0.0	0.0	66.7	16.7	0.0	100.1
Other political	(41)	(1)	(1)	(1)	(9)	(14)	(6)	(2)	(75)
	54.7	1.3	1.3	1.3	12.0	18.7	8.0	2.7	100.0
Yearbook	(17)	(4)	(0)	(2)	(2)	(13)	(4)	(3)	(45)
	37.8	8.9	0.0	4.4	4.4	28.9	8.9	6.7	100.0
Editor in chief	(1)	(0)	(0)	(0)	(0)	(2)	(1)	(0)	(4)
	25.0	0.0	0.0	0.0	0.0	50.0	25.0	0.0	100.0
Other news	(16)	(0)	(1)	(2)	(4)	(9)	(3)	(2)	(37)
	43.2	0.0	2.7	5.4	10.8	24.3	8.1	5.4	99.9
Other publications	(3)	(0)	(1)	(0)	(1)	(2)	(1)	(0)	(8)
	37.5	0.0	12.5	0.0	12.5	25.0	12.5	0.0	100.0
Physics	(22)	(2)	(0)	(0)	(2)	(7)	(3)	(3)	(39)
	56.4	5.1	0.0	0.0	5.1	17.9	7.7	7.7	99.9
Chemistry	(11)	(0)	(0)	(2)	(1)	(9)	(4)	(2)	(29)
	37.9	0.0	0.0	6.9	3.4	31.0	13.8	6.9	99.9
Other science	(21)	(0)	(1)	(0)	(1)	(8)	(3)	(7)	(41)
	51.2	0.0	2.4	0.0	2.4	19.5	7.3	17.1	99.9
Other academic	(79)	(2)	(2)	(3)	(8)	(26)	(9)	(12)	(141)
	56.0	1.4	1.4	2.1	5.7	18.4	6.4	8.5	99.9
Arista	(40)	(1)	(0)	(1)	(3)	(14)	(10)	(4)	(73)
	54.8	1.4	0.0	1.4	4.1	19.2	13.7	5.5	100.1
Glee club	(12)	(2)	(3)	(2)	(0)	(11)	(6)	(1)	(37)
	32.4	5.4	8.1	5.4	0.0	29.7	16.2	2.7	99.9
Orchestra	(15)	(2)	(1)	(0)	(2)	(6)	(2)	(2)	(30)
	50.0	6.7	3.3	0.0	6.7	20.0	6.7	6.7	100.1
Drama	(11)	(0)	(0)	(0)	(1)	(5)	(0)	(0)	(17)
	64.7	0.0	0.0	0.0	5.9	29.4	0.0	0.0	100.0
Social	(19)	(4)	(2)	(1)	(2)	(10)	(8)	(2)	(48)
	39.6	8.3	4.2	2.1	4.2	20.8	16.7	4.2	100.1
Service	(182)	(19)	(19)	(19)	(38)	(91)	(47)	(16)	(431)
	42.2	4.4	4.4	4.4	8.8	21.1	10.9	3.7	99.9
Religious	(0)	(1)	(0)	(0)	(2)	(3)	(1)	(1)	(8)
	0.0	12.5	0.0	0.0	25.0	37.5	12.5	12.5	100.0
Other activity	(69)	(6)	(7)	(5)	(20)	(37)	(19)	(10)	(173)
	39.9	3.5	4.0	2.9	11.6	21.4	11.0	5.8	100.1
Celebrity	(9)	(1)	(0)	(0)	(0)	(10)	(2)	(0)	(22)
	40.9	4.5	0.0	0.0	0.0	45.5	9.1	0.0	100.0
# in each group % male population	(371)	(38)	(44)	(45)	(92)	(199)	(102)	(55)	(946)
	39.2	4.0	4.7	4.8	9.7	21.0	10.8	5.8	100.0

Note: The figure in parentheses gives the number of men of each ethnic group involved in the designated activity. The percentage represents the proportion of all male students involved in the designated activity who were members of each ethnic group.

TABLE D. Extracurricular Participation by Women by Ethnic Group and Activity

GEORGE WASHINGTON HIGH SCHOOL

	Jewish	*Italian*	*Black*	*Irish*	*German*	*Native*	*Other*	*Undecided*	*Total*
Some activity	(425)	(40)	(53)	(26)	(69)	(208)	(131)	(66)	(1018)
	41.7	3.9	5.2	2.6	6.8	20.4	12.9	6.5	100.0
Football	—	—	—	—	—	—	—	—	—
Basketball	(17)	(1)	(8)	(3)	(1)	(1)	(6)	(4)	(41)
	41.5	2.4	19.5	7.3	2.4	2.4	14.6	9.8	99.9
Track	—	—	—	—	—	—	—	—	—
Other sport	(82)	(6)	(10)	(3)	(11)	(49)	(22)	(18)	(201)
	40.8	3.0	5.0	1.5	5.5	24.3	10.9	9.0	100.0
President	(1)	(0)	(0)	(0)	(0)	(0)	(0)	(0)	(1)
	100.0	0.0	0.0	0.0	0.0	0.0	0.0	0.0	100.0
Other political	(41)	(3)	(3)	(6)	(9)	(31)	(9)	(5)	(107)
	38.3	2.8	2.8	5.6	8.4	29.0	8.4	4.7	100.0
Yearbook	(9)	(2)	(1)	(4)	(8)	(12)	(5)	(4)	(45)
	20.0	4.4	2.2	8.9	17.8	26.7	11.1	8.9	100.0
Editor in chief	(2)	(0)	(0)	(0)	(0)	(0)	(0)	(0)	(2)
	100.0	0.0	0.0	0.0	0.0	0.0	0.0	0.0	100.0
Other news	(19)	(1)	(1)	(1)	(9)	(15)	(2)	(1)	(49)
	38.8	2.0	2.0	2.0	18.4	30.6	4.1	2.0	99.9
Other publications	(4)	(0)	(0)	(0)	(0)	(1)	(0)	(0)	(5)
	80.0	0.0	0.0	0.0	0.0	20.0	0.0	0.0	100.0
Physics	(1)	(0)	(0)	(0)	(0)	(0)	(1)	(0)	(2)
	50.0	0.0	0.0	0.0	0.0	0.0	50.0	0.0	100.0
Chemistry	(2)	(0)	(1)	(0)	(0)	(1)	(1)	(0)	(5)
	40.0	0.0	20.0	0.0	0.0	20.0	20.0	0.0	100.0
Other science	(21)	(1)	(2)	(1)	(3)	(3)	(8)	(3)	(42)
	50.0	2.4	4.8	2.4	7.1	7.1	19.0	7.1	99.9
Other academic	(122)	(3)	(7)	(4)	(18)	(35)	(38)	(12)	(239)
	51.0	1.3	2.9	1.7	7.5	14.6	15.9	5.0	99.9
Arista	(86)	(5)	(2)	(8)	(11)	(27)	(19)	(10)	(168)
	51.2	3.0	1.2	4.8	6.5	16.1	11.3	6.0	100.1
Glee club	(34)	(1)	(17)	(2)	(3)	(20)	(15)	(8)	(100)
	34.0	1.0	17.0	2.0	3.0	20.0	15.0	8.0	100.0
Orchestra	(8)	(0)	(0)	(0)	(0)	(0)	(2)	(1)	(11)
	72.7	0.0	0.0	0.0	0.0	0.0	18.2	9.1	100.0
Drama	(20)	(1)	(1)	(0)	(2)	(9)	(6)	(2)	(41)
	48.8	2.4	2.4	0.0	4.9	22.0	14.6	4.9	100.0
Social	(74)	(8)	(4)	(6)	(13)	(43)	(22)	(13)	(183)
	40.4	4.4	2.2	3.3	7.1	23.5	12.0	7.1	100.0
Service	(289)	(33)	(38)	(20)	(46)	(133)	(94)	(45)	(698)
	41.4	4.7	5.4	2.9	6.6	19.1	13.5	6.4	100.0
Religious	(0)	(2)	(1)	(4)	(0)	(2)	(4)	(0)	(13)
	0.0	15.4	7.7	30.8	0.0	15.4	30.8	0.0	100.1
Other activity	(107)	(10)	(12)	(2)	(20)	(51)	(32)	(18)	(252)
	42.5	4.0	4.8	0.8	8.0	20.2	12.7	7.1	100.1
Celebrity	(6)	(2)	(0)	(5)	(2)	(4)	(0)	(3)	(22)
	27.3	9.1	0.0	22.7	9.1	18.2	0.0	13.6	100.0
# in each group % female population	(525)	(56)	(76)	(36)	(93)	(262)	(164)	(89)	(1301)
	40.4	4.3	5.8	2.8	7.1	20.1	12.6	6.8	99.9

Note: The figure in parentheses gives the number of women of each ethnic group involved in the designated activity. The percentage represents the proportion of all female students involved in the designated activity who were members of each ethnic group.

TABLE E. Extracurricular Participation by Men by Ethnic Group and Activity
EVANDER CHILDS HIGH SCHOOL

	Jewish	Italian	Black	Irish	German	Native	Other	Undecided	Total
Some activity	(382)	(171)	(11)	(42)	(61)	(154)	(67)	(34)	(922)
	41.4	18.5	1.2	4.6	6.6	16.7	7.3	3.7	100.0
Football	(9)	(6)	(0)	(3)	(3)	(12)	(1)	(1)	(35)
	25.7	17.1	0.0	8.6	8.6	34.3	2.9	2.9	100.1
Basketball	(17)	(2)	(0)	(2)	(3)	(4)	(3)	(3)	(34)
	50.0	5.9	0.0	5.9	8.8	11.8	8.8	8.8	100.0
Track	(33)	(22)	(6)	(5)	(9)	(7)	(6)	(3)	(91)
	36.3	24.2	6.6	5.5	9.9	7.7	6.6	3.3	100.1
Other sport	(66)	(14)	(0)	(5)	(8)	(26)	(10)	(2)	(131)
	50.4	10.6	0.0	3.8	6.1	19.8	7.6	1.5	99.8
President	(1)	(1)	(0)	(0)	(2)	(2)	(0)	(0)	(6)
	16.7	16.7	0.0	0.0	33.3	33.3	0.0	0.0	100.0
Other political	(7)	(2)	(0)	(2)	(2)	(9)	(0)	(1)	(23)
	30.4	8.7	0.0	8.7	8.7	39.1	0.0	4.4	100.0
Yearbook	(8)	(4)	(0)	(0)	(1)	(4)	(1)	(0)	(18)
	44.4	22.2	0.0	0.0	5.6	22.2	5.6	0.0	100.0
Editor in chief	(1)	(1)	(0)	(0)	(0)	(1)	(0)	(0)	(3)
	33.3	33.3	0.0	0.0	0.0	33.3	0.0	0.0	99.9
Other news	(24)	(1)	(0)	(0)	(0)	(8)	(0)	(2)	(35)
	68.6	2.8	0.0	0.0	0.0	22.9	0.0	5.7	100.0
Other publications	(14)	(4)	(0)	(1)	(0)	(5)	(0)	(0)	(24)
	58.3	16.7	0.0	4.2	0.0	20.8	0.0	0.0	100.0
Physics	(0)	(1)	(0)	(0)	(0)	(0)	(0)	(0)	(1)
	0.0	100.0	0.0	0.0	0.0	0.0	0.0	0.0	100.0
Chemistry	(9)	(1)	(0)	(0)	(0)	(1)	(0)	(1)	(12)
	75.0	8.3	0.0	0.0	0.0	8.3	0.0	8.3	99.9
Other science	(5)	(0)	(0)	(0)	(0)	(1)	(2)	(1)	(9)
	55.6	0.0	0.0	0.0	0.0	11.1	22.2	11.1	100.0
Other academic	(21)	(11)	(0)	(0)	(1)	(9)	(5)	(0)	(47)
	44.7	23.4	0.0	0.0	2.1	19.1	10.6	0.0	99.9
Arista	(33)	(5)	(0)	(0)	(0)	(11)	(1)	(3)	(53)
	62.3	9.4	0.0	0.0	0.0	20.8	1.9	5.7	100.1
Glee club	(17)	(6)	(0)	(0)	(1)	(5)	(1)	(1)	(31)
	54.8	19.4	0.0	0.0	3.2	16.1	3.2	3.2	99.9
Orchestra	(14)	(2)	(1)	(0)	(1)	(1)	(2)	(1)	(22)
	63.6	9.1	4.5	0.0	4.5	4.5	9.1	4.5	99.8
Drama	(17)	(4)	(0)	(0)	(0)	(1)	(0)	(0)	(22)
	77.3	18.2	0.0	0.0	0.0	4.5	0.0	0.0	100.0
Social	(23)	(13)	(0)	(2)	(1)	(11)	(5)	(5)	(60)
	38.3	21.7	0.0	3.3	1.7	18.3	8.3	8.3	99.9
Service	(325)	(143)	(9)	(37)	(52)	(126)	(57)	(29)	(778)
	41.8	18.4	1.2	4.8	6.7	16.2	7.3	3.7	100.1
Religious	(1)	(0)	(0)	(0)	(0)	(0)	(0)	(0)	(1)
	100.0	0.0	0.0	0.0	0.0	0.0	0.0	0.0	100.0
Other activity	(40)	(15)	(2)	(5)	(5)	(18)	(5)	(6)	(96)
	41.7	15.6	2.1	5.2	5.2	18.7	5.2	6.2	99.9
Celebrity	(2)	(0)	(1)	(0)	(1)	(0)	(0)	(0)	(4)
	50.0	0.0	25.0	0.0	25.0	0.0	0.0	0.0	100.0
# in each group	(430)	(211)	(12)	(52)	(77)	(181)	(73)	(46)	(1082)
% male population	39.7	19.5	1.1	4.8	7.1	16.7	6.7	4.3	99.9

Note: The figure in parentheses gives the number of men of each ethnic group involved in the designated activity. The percentage represents the proportion of all male students involved in the designated activity who were members of each ethnic group.

TABLE F. Extracurricular Participation by Women by Ethnic Group and Activity
EVANDER CHILDS HIGH SCHOOL

	Jewish	Italian	Black	Irish	German	Native	Other	Undecided	Total
Some activity	(786)	(229)	(21)	(23)	(101)	(336)	(71)	(70)	(1637)
	48.0	14.0	1.3	1.4	6.2	20.5	4.3	4.3	100.0
Football	—	—	—	—	—	—	—	—	—
Basketball	(2)	(1)	(1)	(1)	(0)	(1)	(0)	(0)	(6)
	33.3	16.7	16.7	16.7	0.0	16.7	0.0	0.0	100.1
Track	(1)	(0)	(0)	(0)	(1)	(0)	(0)	(0)	(2)
	50.0	0.0	0.0	0.0	50.0	0.0	0.0	0.0	100.0
Other sport	(11)	(4)	(0)	(2)	(1)	(8)	(1)	(0)	(27)
	40.7	14.8	0.0	7.4	3.7	29.6	3.7	0.0	99.9
President	—	—	—	—	—	—	—	—	—
Other political	(10)	(1)	(0)	(0)	(0)	(15)	(0)	(1)	(27)
	37.0	3.7	0.0	0.0	0.0	55.5	0.0	3.7	99.9
Yearbook	(31)	(5)	(0)	(2)	(2)	(11)	(1)	(2)	(54)
	57.4	9.3	0.0	3.7	3.7	20.4	1.9	3.7	100.1
Editor in chief	(2)	(0)	(0)	(0)	(1)	(0)	(0)	(0)	(3)
	66.7	0.0	0.0	0.0	33.3	0.0	0.0	0.0	100.0
Other news	(23)	(1)	(0)	(1)	(0)	(10)	(3)	(2)	(40)
	57.5	2.5	0.0	2.5	0.0	25.0	7.5	5.0	100.0
Other publications	(20)	(3)	(0)	(0)	(2)	(8)	(0)	(1)	(34)
	58.8	8.8	0.0	0.0	5.9	23.5	0.0	2.9	99.9
Physics	(1)	(0)	(0)	(0)	(0)	(0)	(0)	(0)	(1)
	100.0	0.0	0.0	0.0	0.0	0.0	0.0	0.0	100.0
Chemistry	(1)	(0)	(0)	(0)	(0)	(0)	(0)	(0)	(1)
	100.0	0.0	0.0	0.0	0.0	0.0	0.0	0.0	100.0
Other science	(13)	(3)	(0)	(0)	(1)	(4)	(1)	(0)	(22)
	59.1	13.6	0.0	0.0	4.5	18.2	4.5	0.0	99.9
Other academic	(122)	(41)	(0)	(1)	(11)	(32)	(7)	(5)	(219)
	55.7	18.7	0.0	0.4	5.0	14.6	3.2	2.3	99.9
Arista	(162)	(16)	(0)	(2)	(7)	(59)	(12)	(7)	(265)
	61.1	6.0	0.0	0.7	2.6	22.3	4.6	2.6	99.9
Glee club	(72)	(16)	(1)	(3)	(4)	(35)	(3)	(5)	(139)
	51.8	11.5	0.7	2.2	2.9	25.2	2.2	3.6	100.1
Orchestra	(6)	(3)	(0)	(0)	(2)	(6)	(2)	(1)	(20)
	30.0	15.0	0.0	0.0	10.0	30.0	10.0	5.0	100.0
Drama	(14)	(0)	(0)	(1)	(2)	(10)	(0)	(2)	(29)
	48.3	0.0	0.0	3.4	6.9	34.5	0.0	6.9	100.0
Social	(89)	(11)	(0)	(2)	(6)	(29)	(2)	(7)	(146)
	61.0	7.5	0.0	1.4	4.1	19.9	1.4	4.8	100.1
Service	(689)	(196)	(17)	(19)	(91)	(303)	(61)	(61)	(1437)
	47.9	13.6	1.2	1.3	6.3	21.1	4.2	4.2	99.8
Religious	(4)	(5)	(0)	(0)	(0)	(0)	(0)	(0)	(9)
	44.4	55.6	0.0	0.0	0.0	0.0	0.0	0.0	100.0
Other activity	(162)	(28)	(6)	(6)	(20)	(57)	(12)	(17)	(308)
	52.6	9.1	1.9	1.9	6.5	18.5	3.9	5.5	99.9
Celebrity	(1)	(0)	(0)	(0)	(0)	(4)	(2)	(0)	(7)
	14.3	0.0	0.0	0.0	0.0	57.1	28.6	0.0	100.0
# in each group % female population	(858)	(276)	(24)	(27)	(122)	(366)	(79)	(80)	(1832)
	46.8	15.1	1.3	1.5	6.7	20.0	4.3	4.4	100.1

Note: The figure in parentheses gives the number of women of each ethnic group involved in the designated activity. The percentage represents the proportion of all female students involved in the designated activity who were members of each ethnic group.

TABLE G. Extracurricular Participation by Men by Ethnic Group and Activity
SEWARD PARK HIGH SCHOOL

	Jewish	Italian	Black	Irish	German	Native	Other	Undecided	Total
Some activity	(626)	(106)	(11)	(2)	(9)	(35)	(60)	(25)	(874)
	71.6	12.1	1.3	0.2	1.0	4.0	6.9	2.9	100.0
Football	(21)	(7)	(1)	(0)	(0)	(1)	(3)	(1)	(34)
	61.8	20.6	2.9	0.0	0.0	2.9	8.8	2.9	99.9
Basketball	(36)	(1)	(3)	(0)	(1)	(2)	(3)	(2)	(48)
	75.0	2.1	6.2	0.0	2.1	4.2	6.2	4.2	100.0
Track	(22)	(1)	(3)	(0)	(1)	(1)	(3)	(0)	(31)
	71.0	3.2	9.7	0.0	3.2	3.2	9.7	0.0	100.0
Other sport	(76)	(8)	(0)	(0)	(1)	(8)	(12)	(3)	(108)
	70.4	7.4	0.0	0.0	1.0	7.4	11.1	2.8	100.1
President	(4)	(0)	(0)	(0)	(0)	(0)	(0)	(0)	(4)
	100.0	0.0	0.0	0.0	0.0	0.0	0.0	0.0	100.0
Other political	(117)	(17)	(1)	(0)	(0)	(4)	(11)	(5)	(155)
	75.5	11.0	0.6	0.0	0.0	2.5	7.1	3.2	99.9
Yearbook	(16)	(0)	(0)	(0)	(1)	(2)	(1)	(1)	(21)
	76.2	0.0	0.0	0.0	4.8	9.5	4.8	4.8	100.1
Editor in chief	(2)	(0)	(0)	(0)	(0)	(1)	(0)	(0)	(3)
	66.7	0.0	0.0	0.0	0.0	33.3	0.0	0.0	100.0
Other news	(26)	(0)	(0)	(0)	(2)	(2)	(1)	(1)	(32)
	81.3	0.0	0.0	0.0	6.3	6.3	3.1	3.1	100.1
Other publications	(26)	(2)	(0)	(0)	(1)	(1)	(2)	(1)	(33)
	78.8	6.1	0.0	0.0	3.0	3.0	6.1	3.0	100.0
Physics	(8)	(0)	(0)	(0)	(0)	(1)	(1)	(0)	(10)
	80.0	0.0	0.0	0.0	0.0	10.0	10.0	0.0	100.0
Chemistry	(12)	(0)	(0)	(0)	(0)	(0)	(0)	(0)	(12)
	100.0	0.0	0.0	0.0	0.0	0.0	0.0	0.0	100.0
Other science	(5)	(2)	(0)	(0)	(0)	(1)	(0)	(1)	(9)
	55.6	22.2	0.0	0.0	0.0	11.1	0.0	11.1	100.0
Other academic	(75)	(13)	(2)	(1)	(0)	(4)	(6)	(1)	(102)
	73.5	12.7	2.0	1.0	0.0	4.0	5.9	1.0	100.1
Arista	(57)	(1)	(0)	(0)	(1)	(4)	(3)	(1)	(67)
	85.1	1.5	0.0	0.0	1.5	6.0	4.5	1.5	100.1
Glee club	(5)	(1)	(0)	(0)	(0)	(0)	(0)	(0)	(6)
	83.3	16.7	0.0	0.0	0.0	0.0	0.0	0.0	100.0
Orchestra	(22)	(5)	(0)	(0)	(1)	(1)	(2)	(0)	(31)
	71.0	16.1	0.0	0.0	3.2	3.2	6.5	0.0	100.0
Drama	(15)	(3)	(0)	(0)	(0)	(0)	(0)	(2)	(20)
	75.0	15.0	0.0	0.0	0.0	0.0	0.0	10.0	100.0
Social	(80)	(10)	(1)	(1)	(1)	(5)	(6)	(0)	(104)
	76.9	9.6	1.0	1.0	1.0	4.8	5.8	0.0	100.1
Service	(443)	(86)	(8)	(2)	(6)	(25)	(41)	(17)	(628)
	70.5	13.7	1.3	0.3	0.9	4.0	6.5	2.7	99.9
Religious	(5)	(1)	(0)	(0)	(0)	(0)	(0)	(0)	(6)
	83.3	16.7	0.0	0.0	0.0	0.0	0.0	0.0	100.0
Other activity	(76)	(9)	(4)	(0)	(0)	(3)	(5)	(2)	(99)
	76.8	9.1	4.0	0.0	0.0	3.0	5.0	2.0	99.9
Celebrity	—	—	—	—	—	—	—	—	—
# in each group	(810)	(147)	(13)	(3)	(10)	(41)	(75)	(31)	(1130)
% male population	71.7	13.0	1.2	0.3	0.9	3.6	6.6	2.7	100.0

Note: The figure in parentheses gives the number of men of each ethnic group involved in the designated activity. The percentage represents the proportion of all male students involved in the designated activity who were members of each ethnic group.

Table H. Extracurricular Participation by Women by Ethnic Group and Activity
SEWARD PARK HIGH SCHOOL

	Jewish	Italian	Black	*Irish	*German	Native	Other	Undecided	Total
Some activity	(658)	(52)	(32)	(1)	(10)	(39)	(50)	(23)	(865)
	76.1	6.0	3.7	0.1	1.2	4.5	5.8	2.6	100.0
Football	—	—	—	—	—	—	—	—	—
Basketball	(35)	(4)	(1)	(0)	(0)	(1)	(8)	(2)	(51)
	68.6	7.8	2.0	0.0	0.0	2.0	15.7	3.9	100.0
Track	—	—	—	—	—	—	—	—	—
Other sport	(111)	(9)	(5)	(0)	(1)	(11)	(13)	(4)	(154)
	72.1	5.8	3.2	0.0	0.6	7.1	8.4	2.6	99.8
President	(5)	(0)	(0)	(0)	(0)	(0)	(0)	(0)	(5)
	100.0	0.0	0.0	0.0	0.0	0.0	0.0	0.0	100.0
Other political	(149)	(12)	(12)	(0)	(0)	(9)	(10)	(3)	(195)
	76.4	6.1	6.1	0.0	0.0	4.6	5.1	1.6	99.9
Yearbook	(20)	(1)	(1)	(0)	(1)	(1)	(1)	(1)	(26)
	76.9	3.8	3.8	0.0	3.8	3.8	3.8	3.8	99.7
Editor in chief	—	—	—	—	—	—	—	—	—
Other news	(29)	(0)	(0)	(0)	(1)	(3)	(1)	(0)	(34)
	85.3	0.0	0.0	0.0	2.9	8.8	2.9	0.0	99.9
Other publications	(25)	(1)	(1)	(0)	(0)	(0)	(0)	(2)	(29)
	86.2	3.4	3.4	0.0	0.0	0.0	0.0	6.9	99.9
Physics	—	—	—	—	—	—	—	—	—
Chemistry	(5)	(0)	(0)	(0)	(0)	(1)	(0)	(1)	(7)
	71.4	0.0	0.0	0.0	0.0	14.3	0.0	14.3	100.0
Other science	(12)	(0)	(0)	(0)	(0)	(0)	(2)	(2)	(16)
	75.0	0.0	0.0	0.0	0.0	0.0	12.5	12.5	100.0
Other academic	(99)	(17)	(3)	(0)	(1)	(10)	(8)	(4)	(142)
	69.7	12.0	2.1	0.0	0.7	7.0	5.6	2.8	99.9
Arista	(67)	(1)	(1)	(0)	(1)	(6)	(3)	(0)	(79)
	84.8	1.3	1.3	0.0	1.3	7.6	3.8	0.0	100.1
Glee club	(18)	(1)	(1)	(0)	(0)	(0)	(0)	(0)	(20)
	90.0	5.0	5.0	0.0	0.0	0.0	0.0	0.0	100.0
Orchestra	(14)	(1)	(4)	(0)	(0)	(0)	(1)	(1)	(21)
	66.7	4.8	19.0	0.0	0.0	0.0	4.8	4.8	100.1
Drama	(20)	(1)	(0)	(0)	(0)	(2)	(1)	(1)	(25)
	80.0	4.0	0.0	0.0	0.0	8.0	4.0	4.0	100.0
Social	(143)	(13)	(5)	(0)	(4)	(7)	(8)	(1)	(181)
	79.0	7.2	2.8	0.0	2.2	3.9	4.4	0.6	100.1
Service	(478)	(37)	(21)	(1)	(8)	(27)	(34)	(17)	(623)
	76.7	5.9	3.4	0.2	1.3	4.3	5.4	2.7	99.9
Religious	(7)	(0)	(0)	(0)	(0)	(0)	(0)	(0)	(7)
	100.0	0.0	0.0	0.0	0.0	0.0	0.0	0.0	100.0
Other activity	(232)	(18)	(4)	(1)	(3)	(14)	(15)	(10)	(297)
	78.1	6.1	1.3	0.3	1.0	4.7	5.0	3.4	99.9
Celebrity	—	—	—	—	—	—	—	—	—
# in each group	(770)	(62)	(35)	(2)	(12)	(46)	(59)	(28)	(1014)
% female population	75.9	6.1	3.5	0.2	1.2	4.5	5.8	2.8	100.0

*Numbers too small to be meaningful.

Note: The figure in parentheses gives the number of women of each ethnic group involved in the designated activity. The percentage represents the proportion of all female students involved in the designated activity who were members of each ethnic group.

TABLE I. Extracurricular Participation by Men by Ethnic Group and Activity
THEODORE ROOSEVELT HIGH SCHOOL

	Jewish	Italian	Black	Irish	German	Native	Other	Undecided	Total
Some activity	(116)	(68)	(2)	(20)	(26)	(68)	(30)	(16)	(346)
	33.5	19.7	0.6	5.8	7.5	19.7	8.7	4.6	100.1
Football	(1)	(2)	(0)	(0)	(1)	(3)	(0)	(1)	(8)
	12.5	25.0	0.0	0.0	12.5	37.5	0.0	12.5	100.0
Basketball	(3)	(1)	(0)	(0)	(0)	(3)	(1)	(2)	(10)
	30.0	10.0	0.0	0.0	0.0	30.0	10.0	20.0	100.0
Track	(4)	(1)	(0)	(2)	(1)	(4)	(0)	(0)	(12)
	33.3	8.3	0.0	16.7	8.3	33.3	0.0	0.0	99.9
Other sport	(11)	(4)	(0)	(4)	(4)	(16)	(1)	(1)	(41)
	26.8	9.8	0.0	9.8	9.8	39.0	2.4	2.4	100.0
President	(1)	(1)	(0)	(1)	(0)	(3)	(0)	(0)	(6)
	16.7	16.7	0.0	16.7	0.0	50.0	0.0	0.0	100.1
Other political	(19)	(8)	(0)	(4)	(2)	(17)	(3)	(3)	(56)
	33.9	14.3	0.0	7.1	3.6	30.4	5.3	5.3	99.9
Yearbook	(8)	(4)	(0)	(2)	(1)	(7)	(1)	(1)	(24)
	33.3	16.7	0.0	8.3	4.2	29.2	4.2	4.2	100.1
Editor in chief	(2)	(0)	(0)	(0)	(1)	(0)	(0)	(0)	(3)
	66.7	0.0	0.0	0.0	33.3	0.0	0.0	0.0	100.0
Other news	(2)	(2)	(0)	(1)	(1)	(2)	(2)	(0)	(10)
	20.0	20.0	0.0	10.0	10.0	20.0	20.0	0.0	100.0
Other publications	(0)	(0)	(0)	(0)	(0)	(1)	(0)	(0)	(1)
	0.0	0.0	0.0	0.0	0.0	100.0	0.0	0.0	100.0
Physics	—	—	—	—	—	—	—	—	—
Chemistry	(8)	(2)	(0)	(0)	(3)	(2)	(1)	(1)	(17)
	47.1	11.8	0.0	0.0	17.6	11.8	5.9	5.9	100.1
Other science	(8)	(1)	(0)	(0)	(4)	(2)	(0)	(0)	(15)
	53.3	6.7	0.0	0.0	26.7	13.3	0.0	0.0	100.0
Other academic	(17)	(15)	(0)	(1)	(4)	(8)	(5)	(2)	(52)
	32.7	28.9	0.0	2.0	7.7	15.4	9.6	3.8	100.1
Arista	(16)	(14)	(1)	(1)	(2)	(5)	(6)	(1)	(46)
	34.8	30.4	2.2	2.2	4.3	10.9	13.0	2.2	100.0
Glee club	(1)	(0)	(0)	(0)	(1)	(2)	(0)	(0)	(4)
	25.0	0.0	0.0	0.0	25.0	50.0	0.0	0.0	100.0
Orchestra	(4)	(1)	(0)	(0)	(1)	(0)	(0)	(0)	(6)
	66.7	16.7	0.0	0.0	16.7	0.0	0.0	0.0	100.1
Drama	(2)	(4)	(0)	(2)	(2)	(3)	(0)	(0)	(13)
	15.4	30.8	0.0	15.4	15.4	23.1	0.0	0.0	100.1
Social	(19)	(3)	(0)	(2)	(5)	(12)	(0)	(2)	(43)
	44.2	7.0	0.0	4.7	11.6	27.9	0.0	4.7	100.1
Service	(69)	(33)	(1)	(11)	(16)	(42)	(20)	(9)	(201)
	34.3	16.4	0.4	5.5	8.0	20.9	9.9	4.5	99.9
Religious	(2)	(2)	(0)	(8)	(2)	(13)	(6)	(0)	(33)
	6.1	6.1	0.0	24.2	6.1	39.4	18.2	0.0	100.1
Other activity	(53)	(32)	(1)	(4)	(16)	(38)	(19)	(2)	(165)
	32.1	19.4	0.6	2.4	9.7	23.0	11.5	1.2	99.9
Celebrity	(7)	(3)	(0)	(1)	(1)	(10)	(0)	(0)	(22)
	31.8	13.6	0.0	4.5	4.5	45.5	0.0	0.0	99.9
# in each group	(236)	(167)	(3)	(38)	(52)	(124)	(62)	(35)	(717)
% male population	32.9	23.3	0.4	5.3	7.3	17.3	8.6	4.9	100.0

Note: The figure in parentheses gives the number of men of each ethnic group involved in the designated activity. The percentage represents the proportion of all male students involved in the designated activity who were members of each ethnic group.

TABLE J. Extracurricular Participation by Women by Ethnic Group and Activity

THEODORE ROOSEVELT HIGH SCHOOL

	Jewish	Italian	Black	Irish	German	Native	Other	Undecided	Total
Some activity	(379)	(143)	(4)	(25)	(33)	(120)	(53)	(48)	(805)
	47.1	17.8	0.5	3.1	4.1	14.9	6.6	6.0	100.1
Football	—	—	—	—	—	—	—	—	—
Basketball	(15)	(5)	(1)	(4)	(3)	(7)	(4)	(3)	(42)
	35.7	11.9	2.4	9.5	7.1	16.7	9.5	7.1	99.9
Track	(0)	(0)	(1)	(0)	(0)	(0)	(0)	(0)	(1)
	0.0	0.0	100.0	0.0	0.0	0.0	0.0	0.0	100.0
Other sport	(41)	(17)	(0)	(3)	(7)	(15)	(9)	(7)	(99)
	41.4	17.2	0.0	3.0	7.1	15.2	9.1	7.1	100.1
President	(0)	(0)	(0)	(0)	(0)	(1)	(0)	(0)	(1)
	0.0	0.0	0.0	0.0	0.0	100.0	0.0	0.0	100.0
Other political	(37)	(20)	(1)	(8)	(3)	(14)	(9)	(7)	(99)
	37.4	20.2	1.0	8.1	3.0	14.1	9.1	7.1	100.0
Yearbook	(41)	(10)	(1)	(3)	(2)	(15)	(6)	(4)	(82)
	50.0	12.2	1.2	3.6	2.4	18.3	7.3	4.9	99.9
Editor in chief	(3)	(0)	(0)	(0)	(0)	(0)	(0)	(0)	(3)
	100.0	0.0	0.0	0.0	0.0	0.0	0.0	0.0	100.0
Other news	(21)	(3)	(0)	(0)	(1)	(5)	(2)	(1)	(33)
	63.7	9.1	0.0	0.0	3.0	15.2	6.1	3.0	100.1
Other publications	(0)	(0)	(0)	(0)	(0)	(1)	(0)	(0)	(1)
	0.0	0.0	0.0	0.0	0.0	100.0	0.0	0.0	100.0
Physics	—	—	—	—	—	—	—	—	—
Chemistry	(6)	(2)	(0)	(3)	(1)	(2)	(2)	(2)	(18)
	33.3	11.1	0.0	16.7	5.6	11.1	11.1	11.1	100.0
Other science	(7)	(2)	(0)	(0)	(1)	(5)	(1)	(2)	(18)
	38.9	11.1	0.0	0.0	5.6	27.8	5.6	11.1	100.1
Other academic	(96)	(43)	(2)	(5)	(10)	(35)	(9)	(12)	(212)
	45.3	20.3	0.9	2.4	4.7	16.5	4.2	5.7	100.0
Arista	(41)	(21)	(0)	(2)	(4)	(10)	(7)	(3)	(88)
	46.6	23.9	0.0	2.3	4.5	11.4	8.0	3.4	100.1
Glee club	(7)	(2)	(0)	(2)	(1)	(2)	(2)	(2)	(18)
	38.9	11.1	0.0	11.1	5.6	11.1	11.1	11.1	100.0
Orchestra	(0)	(2)	(0)	(0)	(0)	(1)	(0)	(0)	(3)
	0.0	66.7	0.0	0.0	0.0	33.3	0.0	0.0	100.0
Drama	(19)	(3)	(0)	(1)	(3)	(3)	(2)	(2)	(33)
	57.6	9.1	0.0	3.0	9.1	9.1	6.1	6.1	100.1
Social	(56)	(17)	(0)	(2)	(6)	(19)	(6)	(14)	(120)
	46.7	14.2	0.0	1.7	5.0	15.8	5.0	11.7	100.1
Service	(245)	(84)	(0)	(17)	(19)	(82)	(34)	(29)	(510)
	48.0	16.5	0.0	3.3	3.7	16.1	6.7	5.7	100.0
Religious	(4)	(35)	(0)	(13)	(4)	(25)	(7)	(4)	(92)
	4.3	38.0	0.0	14.1	4.3	27.2	7.6	4.3	99.8
Other activity	(212)	(47)	(2)	(12)	(17)	(58)	(18)	(20)	(386)
	54.9	12.2	0.5	3.1	4.4	15.0	4.7	5.2	100.0
Celebrity	(4)	(1)	(0)	(2)	(3)	(6)	(2)	(2)	(20)
	20.0	5.0	0.0	10.0	15.0	30.0	10.0	10.0	100.0
# in each group % female population	(575)	(271)	(11)	(39)	(51)	(192)	(102)	(83)	(1324)
	43.4	20.5	0.8	2.9	3.9	14.5	7.7	6.3	100.0

Note: The figure in parentheses gives the number of women of each ethnic group involved in the designated activity. The percentage represents the proportion of all female students involved in the designated activity who were members of each ethnic group.

TABLE K. Extracurricular Participation by Men by Ethnic Group and Activity
NEW UTRECHT HIGH SCHOOL

	Jewish	Italian	Black	Irish	German	Native	Other	Undecided	Total
Some activity	(747)	(296)	—	(16)	(31)	(95)	(63)	(49)	(1297)
	57.6	22.8		1.2	2.4	7.3	4.9	3.8	100.0
Football	(25)	(15)	—	(0)	(3)	(2)	(5)	(1)	(51)
	49.0	29.4		0.0	5.9	3.9	9.8	2.0	100.0
Basketball	(22)	(0)	—	(1)	(0)	(4)	(1)	(0)	(28)
	78.6	0.0		3.6	0.0	14.3	3.6	0.0	100.1
Track	(63)	(14)	—	(1)	(2)	(5)	(8)	(4)	(97)
	64.9	14.4		1.0	2.1	5.2	8.2	4.1	99.9
Other sport	(102)	(34)	—	(4)	(8)	(13)	(15)	(8)	(184)
	55.4	18.5		2.2	4.3	7.1	8.1	4.3	99.9
President	(7)	(1)	—	(0)	(0)	(0)	(0)	(0)	(8)
	87.5	12.5		0.0	0.0	0.0	0.0	0.0	100.0
Other political	(62)	(24)	—	(1)	(4)	(5)	(6)	(6)	(108)
	57.4	22.2		0.9	3.7	4.6	5.6	5.6	100.0
Yearbook	(40)	(6)	—	(1)	(2)	(7)	(1)	(2)	(59)
	67.8	10.2		1.7	3.4	11.9	1.7	3.4	100.1
Editor in chief	(5)	(0)	—	(0)	(0)	(0)	(0)	(0)	(5)
	100.0	0.0		0.0	0.0	0.0	0.0	0.0	100.0
Other news	(48)	(3)	—	(0)	(4)	(7)	(1)	(2)	(65)
	73.8	4.6		0.0	6.2	10.8	1.5	3.1	100.0
Other publications	(16)	(7)	—	(1)	(2)	(5)	(2)	(3)	(36)
	44.4	19.4		2.8	5.6	13.9	5.6	8.3	100.0
Physics	—	—	—	—	—	—	—	—	—
Chemistry	(6)	(3)	—	(0)	(0)	(2)	(0)	(2)	(13)
	46.2	23.1		0.0	0.0	15.4	0.0	15.4	100.1
Other science	(18)	(5)	—	(0)	(0)	(3)	(1)	(2)	(29)
	62.1	17.2		0.0	0.0	10.3	3.5	6.9	100.0
Other academic	(61)	(27)	—	(2)	(4)	(7)	(6)	(1)	(108)
	56.5	25.0		1.9	3.7	6.5	5.6	0.9	100.1
Arista	(93)	(21)	—	(0)	(3)	(6)	(6)	(5)	(134)
	69.4	15.7		0.0	2.2	4.5	4.5	3.7	100.0
Glee club	(15)	(16)	—	(1)	(1)	(3)	(1)	(0)	(37)
	40.5	43.2		2.7	2.7	8.1	2.7	0.0	99.9
Orchestra	(20)	(9)	—	(0)	(0)	(3)	(1)	(2)	(35)
	57.1	25.7		0.0	0.0	8.6	2.9	5.7	100.0
Drama	(17)	(2)	—	(0)	(2)	(5)	(1)	(2)	(29)
	58.6	6.9		0.0	6.9	17.2	3.4	6.9	99.9
Social	(4)	(0)	—	(0)	(1)	(1)	(0)	(0)	(6)
	66.7	0.0		0.0	16.7	16.7	0.0	0.0	100.1
Service	(582)	(228)	—	(10)	(22)	(64)	(43)	(38)	(987)
	59.0	23.1		1.0	2.2	6.5	4.3	3.8	99.9
Religious	(1)	(11)	—	(2)	(1)	(3)	(4)	(0)	(22)
	4.5	50.0		9.1	4.5	13.6	18.2	0.0	99.9
Other activity	(144)	(60)	—	(2)	(4)	(17)	(10)	(13)	(250)
	57.6	24.0		0.8	1.6	6.8	4.0	5.2	100.0
Celebrity	(14)	(4)	—	(0)	(0)	(0)	(0)	(1)	(19)
	73.7	21.1		0.0	0.0	0.0	0.0	5.3	100.1
# in each group	(870)	(376)	—	(18)	(40)	(121)	(78)	(52)	(1555)
% male population	55.9	24.2	—	1.2	2.6	7.8	5.0	3.3	100.0

Note: The figure in parentheses gives the number of men of each ethnic group involved in the designated activity. The percentage represents the proportion of all male students involved in the designated activity who were members of each ethnic group.

TABLE L. Extracurricular Participation by Women by Ethnic Group and Activity
NEW UTRECHT HIGH SCHOOL

	Jewish	Italian	Black	Irish	German	Native	Other	Undecided	Total
Some activity	(961)	(227)	—	(12)	(31)	(160)	(106)	(60)	(1557)
	61.7	14.6		0.8	2.0	10.3	6.8	3.9	100.1
Football	—	—	—	—	—	—	—	—	—
Basketball	(45)	(12)	—	(0)	(1)	(6)	(4)	(3)	(71)
	63.4	16.9		0.0	1.4	8.5	5.6	4.2	100.0
Track	(1)	(1)	—	(0)	(0)	(0)	(0)	(0)	(2)
	50.0	50.0		0.0	0.0	0.0	0.0	0.0	100.0
Other sport	(96)	(46)	—	(1)	(1)	(22)	(15)	(0)	(181)
	53.0	25.4		0.6	0.6	12.1	8.3	0.0	100.0
President	(3)	(1)	—	(0)	(0)	(0)	(0)	(0)	(4)
	75.0	25.0		0.0	0.0	0.0	0.0	0.0	100.0
Other political	(86)	(18)	—	(0)	(3)	(17)	(7)	(7)	(138)
	62.3	13.0		0.0	2.2	12.3	5.1	5.1	100.0
Yearbook	(55)	(3)	—	(0)	(1)	(7)	(4)	(2)	(72)
	76.4	4.2		0.0	1.4	9.7	5.6	2.8	100.1
Editor in chief	(1)	(0)	—	(0)	(0)	(0)	(0)	(0)	(1)
	100.0	0.0		0.0	0.0	0.0	0.0	0.0	100.0
Other news	(73)	(2)	—	(0)	(2)	(9)	(1)	(5)	(92)
	79.3	2.2		0.0	2.2	9.8	1.1	5.4	100.0
Other publications	(33)	(2)	—	(0)	(2)	(4)	(3)	(1)	(45)
	73.3	4.4		0.0	4.4	8.9	6.7	2.2	99.9
Physics	—	—	—	—	—	—	—	—	—
Chemistry	(0)	(0)	—	(0)	(0)	(0)	(1)	(0)	(1)
	0.0	0.0		0.0	0.0	0.0	100.0	0.0	100.0
Other science	(9)	(2)	—	(0)	(0)	(0)	(1)	(2)	(14)
	64.3	14.3		0.0	0.0	0.0	7.1	14.3	100.0
Other academic	(113)	(56)	—	(0)	(1)	(14)	(15)	(8)	(207)
	54.6	27.1		0.0	0.5	6.8	7.2	3.9	100.1
Arista	(148)	(15)	—	(1)	(3)	(18)	(13)	(7)	(205)
	72.2	7.3		0.5	1.5	8.8	6.3	3.4	100.0
Glee club	(35)	(13)	—	(0)	(0)	(9)	(9)	(0)	(66)
	53.0	19.7		0.0	0.0	13.6	13.6	0.0	99.9
Orchestra	(12)	(1)	—	(0)	(0)	(2)	(1)	(0)	(16)
	75.0	6.3		0.0	0.0	12.5	6.3	0.0	100.1
Drama	(48)	(6)	—	(0)	(1)	(8)	(5)	(4)	(72)
	66.7	8.3		0.0	1.4	11.1	6.9	5.5	99.9
Social	(50)	(7)	—	(0)	(1)	(12)	(7)	(3)	(80)
	62.5	8.7		0.0	1.2	15.0	8.7	3.8	99.9
Service	(744)	(168)	—	(12)	(25)	(131)	(77)	(49)	(1206)
	61.7	13.9		1.0	2.1	10.9	6.4	4.1	100.1
Religious	(2)	(25)	—	(1)	(0)	(4)	(0)	(0)	(32)
	6.3	78.1		3.1	0.0	12.5	0.0	0.0	100.0
Other activity	(182)	(62)	—	(2)	(6)	(32)	(35)	(12)	(331)
	55.0	18.7		0.6	1.8	9.7	10.6	3.6	100.0
Celebrity	(9)	(2)	—	(0)	(0)	(6)	(1)	(1)	(19)
	47.4	10.5		0.0	0.0	31.6	5.3	5.3	100.1
# in each group % female	(1026)	(265)	—	(12)	(34)	(168)	(119)	(66)	(1690)
population	60.7	15.7		0.7	2.0	9.9	7.0	3.9	99.9

Note: The figure in parentheses gives the number of women of each ethnic group involved in the designated activity. The percentage represents the proportion of all female students involved in the designated activity who were members of each ethnic group.

TABLE M. Extracurricular Participation by Men by Ethnic Group and Activity

HIGH SCHOOL OF COMMERCE

	Jewish	Italian	Black	Irish	German	Native	Other	Undecided	Total
Some activity	(209)	(166)	(47)	(108)	(61)	(179)	(139)	(48)	(957)
	21.8	17.3	4.9	11.3	6.4	18.7	14.5	5.0	99.9
Football	(4)	(8)	(0)	(3)	(3)	(6)	(2)	(0)	(26)
	15.4	30.8	0.0	11.5	11.5	23.1	7.7	0.0	100.0
Basketball	(6)	(5)	(2)	(6)	(0)	(8)	(7)	(1)	(35)
	17.1	14.3	5.7	17.1	0.0	22.9	20.0	2.9	100.0
Track	(9)	(13)	(10)	(13)	(6)	(6)	(12)	(3)	(72)
	12.5	18.1	13.9	18.1	8.3	8.3	16.7	4.2	100.1
Other sport	(25)	(24)	(3)	(19)	(12)	(22)	(22)	(7)	(134)
	18.7	17.9	2.2	14.2	8.9	16.4	16.4	5.2	99.9
President	(2)	(0)	(0)	(3)	(2)	(0)	(2)	(0)	(9)
	22.2	0.0	0.0	33.3	22.2	0.0	22.2	0.0	99.9
Other political	(64)	(45)	(8)	(35)	(20)	(44)	(28)	(10)	(254)
	25.2	17.7	3.1	13.8	7.9	17.3	11.0	3.9	99.9
Yearbook	(4)	(5)	(0)	(5)	(1)	(6)	(0)	(0)	(21)
	19.0	23.8	0.0	23.8	4.8	28.6	0.0	0.0	100.0
Editor in chief	(2)	(2)	(0)	(0)	(0)	(0)	(0)	(0)	(4)
	50.0	50.0	0.0	0.0	0.0	0.0	0.0	0.0	100.0
Other news	(11)	(9)	(1)	(5)	(2)	(7)	(2)	(4)	(41)
	26.8	21.9	2.4	12.2	4.9	17.1	4.9	9.7	99.9
Other publications	(26)	(7)	(0)	(4)	(4)	(12)	(2)	(1)	(56)
	46.4	12.5	0.0	7.1	7.1	21.4	3.6	1.9	100.0
Physics	—	—	—	—	—	—	—	—	—
Chemistry	—	—	—	—	—	—	—	—	—
Other science	(0)	(3)	(0)	(0)	(0)	(2)	(1)	(0)	(6)
	0.0	50.0	0.0	0.0	0.0	33.3	16.7	0.0	100.0
Other academic	(35)	(13)	(0)	(3)	(5)	(18)	(14)	(6)	(94)
	37.2	13.8	0.0	3.2	5.3	19.1	14.9	6.4	99.9
Arista	(34)	(17)	(2)	(9)	(7)	(22)	(16)	(4)	(111)
	30.6	15.3	1.8	8.1	6.3	19.8	14.4	3.6	99.9
Glee club	(19)	(10)	(2)	(7)	(6)	(17)	(7)	(0)	(68)
	27.9	14.7	2.9	10.3	8.8	25.0	10.3	0.0	99.9
Orchestra	(1)	(5)	(1)	(1)	(0)	(5)	(3)	(1)	(17)
	5.9	29.4	5.9	5.9	0.0	29.4	17.6	5.9	100.0
Drama	(2)	(0)	(0)	(0)	(2)	(1)	(1)	(1)	(7)
	28.6	0.0	0.0	0.0	28.6	14.3	14.3	14.3	100.1
Social	(5)	(4)	(1)	(4)	(3)	(7)	(4)	(1)	(29)
	17.2	13.8	3.4	13.8	10.3	24.1	13.8	3.4	99.8
Service	(185)	(140)	(40)	(90)	(49)	(153)	(120)	(42)	(819)
	22.6	17.1	4.9	11.0	6.0	18.7	14.6	5.1	100.0
Religious	(0)	(4)	(0)	(5)	(1)	(7)	(1)	(0)	(18)
	0.0	22.2	0.0	27.8	5.6	38.9	5.6	0.0	100.1
Other activity	(71)	(51)	(6)	(33)	(23)	(61)	(38)	(12)	(295)
	24.1	17.3	2.0	11.2	7.8	20.7	12.9	4.1	100.1
Celebrity	(12)	(6)	(1)	(1)	(5)	(12)	(3)	(1)	(41)
	29.3	14.6	2.4	2.4	12.2	29.3	7.3	2.4	99.9
# in each group % male	(230)	(202)	(60)	(133)	(78)	(212)	(162)	(61)	(1138)
population	20.2	17.8	5.3	11.7	6.9	18.6	14.2	5.4	100.1

Note: The figure in parentheses gives the number of men of each ethnic group involved in the designated activity. The percentage represents the proportion of all male students involved in the designated activity who were members of each ethnic group.

TABLE N. Extracurricular Participation by Women by Ethnic Group and Activity
BAY RIDGE HIGH SCHOOL

	Jewish	Italian	*Black	Irish	German	Native	Other	Undecided	Total
Some activity	(88)	(424)	(4)	(109)	(73)	(358)	(234)	(79)	(1369)
	6.4	31.0	0.3	8.0	5.3	26.2	17.1	5.8	100.1
Football	—	—	—	—	—	—	—	—	—
Basketball	(7)	(43)	(1)	(13)	(10)	(40)	(23)	(8)	(145)
	4.8	29.7	0.7	9.0	6.9	27.6	15.9	5.5	100.1
Track	—	—	—	—	—	—	—	—	—
Other sport	(28)	(111)	(1)	(30)	(23)	(98)	(59)	(17)	(367)
	7.7	30.2	0.3	8.2	6.3	26.7	16.1	4.6	100.1
President	(0)	(1)	(0)	(0)	(0)	(2)	(0)	(1)	(4)
	0.0	25.0	0.0	0.0	0.0	50.0	0.0	25.0	100.0
Other political	(8)	(53)	(0)	(11)	(5)	(29)	(21)	(10)	(137)
	5.8	38.7	0.0	8.0	3.6	21.2	15.3	7.3	99.9
Yearbook	(15)	(31)	(0)	(9)	(8)	(53)	(30)	(8)	(154)
	9.7	20.1	0.0	5.8	5.2	34.4	19.5	5.2	99.9
Editor in chief	(1)	(2)	(0)	(1)	(0)	(2)	(0)	(0)	(6)
	16.7	33.3	0.0	16.7	0.0	33.3	0.0	0.0	100.0
Other news	(8)	(15)	(0)	(2)	(5)	(15)	(13)	(2)	(60)
	13.3	25.0	0.0	3.3	8.3	25.0	21.7	3.3	99.9
Other publications	(4)	(4)	(0)	(2)	(0)	(10)	(1)	(1)	(22)
	18.2	18.2	0.0	9.1	0.0	45.4	4.5	4.5	99.9
Physics	(0)	(3)	(0)	(0)	(0)	(0)	(1)	(0)	(4)
	0.0	75.0	0.0	0.0	0.0	0.0	25.0	0.0	100.0
Chemistry	(2)	(5)	(0)	(3)	(2)	(12)	(5)	(3)	(32)
	6.3	15.6	0.0	9.4	6.3	37.5	15.6	9.4	100.1
Other science	(14)	(29)	(0)	(8)	(2)	(28)	(13)	(12)	(106)
	13.2	27.3	0.0	7.5	1.9	26.4	12.3	11.3	99.9
Other academic	(20)	(86)	(2)	(19)	(14)	(72)	(50)	(8)	(271)
	7.4	31.7	0.7	7.0	5.2	26.6	18.5	2.9	100.0
Arista	(11)	(30)	(0)	(9)	(6)	(37)	(26)	(9)	(128)
	8.6	23.4	0.0	7.0	4.7	28.9	20.3	7.0	99.9
Glee club	(9)	(22)	(1)	(5)	(4)	(28)	(19)	(5)	(93)
	9.7	23.7	1.1	5.4	4.3	30.1	20.4	5.4	100.1
Orchestra	(4)	(18)	(1)	(2)	(1)	(19)	(8)	(6)	(59)
	6.8	30.5	1.7	3.4	1.7	32.2	13.6	10.2	100.1
Drama	(12)	(9)	(0)	(5)	(6)	(32)	(15)	(5)	(84)
	14.3	10.7	0.0	6.0	7.1	38.1	17.9	6.0	100.1
Social	(18)	(155)	(3)	(35)	(16)	(89)	(64)	(19)	(399)
	4.5	38.9	0.7	8.8	4.0	22.3	16.0	4.8	100.0
Service	(38)	(264)	(4)	(54)	(32)	(186)	(122)	(39)	(739)
	5.1	35.7	0.5	7.3	4.3	25.2	16.5	5.3	99.9
Religious	—	—	—	—	—	—	—	—	—
Other activity	(80)	(416)	(4)	(106)	(71)	(339)	(225)	(78)	(1319)
	6.1	31.5	0.3	8.0	5.4	25.7	17.0	5.9	99.9
Celebrity	—	—	—	—	—	—	—	—	—
# in each group	(88)	(427)	(4)	(109)	(74)	(365)	(235)	(80)	(1382)
% female population	6.4	30.9	0.3	7.9	5.4	26.4	17.0	5.8	100.1

* Number too small to be meaningful.

Note: The figure in parentheses gives the number of women of each ethnic group involved in the designated activity. The percentage represents the proportion of all female students involved in the designated activity who were members of each ethnic group.

Notes

1: THE PROGRESSIVE, THE IMMIGRANT, AND THE SCHOOL

1. The quote is taken from Carl F. Kaestle, *Pillars of the Republic: Common Schools and American Society, 1780–1860* (New York: Hill and Wang, 1983), 98. Kaestle's book is a model of historical synthesis and explication. See also Lawrence A. Cremin, *American Education: The National Experience 1783–1876* (New York: Harper and Row, 1980), especially 103–245.

2. The historiography of progressive education has gone through several revisionist assaults. It began with the view that progressive education was an enlightened and laudable response to the problems facing American education and society. Cremin's book, still the best introduction to progressive education, was very much an expression of the early dominant optimism; see *The Transformation of the School: Progressivism in American Education, 1876–1957* (New York: Alfred A. Knopf, 1961). Also within this tradition and still useful is Sol Cohen, *Progressives and Urban School Reform: The Public Education Association of New York City, 1895–1954* (New York: Bureau of Publications, Teachers College, Columbia University, 1964); Timothy Smith, "Progressivism in American Education, 1880–1900," *Harvard Educational Review*, 31 (Spring 1961): 168–193; Oscar Handlin, *John Dewey's Challenge to Education: Historical Perspectives on the Cultural Context* (Westport, Connecticut: Greenwood Press, 1959); Rush Welter, *Public Education and Democratic Thought in America* (New York: Columbia University Press, 1962).

By the late 1960s and early '70s, progressive education was beginning to elicit a very different response from historians, as they became more critical of the motives of reformers and the effects of the reforms. The most important reevaluations are Marvin Lazerson, *Origins of the Urban School: Public Education in Massachusetts, 1870–1915* (Cambridge: Harvard University Press, 1971); David B. Tyack, *The One Best System: A History of American Urban Education* (Cambridge: Harvard University Press, 1974); Patricia A. Graham, *Community and Class in Urban*

Education (New York: John Wiley & Sons, 1974). These provide critical but judicious reappraisals.

Progressive education also came in for far more corrosive criticism which condemned the whole progressive experiment and the reformers as consciously eager to use the schools in the interest of social and class control. See Colin Greer, *The Great School Legend: A Revisionist Interpretation of American Public Education* (New York: Basic Books, 1972); Clarence Karier, *Shaping the American Educational State* (New York: Free Press, 1975); Paul Violas, *The Training of the Urban Working Class* (Chicago: Rand McNally, 1978); Joel Spring, *Education and the Rise of the Corporate Liberal State* (Boston: Beacon Press, 1972); Samuel Bowles and Herbert Gintis, *Schooling in Capitalist America: Educational Reform and the Contradictions of Economic Life* (New York: Basic Books, 1976). The most recent addition to this tradition, though in somewhat modified form, is David John Hogan, *Class and Reform: School and Society in Chicago, 1880–1930* (Philadelphia: University of Pennsylvania Press, 1985).

In the 1980s, progressive education is once more being reappraised. The historiography has moved beyond social control to a new appreciation of the complexity of different forces operating on the schools and a new awareness of how various parts of the democratic constituency contributed to reform, each for different motives and reasons. Among the best of the recent studies is William J. Reese, *Power and the Promise of School Reform: Grassroots Movements during the Progressive Era* (London: Routledge and Kegan Paul, 1985). In Reese's account, progressive education once more becomes a popular movement to wrest the schools from elite control. See also Paul Peterson, *The Politics of School Reform, 1870–1940* (Chicago: University of Chicago Press, 1985); Julia Wrigley, *Class, Politics and Public Schools 1900–1930* (New Brunswick, N.J.: Rutgers University Press, 1982). Ira Katznelson and Margaret Weir, *Schooling for All: Class, Race and the Decline of the Democratic Ideal* (New York: Basic Books, 1985), although critical of the progressives, also adopt a similar model for American educational development. While far more sophisticated than the examination based on social and class control, these books often fail to capture the cultural construct that progressive education represented. What is most often missing from the most recent reappraisals—but amply present in both the older positive views and the seventies' critiques—is an understanding of the vital link between reform values for the society as a whole and the reform of the schools. Reese is an exception since he places the various groups in the progressive coalition in a rich context of ideology and social objectives.

But the anti-progressive views continue. Perhaps their most articulate and intelligent spokesperson has been Diane Ravitch, whose views are very different than those who have adopted a class or social control perspective. See *The Troubled Crusade: American Education, 1945–1980* (New York: Basic Books, 1983). Ravitch's perspective is perhaps the most faithful to the view adopted by Richard Hofstadter in *Anti-Intellectualism in American Life* (New York: Alfred A. Knopf, 1962). Hofstadter was the first to challenge effectively the optimistic progressivist values which informed the early historiography. Like Hofstadter, Ravitch challenges the educational and intellectual, rather than the social, implications of progressive education.

For John Dewey's views see *School and Society* (Chicago: University of Chicago Press, 1915; reprinted 1971); *The Child and the Curriculum* (Chicago: University of Chicago Press, 1902; reprinted 1971); and especially *Democracy and Education: An Introduction to the Philosophy of Education* (New York: Macmillan, 1916) in which Dewey most completely set forth his views. For the nature of Dewey's philosophy, see Morton White, *Social Thought in America: The Revolt Against Formalism* (New York: Viking, 1949). Cremin dealt sensitively with Dewey throughout *The Transformation of the School* and addressed the question of his influence there and in "John Dewey and the Progressive Education Movement, 1915–1952," *School Review*, 67(Summer 1959): 160–173. Richard Hofstadter's discussion of progressive education and Dewey in *Anti-Intellectualism in American Life* is bruising; see pp. 323–390.

3. In *Origins of the Urban School*, Marvin Lazerson presents an important analysis of this dual source of reform. Another effort to capture the complex and contradictory nature of reform is made by Ronald D. Cohen and Raymond A. Mohl in *The Paradox of Progressive Education: The Gary Plan and Urban Schooling* (Port Washington, N.Y.: Kennikat Press, 1979). For a discussion of the contemporary issues involved in discussions of school history, see Michael B. Katz, *Reconstructing American Education* (Cambridge: Harvard University Press, 1987), 111–159.

4. Niles Carpenter, *Immigrants and Their Children, 1920*, Census Monograph VII (Washington, D.C.: U.S. Government Printing Office, 1927). I have used the Arno Press reprint (New York, 1969), 24–27.

For a discussion of some of the industrial problems introduced by continuous immigration, see Herbert H. Gutman, "Work, Culture and Society in Industrializing America, 1815–1919," in his *Work, Culture and Society in Industrializing America* (New York: Vintage; 1977), 3–78. The general problem of how immigrants were perceived is discussed in John Higham's classic study of nativism, *Strangers in the Land: Patterns of American Nativism, 1860–1925* (New Brunswick, N.J.: Rutgers University Press, 1955).

5. Bernard J. Weiss gives a good introduction to the social context within which the education of immigrants was seen as a social necessity in, Bernard J. Weiss (ed.), *American Education and the European Immigrant: 1840–1940* (Urbana: University of Illinois Press, 1982), xi–xxviii.

6. Robert H. Bremner, *From the Depths: The Discovery of Poverty in America* (New York: New York University Press, 1956) remains a good introduction to the broad response to poverty and industrialization.

One of the very best discussions of how industrialization threatened American ideas and values is Daniel T. Rodgers, *The Work Ethic in Industrial America, 1850–1920* (Chicago: University of Chicago Press, 1978).

7. For American ideology in the middle of the nineteenth century, see Eric Foner, *Free Labor, Free Soil, Free Men: The Ideology of the Republican Party Before the Civil War* (New York: Oxford University Press, 1970). For the specific issue of how immigrants were envisaged in American ideology, see also the illuminating discussion in Arthur Mann, *The One and the Many: Reflections on the American Identity* (Chicago: University of Chicago Press, 1979), especially 46–96.

8. John R. Commons, *Races and Immigrants in America* (New York: Macmillan, 1930), 167–168. A similar, though more romantic, view was adopted by Jacob Riis. "The ideal, always in my mind, is that of a man with his feet upon the soil, and his children growing up there. So it seems to me, we should have responsible citizenship by the surest road." *Peril and Preservation of the Home* (London: Alexander Moring, 1903), 24. An excellent discussion of Riis's romanticism can

be found in Roy Lubove, *The Progressives and the Slums* (Pittsburgh: University of Pittsburgh Press, 1962), 66–80. For criticism of the immigrants' failure to seek out the farms, see Jeremiah W. Jencks and W. Jett Lauck, *The Immigration Problem: A Study of American Immigration Conditions and Needs* (New York: Funk and Wagnals, 1913), 81–103.

Ronald Cohen and Raymond Mohl discuss William Wirt's rural background and commitments in *The Paradox of Progressive Education,* pp. 11–22. Wirt was the principle architect of the model progressive school system of Gary, Indiana, and an influential proponent of progressive education. In fact, this rural romanticism and its firm links with traditional values and ideology pervades progressive social thinkers' sentiments.

9. Mary Kingsbury Simkovitch, *The City Worker's World in America* (New York: Macmillan, 1917), 53.

10. Margaret Byington, *Homestead: The Households of a Mill Town* (1910; reprinted by University Center for International Studies, University of Pittsburgh, 1974). See Samuel P. Hayes's excellent introduction in this edition called "Homestead Revisited," xvii–xxxiv.

11. Simkovitch, *City Worker's,* pp. 81–82.

12. Felix Adler, *Our Part in this World* (New York: King's Crown Press, 1946), 27. For Adler and progressivism, see Robert H. Beck, "Progressive Education and American Progressivism: Felix Adler," *Teachers College Record,* 60(November 1958): 77–89.

13. David J. Rothman, "The State as Parent," in, Willard Gaylin, Ira Glasser, Steven Marcus, and David J. Rothman, *Doing Good: The Limits of Benevolence* (New York: Pantheon, 1978); see also Paul Boyer, *Urban Masses and Moral Order in America, 1820–1920* (Cambridge: Harvard University Press, 1978).

14. Jacob A. Riis, *The Children of the Poor* (New York: Charles Scribner's Sons, 1892; reprinted by Arno Press, 1971), 1.

15. Frank T. Carlton, "School as a Factor in Industrial and Social Problems," *Education,* 24(September 1903–June 1904): 74.

16. For a discussion of the theories about family change and industrialization see Paula S. Fass, *The Damned and the Beautiful: American Youth in the 1920's* (New York: Oxford University Press, 1977), 95–118.

17. Sophonisba P. Breckinridge and Edith Abbott, *The Delinquent Child and the Home* (New York: Survey Associates, 1912), 176–177.

18. David J. Rothman, *Conscience and Convenience: The Asylum and Its Alternatives in Progressive America* (Boston: Little Brown, 1980), 46–53.

19. David Tyack, "Education and Social Unrest, 1873–1878," *Harvard Educational Review,* 31(Spring 1961): 194–212.

For the broad-based support for schools, see especially Reese, *The Promise of School Reform,* passim; Peterson, *Politics of School Reform,* pp. 7–51. Katznelson and Weir have recently argued for the important ways in which schooling in America was democratized because of the wide-based suffrage; *Schooling for All,* especially pp. 28–85.

20. Peterson, *Politics of School Reform,* pp. 11–12. The high-school enrollment figures are from Leonard V. Koos, *The American Secondary School* (Boston: Ginn and Co., 1927), 5, 10, 33.

21. Robert Hunter, *Poverty* (New York: Macmillan, 1904), 261; Riis, *Children of the Poor,* p. 2; Simkovitch, *City Worker's,* p. 5.

22. See Dorothy Ross, *G. Stanley Hall: The Psychologist as Prophet* (Chicago: University of Chicago Press, 1972).

23. See, for example, Kaestle, *Pillars of the Republic,* pp. 136–181; Reese, *Promise of School Reform,* pp. 12–18; Peterson, *Politics of School Reform,* pp. 52–71.

24. See especially Higham, *Strangers in the Land,* pp. 131–157; also, Barbara Miller Solomon, *Ancestors and Immigrants: A Changing New England Tradition* (Cambridge: Harvard University Press, 1956).

25. For the political dangers inherent in portrayals of immigrant separateness, see Frances Kellor, *Immigration and the Future* (New York: George H. Doran, 1920). See also Higham's brief biography of Kellor in *Strangers in the Land,* pp. 239–249.

26. This confusion is especially apparent in the case of Robert Hunter, whose attempts to distinguish pauperism, "a disease of character," from poverty, the result of economic exploitation, was extremely confusing. Pauperism, Hunter argued, was "analogous to parasitism in biological science," (p. 69) and was at once inheritable and socially induced. His confusion was not helped by his constant reference to the infamous Jukes family and the Tribe of Ishmael, cases that were used at the time to demonstrate the effects of bad heredity in charting generations of crime, feeblemindedness, and immorality. See Hunter, *Poverty,* pp. 69, 72, 80, 327.

John R. Commons's study, *Races and Immigrants in America,* also suffers from these confusions. Commons believed that immigrants were genetically inferior to native Anglo-Saxons whose mission was to conquer the world, an explicitly racist theory. But Commons rejected the idea of race purity and many of the stereotypes of Chinese and black workers common at the time. Moreover, Commons's views may be best summarized in his statement that "Race and heredity may be beyond our organized control, but the instrument of a common language is at hand for conscious improvement through education and social environment." (p. 21) That statement perhaps best represents the consensus of progressive reformers' beliefs at a time when heredity and environment were tangled issues in immigration discussions.

For an informed discussion of the heredity-environment controversy, see Hamilton Cravens, *The Triumph of Evolution: American Scientists and the Heredity-Environment Controversy, 1900–1941* (Philadelphia: University of Pennsylvania Press, 1978).

27. Simkovitch, *City Worker's,* p. 7. For racial stereotypes, see especially Robert A. Woods (ed.), *Americans in Process: A Settlement Study by the Residents of the South End House* (Boston: Houghton Mifflin, 1902); Jacob A. Riis, *How the Other Half Lives* (New York: Hill and Wang, 1957; orignally published in 1890).

28. Hunter, *Poverty,* p. 264; Kellor, *Immigration and the Future,* p. 65; Riis, *Children of the Poor,* p. 173; Edward Hale Bierstadt, *Aspects of Americanization* (Cincinnati: Steward Kidd, 1922), 30.

For the Americanization attempts that began with the outbreak of World War I see John McClymer, "The Americanization Movement," in Weiss (ed.), *American Education and the European Immigrant,* pp. 96–116. For some contemporary definitions of Americanization, see Howard C. Hill, "The Americanization Movement," *American Journal of Sociology,* 24(May 1919): 609–642; Carol Aronovici, "Americanization: Its Meaning and Function," *American Journal of Sociology,* 25(May 1920): 695–730. A dated, but still useful, history is Edward George Hartmann, *The Movement to Americanize the Immigrant* (New York: Columbia University Press, 1948).

29. Hunter, *Poverty,* p. 219. For Dewey, see Robert L. McCaul, "Dewey's Chicago," *The School Review,* 67(1959): 258–280.

30. Hunter, *Poverty,* p. 201.

31. Many reformers understood the strengths provided by ethnic culture, strengths that were especially necessary given the atomization of industrial society. See, for example, John Daniels, *America via the Neighborhoods* (New York: Harper and Brothers, 1920); Grace Abbott, *The Immigrant and the Community* (New York: Century Co., 1917); and many of the articles in *The Immigrants in America Review,* vol. I (March 1915–March 1916) published by the Committee for Immigrants in America.

The efforts to maintain immigrant culture and community integrity continued after the war. See Raymond Mohl, "The International Institutes and Immigrant Education, 1910–1940," in Weiss (ed.), *American Education and the European Immigrant,* pp. 117–141; also Nicholas V. Montalto, "The Intercultural Education Movement, 1924–41: The Growth of Tolerance as a Form of Intolerance," pp. 142–160 in the same volume.

32. Dewey, *School and Society,* p. 29.

33. Charles Horton Cooley, "A Primary Culture for Democracy," *Publications of the American Sociological Society,* 13(1918): 2; Simkovitch, *City Worker's,* p. 50. See also Mary Simkovitch, "The Enlarged Function of the Public School," *Proceedings of the National Conference of Charities and Correction, 31st Annual Session, 1904.* For many reformers, the schools were to function like settlements. See Morris Isaiah Berger, "The Settlement, The Immigrant and the Public School," (Ph.D. dissertation, Teachers College, Columbia University, 1959).

34. Marvin Lazerson discusses this in *Origins of the Urban School,* pp. 77–80 and passim.

35. Simkovitch, *City Worker's,* pp. 50, 67.

36. Commons, *Race and Immigrants,* p. 38.

37. Lillian D. Wald, "Qualifications and Training for Service with Children in a Crowded City Neighborhood," in Sophonisba P. Breckinridge (ed.), *The Child in the City* (Chicago: The Hollister Press, 1912), 256; Henry Moskowitz, "The Place of the Immigrant Child in the Social Program," in the same volume, p. 265.

38. Simkovitch, *City Worker's,* p. 63; Edward Divine, "Discussion," *Publications of the American Sociological Society,* 12(1918): 105.

39. Abbott, *Immigrant and the Community,* pp. 224, 225; Simkovitch, *City Worker's,* p. 61. For a discussion of the limitations on the attendance of immigrant children, see Selma Berrol, "Public Schools and Immigrants," in Weiss (ed.), *American Education and the European Immigrant,* pp. 39–41.

40. Wald, "Children in a Crowded City Neighborhood," p. 253; Moskowitz, "Immigrant Child," p. 257.

41. Simkovitch, *City Worker's,* p. 64; Abbott, *Immigrant and the Community,* p. 236.

42. On this matter, see Lazerson, *Origins of the Urban School,* pp. 76–154.

43. Hunter, *Poverty,* pp. 211, 213.

44. Hunter, *Poverty,* p. 214.

45. Robert A. Woods, "The Basis of an Efficient Education, Culture or Vocation," *The School Review,* 15(May 1907): 337–338.

46. See, for example, Hutchins Hapgood's classic series on Jews in New York, reprinted as *The Spirit of the Ghetto: Studies of the Jewish Quarter of New York* (New York: Schocken Books, 1966).

2: EDUCATION, DEMOCRACY AND THE SCIENCE OF INDIVIDUAL DIFFERENCES

1. W. I. Thomas, "Race Psychology: Standpoint and Questionnaire with Particular Reference to the Immigrant and the Negro," *American Journal of Sociology*, *17*(May 1912): 725.

2. Hollis L. Caswell and Doak S. Campbell, *Curriculum Development* (New York: American Book Co., 1935), 35.

3. Thomas, "Race Psychology," p. 753.

4. Newton Edwards and Herman G. Richey, *The School in the American Social Order: The Dynamics of American Education* (Boston: Houghton Mifflin, 1947), 682.

5. Historians who discuss the various groups and conflicts which marked nineteenth-century educational development include: Michael B. Katz, *The Irony of Early School Reform: Educational Innovation in Mid-Nineteenth Century Massachusetts* (Boston: Beacon Press, 1968); Diane Ravitch, *The Great School Wars, New York City, 1805–1973: A History of the Public Schools As Battlefields of Social Change* (New York: Basic Books, 1974); Selwyn K. Troen, *The Public and the Schools: Shaping the St. Louis System, 1838–1920* (Columbia: University of Missouri Press, 1975); Patricia Albjerg Graham, *Community and Class in American Education, 1865–1918* (New York: John Wiley & Sons, 1974); Marvin Lazerson, *Origins of the Urban School: Public Education in Massachusetts, 1870–1915* (Cambridge: Harvard University Press, 1971); David John Hogan, *Class and Reform: School and Society in Chicago, 1880–1930* (Philadelphia: University of Pennsylvania Press, 1985); Carl F. Kaestle, *Pillars of the Republic: Common Schools and American Society, 1780–1860* (New York: Hill and Wang, 1983), especially Chapter 7; Paul Peterson, *The Politics of School Reform, 1870–1940* (Chicago: University of Chicago Press, 1985); William Reese, *Power and the Promise of School Reform: Grassroots Movements during the Progressive Era* (London: Routledge and Kegan Paul, 1986).

6. David B. Tyack, *The One Best System: A History of American Urban Education* (Cambridge: Harvard University Press, 1974). For the centralization of the New York City school system, see David Hammack, "The Centralization of New York City's Public School System, 1896: A Social Analysis of a Decision," (Masters thesis, Columbia University, 1969).

For school surveys, see, for example, James H. Van Sickle, Leonard P. Ayres, Calvin N. Kendall, and William H. Maxwell, "An Investigation of the Efficiency of Schools and School Systems," National Education Association, *Proceedings, 1915* (Bloomington, Ill.: Public School Publishing Co., 1915), 379–402. For a discussion of the school survey movement, see Hollis L. Caswell, *City School Surveys: An Interpretation and Appraisal*, Contributions to Education, Teachers College (New York: Columbia University, 1929).

7. Caswell, *City School Surveys*, p. 20.

8. Graham, *Community and Class*, pp. 15–17, discusses age segregation. The

quote is from Frank Forest Bunker, "Reorganization of the Public School System," *U.S. Office of Education Bulletin 1916 #8* (Washington, D.C.: U.S. Government Printing Office, 1916), 116.

9. In 1870, 20 percent of those ten years of age or older were illiterate; see Graham, *Community and Class,* p. 13. This, of course, included immigrants. By 1940, 3.7 percent of the whole population reported having completed no years of schooling and can be considered illiterate; see, Henry S. Shyrock, Jr., "1940 Census Data on Number of Years of School Completed," *Milbank Memorial Fund Quarterly,* 20(October 1942): 368. For the problem of illiteracy and the imprecision of its meanings, see Patricia Albjerg Graham, "Literacy: A Goal for Secondary Schools," *Daedelus,* 110(Summer 1981): 119–134.

10. For the community context of mobility, see Paul E. Johnson, *A Shopkeeper's Millenium: Society and Revivals in Rochester, New York, 1815–1837* (New York: Hill and Wang, 1978); and Mary Ryan, *Cradle of the Middle Class: The Family in Oneida County, New York, 1790–1865* (Cambridge, England: Cambridge University Press, 1981).

The best general introduction to the nature of adolescence in the nineteenth century is Joseph Kett, *Rites of Passage: Adolescence in America, 1790 to the Present* (New York: Basic Books, 1977), especially 11–108.

11. For the high school in the early twentieth century, see Theodore R. Sizer, *Secondary Schools at the Turn of the Century* (New Haven: Yale University Press, 1964); Edward A. Krug, *The Shaping of the American High School,* 2 vols. (Madison, Wis., University of Wisconsin Press, 1969, 1972); Alexander J. Inglis, "Secondary Education," in I. L. Kandel (ed.), *Twenty-Five Years of American Education* (New York: Macmillan, 1924), 251–269; John Elbert Stout, *The Development of High School Curricula in the North Central States from 1860 to 1918* (Chicago: University of Chicago Press, 1921).

12. U.S. Immigration Commission (Dillingham Commission), "The Children of Immigrants in Schools," *Reports of the Immigration Commission,* vol. I (Washington, D.C.: U.S. Government Printing Office, 1911), 15, 23, 47, 95.

13. For the significance of child-labor laws and progressive reform, see Robert Bremner, *From the Depths: The Discovery of Poverty in the United States* (New York: New York University Press, 1956), 212–229.

My statistics on school enrollment are drawn from Edwards and Richey, *The School in the American Social Order,* pp. 671, 683. See also Michael Olneck and Marvin Lazerson, "The School Achievement of Immigrant Children, 1900–1930," *History of Education Quarterly,* 14(Winter 1974): 453–482.

By 1920, 94.1 percent of the children, seven to thirteen years of age, whose parents were foreign-born, were in school, while only 92.2 percent of those whose parents were native-born were in school. Only the children who were themselves immigrants lagged behind with 84.1 percent in school. By 1930, 97.5 percent of even this group was in school. In 1920, of those fourteen and fifteen years of age, the children of native parents had the edge over those whose parents were foreign-born. Eighty-three percent of the former and 77.9 percent of the latter were in school in 1920. By 1930, the difference between those of native parentage and foreign parentage who were fourteen and fifteen years of age had disappeared. See, T. J. Woofter, Jr., *Races and Ethnic Groups in American Life* (New York: McGraw Hill, 1933), 166.

14. For progressivism, science, and expertise, see Robert H. Wiebe, *The Search*

for Order, 1877–1920 (New York: Hill and Wang, 1967), 111–195; David J. Rothman, *Conscience and Convenience: The Asylum and Its Alternatives in Progressive America* (Boston: Little Brown, 1980), 43–61; David Tyack and Elizabeth Hansot, *Managers of Virtue: Public School Leadership in America, 1820–1980* (New York: Basic Books, 1982), 105–167. John Dewey, of course, used scientific method as the basis of his philosophy of knowledge and action.

15. For a discussion of the historical antecedents of IQ testing, see Kimball Young, "The History of Mental Testing," *Pedagogical Seminar,* 31(March 1923): 1–48; Frank N. Freeman, *Mental Tests: Their History, Principles and Applications,* rev. ed. (Boston: Houghton Mifflin, 1939); Joseph Peterson, *Early Conceptions and Tests of Intelligence* (Yonkers, N.Y.: World Book Co., 1925).

16. Peterson, *Early Conceptions,* pp. 78–83, 93–94; Young, "History of Mental Testing," pp. 30–33. The quotes are from Young, "History of Mental Testing," p. 1; and Harlan Cameron Hines, *Measuring Intelligence* (Boston: Houghton Mifflin, 1923), 59.

17. Peterson, *Early Conceptions,* p. 230. Other testers, notably H. H. Goddard, had used the tests as Binet had to measure feeblemindedness; see Peterson, *Early Conceptions,* pp. 226–229.

18. Lewis Terman, *The Measurement of Intelligence* (Boston: Houghton Mifflin, 1916), 41; Terman quoted in Peterson, *Early Conceptions,* p. 231; Cubberly introduction to Terman, *Measurement,* vii.

19. "It cannot be too strongly emphasized that no one, whatever his previous training may have been, can make proper use of the scale unless he is willing to learn the method of procedure and scaling down to the minutest detail"; Terman, *Measurement,* xi. See also Peterson, *Early Conceptions,* pp. 232–233; Hines, *Measuring Intelligence,* p. 70.

20. See Franz Samuelson, "World War I Intelligence Testing and the Development of Psychology," *Journal of the History of the Behavioral Sciences,* 13(July 1977): 274–288; Daniel J. Kevles, "Testing the Army's Intelligence: Psychologists and the Military in World War I," *Journal of American History,* 55(December 1968): 565–581.

21. IQ testing was constantly compared with other "simple," "exact," and "useful" scientific measurements, like the Babcock test for measuring the cream content of milk (Terman, *Measurement,* p. 35) and the blood count (Cubberly in *Measurement,* viii).

22. The quote is from Freeman, *Mental Tests,* p. 3. Lippmann's essays and Terman's reply are reprinted in N. J. Block and Gerald Dworkin (eds.), *IQ Controversy* (New York: Pantheon, 1976), 4–44. On the furor over early test results, see, for example, Freeman, *Mental Tests,* p. 127.

23. See Robert M. Yerkes, "Psychological Examining in the United States Army," *Memoirs of the National Academy of Sciences,* vol. 15 (Washington, D.C.: U.S. Government Printing Office, 1921); C. C. Brigham, *A Study of American Intelligence* (Princeton, N.J.: Princeton University Press, 1923). For discussion of the test results, see Freeman, *Mental Tests,* pp. 404–430.

24. As one tester, Rudolph Pintner, noted, "That there are differences in intelligence and in other characteristics between races has been assumed by many anthropologists and psychologists." He then proceeded to describe tests designed to elucidate and thus confirm those differences. Rudolph Pintner, *Intelligence Testing: Methods and Results,* rev. ed. (New York: Holt & Co., 1931), 448–457.

25. Peterson, *Early Conceptions,* pp. 274–275; E. J. Varon, "Alfred Binet's Concept of Intelligence," *Psychological Review,* 43(1936): 32–58; Young, "History of Mental Testing," pp. 19–24.

26. Throughout Europe, researchers using the Binet tests were finding strong correlations between performance and sociocultural factors—like class—and noted the marked academic inclination of the tests. One group of Italian psychologists particularly attacked the view that Binet's scale could ever measure "pure intelligence" outside of a specific social milieu, and others emphasized that the validity of the test results could never be more than group specific. Americans tended to draw opposite conclusions from the class correlations found by European investigators. Terman, for example, in discussing the European studies both underplays their significance and concludes, "the common opinion that the child from a cultured home does better in tests solely by reason of his superior home advantages is an entirely gratuitous assumption. Practically all of the investigations which have been made of the influence of nature and nurture on mental performance agree in attributing far more to original endowment than to environment. Common observation would itself suggest that the social class to which the family belongs depends less on chance than on the parents' native qualities of intellect and character." (Terman, *Measurement,* p. 115). Terman, despite his "scientific" veneer, often depended on "common observations" which were elevated into principles of nature. C. C. Brigham, in reporting on and analyzing the army results, used precisely the same words to deny the significance of environmental factors (Brigham, *American Intelligence,* p. 182). Even a careful and judicious investigator like Clifford Kirkpatrick, after first noticing that "nature never exists apart from nurture," explained that heredity was a factor that could be separately investigated by careful research designed to eliminate or keep environment constant and that ultimately, "high germ plasm often leads to better results than high per capita school expenditure. Definite limits are set by heredity, and *immigrants of low innate ability cannot by any amount of Americanization be made into intelligent American citizens capable of approximating and advancing a complex culture*" (my emphasis). Clifford Kirkpatrick, *Intelligence and Immigration,* Mental Monographs #2(Baltimore: The Williams and Wilkins Co., 1926), 2. See also Rudolph Pintner and Ruth Keller, "Intelligence Tests of Foreign Children," *Journal of Educational Psychology,* 13(1922): 214–222; Stephen S. Colvin, "Principles Underlying the Construction and Use of Intelligence Tests," *The Twenty-First Yearbook of the National Society for the Study of Education* (Bloomington, Ill.: Public School Publishing, 1922), 11–25. (The Yearbooks of the National Society for the Study of Education will hereafter be cited as *Yearbook, NSSE.*)

27. As Frank Freeman explained, "Whether they be mental tests or educational tests, both are relative. That is, the score which results from the application of the test has significance only in comparison with scores which are made by other individuals. The score serves as a comparatively exact numerical method of indicating the rank of the individual in a group." Freeman, *Mental Tests,* p. 19. Usually the norm of performance was the fact that 75 percent of children of a certain age could be expected to answer various kinds of questions correctly; see Pintner, *Intelligence Testing,* pp. 448–467.

28. For the growing interest in hereditarian explanations in the late nineteenth and early twentieth centuries, see Charles E. Rosenberg, "The Bitter Fruit: Heredity, Disease and Social Thought in 19th Century America," *Perspectives in Amer-*

ican History, 8(1974): 187–235. For eugenics, William H. Haller, *Eugenics: Hereditarian Attitudes in American Thought* (New Brunswick, N.J.: Rutgers University Press, 1963); Kenneth M. Ludmeier, *Genetics and American Society: A Historical Appraisal* (Baltimore: Johns Hopkins University Press, 1972).

29. "The scientific problem," Kimball Young observed, "is that of eliminating from the tests used as measuring instruments those particular tests which demonstrably measure nurture, and to measure with genuine tests of native intelligence random or impartial samples of each race throughout the entire range of its geographical and institutional distribution." Testers not only insisted on the validity of the tests but also sought to measure what they assumed could be measured, nature alone and apart from nurture. See Kimball Young, "Mental Differences in Certain Immigrant Groups," *University of Oregon Publications* (July 1922), 194–195; and Kirkpatrick, *Intelligence and Immigration,* pp. 7–14.

30. Young, "History of Mental Testing," p. 45.

31. For example, Brigham, *American Intelligence,* pp. 110–117, and the forward by Robert M. Yerkes. Brigham also acknowledged his debt to the two most potent racial analysts of the period, Madison Grant and William Z. Ripley (pp. xvii–xviii). See also Young, "Mental Differences," pp. 3, 72–84; Pintner, *Intelligence Testing,* pp. 466–467; Kirkpatrick, *Intelligence and Immigration,* pp. 105–116.

32. *Bulletin of High Points in the Work of the High Schools of New York City,* vol. I, November 1919, February 1920, April 1920, November 1920, and the December 1920 review of Terman's *The Intelligence of School Children.* For Detroit, see Warren K. Layton, "Group Intelligence Testing Program at the Detroit Public Schools," *Twenty-First Yearbook, NSSE;* for Mt. Vernon, George S. Counts, *The Selective Character of American Secondary Education* (Chicago: University of Chicago Press, 1922), 5. In his autobiography, Leonard Covello notes that this was the period of "intelligence tests insanity"; *The Heart is the Teacher* (New York: McGraw Hill, 1958), 149. See also Paul Chapman, "Schools as Sorters: Testing and Tracking in California, 1910–1925," *Journal of Social History,* 14(Summer 1981): 701–718. The quote is from Layton, "Group Intelligence Testing Program," p. 123.

33. Freeman, *Mental Tests,* p. 3; Guy M. Whipple, "An Annotated List of Group Intelligence Tests," *Twenty-First Yearbook, NSSE,* p. 199.

Thorndike is quoted in Hines, *Measuring Intelligence,* p. 113. For Thorndike, see Geraldine Joncich Clifford, *The Sane Positivist: A Biography of Edward L. Thorndike* (Middleton, Conn.: Wesleyan University Press, 1968). On the careless nature of early tests, see Peterson, *Early Conceptions,* pp. 232–233; Hines, *Measuring Intelligence,* pp. 60, 108; Freeman, *Mental Tests,* pp. 14–16.

34. Josephine Chase, *New York at School: A Description of the Activities and Administration of the Public Schools of New York* (New York: Public Education Association of the City of New York, 1927), 2.

35. Cubberly introduction to Terman, *Measurement,* vii–viii.

Hollis Caswell notes the effect of IQ testing on the school surveys after 1919: "A new note began to creep into the reports on classification and programs. . . . In the Boise, Idaho, survey, for example, 'Individual Differences' among children was added to the report." According to Caswell, the later surveys which could rely on tests were "far more exacting in their demands on school accommodation in curricula and programs to 'individual differences.' " "The testing movement," Caswell concludes, "has contributed in a major way to the survey movement and the survey movement has reciprocated by spreading and encouraging the testing movement." Caswell, *City School Survey,* p. 96. Thus, measurement encouraged measurement as each seemed to promise more precision in the development of the science of education, or as Caswell observes "Statistical methods were everywhere being applied to school problems." (p. 104).

36. Freeman, *Mental Tests,* p. 345.

37. Carleton W. Washburne, *Twenty-Fourth Yearbook, NSSE* (1925), p. xiii.

38. For what was tested by IQ, see, for example, Hines, *Measuring Intelligence,* pp. 53, 111; Pintner, *Intelligence Testing,* pp. 45–71.

39. Young, "Immigrant Groups," 65; Hines, *Measuring Intelligence,* p. 134; but see also Hines's defense of instruction organized to suitable vocational aims, pp. 93–97. For vocational channeling of immigrants, see Marvin Lazerson and W. Norton Grubb (eds.), *American Education and Vocationalism: A Documentary History 1870–1970* (New York: Teachers College Press, 1974), 38; Graham, *Community and Class,* pp. 20, 163.

For the relationship between education and mobility in the twentieth century, see Stanley Lieberson, *A Piece of the Pie: Blacks and White Immigrants Since 1880* (Berkeley: University of California Press, 1980), 214–215, 331–332, 354–359.

40. On overcrowding in New York schools, see Selma Berrol, "Immigrants at School: New York City, 1900–1910," *Urban Education,* 4(October 1969): 220–230. On class size, see P. R. Stevenson, *Class-Size in the Elementary School, Bureau of Educational Research, monograph #3*(Columbus: Ohio State University, 1925).

41. Carleton W. Washburne, "Adapting the Schools to Individual Differences," *Twenty-Fourth Yearbook, NSSE,* p. 31; Allen Raymond, "A Study of New York's Public School System," *New York Herald Tribune,* March 6, 1931 (the series ran from February 23, 1931, to March 6, 1931). See also Charles Judd, *Measuring the Work of the Public Schools,* vol. 10 of The Cleveland School Survey (Cleveland: The Survey Committee of the Cleveland Foundation, 1917), who notes that traditional subjects are wrong for the foreign population (p. 48).

42. Peiser is quoted in Morris Isaiah Berger, "The Settlement, The Immigrant, and the Public School: A Study of the Influence of the Settlement Movement and the New Migration Upon Public Education, 1890–1924" (Ph.D. dissertation, Teachers College, Columbia University, 1959), 120. See Berger generally for a discussion of the influence of the settlements on the public schools. For the incorporation of various services into New York City schools, see Ravitch, *School Wars,* pp. 168–169, 176, 234; also Sol Cohen, *The Progressives and Urban School Reform: The Public Education Association of New York City, 1895–1954* (New York: Bureau of Publications, Teachers College, Columbia University; 1964). For the social center movement, see Reese, *Power and Promise,* pp. 177–208.

43. For a discussion of the growth of services in Chicago high schools, see Thomas W. Gutowski, "The High School as an Adolescent-Raising Institution: An Inner History of Chicago Public Secondary Education, 1856–1940" (Ph.D. dissertation, University of Chicago, 1978), 64–119.

44. Francis T. Spaulding, O. I. Frederick, and Leonard V. Koos, "The Reorganization of Secondary Education," *U.S. Office of Education Bulletin 1932, #17*(Washington, D.C.: U.S. Government Printing Office, 1933): 305–350.

45. Leonard Covello, "A High School and Its Immigrant Community: A Chal-

lenge and an Opportunity," *Journal of Educational Sociology,* 9(February 1936): 331–346; Covello, *Heart is the Teacher,* passim.

46. Angelo Petri, *A Schoolmaster of the Great City* (New York: Macmillan, 1917), 213.

47. Julius John Oppenheimer, *The Visiting Teacher Movement: With Special Reference to Administrative Relationships* (New York: Joint Committee on Methods of Preventing Delinquency, 1925), 20, 26.

48. Oppenheimer, *Visiting Teacher Movement,* pp. xvii, 199, 201. On the effort to use visiting teachers familiar with the neighborhood and culture which they serviced, see Pauline V. Young, "Social Problems in the Education of the Immigrant Child," *American Sociological Review,* 1(1936): 424.

49. William H. Maxwell, *A Quarter Century of Public School Development* (New York: American Book Co., 1912), 190, 192.

50. Frank W. Thompson, *Schooling of the Immigrant* (New York: Harper & Brothers, 1920), 73.

51. See Graham, *Class and Community,* p. 160; Robert A. F. McDonald, *Adjustment of School Organization to Various Population Groups,* Contributions to Education (New York: Teachers College, Columbia University, 1915), 74; Francesco Cordasco, *Immigrant Children in American Schools: A Classified and Annotated Bibliography of Selected Source Documents* (Fairfield, N.J.: A. M. Kelly, 1976), 33–34.

52. New York City *Circular #6* (February 1918), Teachers College Archives, New York; also Frederick Martin, "Defects of Speech," *Bulletin of High Points in the Work of the High Schools of New York City* (May 1919), 22.

53. For the new attention to secondary schooling, see Alexander Inglis, "Secondary Education," pp. 251–269; Leonard V. Koos, *Trends in American Secondary Education* (Cambridge: Harvard University Press, 1927); Charles Judd, "Secondary Education," United States President's Committee on Social Trends, *American Civilization Today: A Summary of Recent Social Trends,* ed. by John T. Greenan (New York: McGraw Hill, 1934), 325–381; Sizer, *Secondary Schools at the Turn of the Century;* Lawrence Cremin, "The Revolution in American Secondary Education, 1893–1918," *Teachers College Record,* 56(March 1955): 295–308. The quote about Jordan is from Spaulding, et al., "Reorganization of Secondary Education," p. 310.

54. Riverda Harding Jordan, *Nationality and School Progress: A Study of Americanization* (Bloomington, Ill.: Public School Publishing Co., 1921), 102, 101; Chase, *New York at School,* p. 61.

55. Covello, *Heart is the Teacher,* p. 181. For the relationship between IQ and curriculum choices, see Mary E. Roberts, "Elimination From the Public High Schools of New Jersey" (Ph.D. dissertation, University of Pennsylvania, 1930), 221–222; Emily G. Palmer, *Pupils Who Leave School,* Division of Vocational Education of the University of California and the State Department of Education, University of California, Part-Time Education Series, no. 17, division bulletin no. 24 (Berkeley, Calif., January 1930), especially 18. See also David K. Cohen, "Immigrants and the Schools," *Review of Educational Research,* 40(February 1970): 13–27.

56. George Sylvester Counts, *The Selective Character of American Secondary Education* (Chicago: University of Chicago Press, 1922); Roberts, *Schools of New Jersey,* pp. 141–151, 165; Palmer, *Pupils Who Leave School,* pp. 18, 32, 42, 25.

57. Everett B. Sackett, *Situations Affecting School Persistence: The Regents' In-*

quiry into the Character and Cost of Public Education in the State of New York (New York: Charles Bruning Co., 1938), 14.

58. According to Edward Sackett, the solution to high-school leaving was for the schools to provide "such variety and vitality that students of every normal ability and interest can find a program giving a school experience which is interesting and satisfying." Sackett, *School Persistence,* p. 15.

59. See, for example, Judd, *Measuring the Work of the Public Schools,* p. 196; Roberts, *Schools of New Jersey,* p. 129. The quote is from Frank W. Thompson, "Commercial High Schools and Commercial Courses in High School," Committee on School Inquiry, *Report on Educational Aspects of the Public School System of the City of New York to the Committee on School Inquiry of the Board of Estimate and Apportionment,* part II, subdivision III, High Schools, section B. (City of New York: 1911–12), 21.

60. Joseph King Van Denburg, *Causes of the Elimination of Students in Public Secondary Schools of New York City,* Contributions to Education (New York: Bureau of Publications, Teachers College, Columbia University, 1911), 35, 80, 47.

61. See Sackett, *School Persistence,* pp. 5, 17.

62. Thompson, "Commercial High Schools," p. 35; John L. Tildsley, "The Reorganization of the High School for the Service of Democracy," address delivered at a meeting of the National Education Association, July 1, 1919, Milwaukee, Wisconsin, reprinted in *Bulletin of High Points in the Work of the High Schools of New York City,* I(October 1919): 13; Sackett, *School Persistence,* p. 16.

63. Alexander Inglis, "The High School in Evolution," *New Republic* (November 7, 1923), 2. For the quality of vocational education, see John D. Russell, et al., *Vocational Education,* prepared for the Advisory Committee on Education, staff study no. 8 (Washington, D.C.: U.S. Government Printing Office, 1938).

64. Sol Cohen, "The Industrial Education Movement, 1906–1917," *American Quarterly,* 20(1968): 95–110; Lazerson, *Origins of the Urban School;* Lazerson and Grubb, "Introduction," pp. 7–10, 26–27.

65. Caswell and Campbell, *Curriculum Development,* p. 5. According to Josephine Chase, "This plan of grouping high school pupils according to their capacities to do varied types of work allows to the slow pupils, for whom academic work is hard, a chance to substitute such subjects as typewriting, shop work, modified science, extra English or even supervised study periods in place of languages or mathematics." Chase, *New York at School,* pp. 61–62.

66. Thompson, "Commercial High Schools," p. 20; Covello, *Heart is the Teacher,* p. 181. See also Judd, *Measuring the Work of the Public Schools,* p. 194.

67. Chase, *New York at School,* pp. 54–55.

68. Palmer, *Pupils Who Leave School,* p. 20. Patricia Graham described the educational innovations in Marquette County, Michigan, in a similar fashion: "The concern for kindergartens, for the child study movement, for geography in the grammar grades, for vocational education in the high school—all were indicative of Marquette's efforts to follow what was considered new and best, or the word that was often used as a synonym for new and best in national pedagogical circles, 'progressive.' " Graham, *Class and Community,* p. 85.

69. Palmer, *Pupils Who Leave School,* p. 49. See Dewey's denunciation of testing and stratification, "Mediocrity and Individuality," *The New Republic* (December 6, 1922), 35–37; "Individuality, Equality, and Superiority," *The New Republic* (December 13, 1922), 61–63.

70. Tildsley, "Reorganization of the High School," pp. 10–11.
71. Inglis, "The High School in Evolution," p. 2.
72. Richard Hofstadter, *Anti-Intellectualism in American Life* (New York: Knopf, 1962), 332–358.

3: "AMERICANIZING" THE HIGH SCHOOLS

1. Among the important assessments of school enrollments, see Bernard D. Karpinos and Herbert J. Sommers, "Educational Attainment of Urban Youth in Various Income Classes," parts I & II, *The Elementary School Journal*, 42(May 1942): 677–687 and (June 1942): 766–774; Michael Olneck and Marvin Lazerson, "The School Achievement of Immigrant Children, 1890–1930," *History of Education Quarterly*, 14(Winter 1974): 453–482; David Hogan, "Education and the Making of the Chicago Working Class, 1880–1930," *History of Education Quarterly* 18(Fall 1978): 227–270; Michael Katz, "Who Went to School?" *History of Education Quarterly*, 12(Fall 1972): 432–454; Carl F. Kaestle and Maris A. Vinovskis, *Education and Social Change in Nineteenth Century Massachusetts* (Cambridge, England: Cambridge University Press, 1980); Carl F. Kaestle, *Pillars of the Republic: Common Schools and American Society, 1780–1860* (New York: Hill and Wang, 1983).

According to the 1960 census, second-generation males from South and East European countries (the largest part of the early twentieth-century migration) who were thirty-five to forty-four years of age (therefore of high-school age in the 1930s and '40s) had achieved the following average number of years of schooling: Austria, 12.2; Poland, 11.5; Czechoslovakia, 11.9; Hungary, 12.0; Yugoslavia, 12.0; Lithuania, 12.3; Finland, 12.3; U.S.S.R., 12.7; Italy, 11.4. The comparable figures for women were: Austria, 12.2; Poland, 11.3; Czechoslovakia, 11.8; Hungary, 12.0; Yugoslavia, 12.1; Lithuania, 12.2; Finland, 12.3; U.S.S.R., 12.4; Italy, 11.3. These figures are adapted from Table 6.4 in Stanley Lieberson, *A Piece of the Pie: Blacks and White Immigrants Since 1880* (Berkeley: University of California Press, 1980), 130. Without exception, this meant that the average child of second-generation Eastern or Southern European immigrants had at least entered high school and in many cases had graduated.

For the educational aspirations of these immigrants, see Timothy L. Smith, "Immigrant Social Aspirations and American Education, 1880–1930," *American Quarterly*, 21(1969): 523–543. For the high achievement that immigrants associated with high-school attendance, see Thomas W. Gutowski, "The High School as an Adolescent-Raising Institution: An Inner History of Chicago Public Secondary Education, 1856–1940" (Ph.D. dissertation, University of Chicago, 1978), 9–18.

The development of the high school is discussed in Edward A. Krug, *The Shaping of the American High School,* 2 vols. (Madison, Wis.: University of Wisconsin Press, 1969, 1972); Theodore R. Sizer, *Secondary Schools at the Turn of the Century* (New Haven: Yale University Press, 1964); Lawrence Cremin, "The Revolution in American Secondary Education, 1893–1918," *Teachers College Record,* 56(March 1955): 295–308; Alexander J. Inglis, "Secondary Education," in I. L. Kandel (ed.), *Twenty-Five Years of American Secondary Education* (New York: Macmillan, 1921).

For how the colleges responded to the new onslaught, see Harold S. Wechsler,

The Qualified Student: A History of Selective College Admissions in America (New York: John Wiley and Sons, 1977); David O. Levine, *The American College and the Culture of Aspiration, 1915–1940* (Ithaca, N.Y.: Cornell University Press, 1986).

Niccolà Sacco's quote appears in Herbert Ehrmann, *The Case That Will Not Die: Commonwealth vs. Sacco and Vanzetti* (Boston: Little Brown, 1969), 315; Leonard Covello (with Guido D'Agostino), *The Heart is the Teacher* (New York: McGraw-Hill, 1958), 39.

2. Elwood Cubberly, editor's introduction to Elbert K. Fretwell, *Extra-Curricular Activities in Secondary Schools* (Boston: Houghton Mifflin, 1931), vi. See also Thomas H. Briggs, *Secondary Education* (New York: Macmillan, 1934), 138–191 and passim.

3. It was common observation in the teens, twenties, and thirties that Jews were the most eager for education. Thus, Alexander Dushkin observed "It has been ascertained that the Russian Jews, in spite of their comparative poverty, send more of their children to the high schools of this city, and permit them to stay there longer than any other ethnic group." Quoted in Robert E. Park and Herbert H. Miller, *Old World Traits Transplanted* (New York: Harper & Brothers, 1921), 37. See also Thomas Kessner, *The Golden Door: Italian and Jewish Immigrant Mobility in New York City 1880–1915* (New York: Oxford University Press, 1977), 95–99; Lazerson and Olneck, "The School Achievement of Immigrant Children"; Ronald H. Bayor, *Neighbors in Conflict, The Irish, Germans, Jews, and Italians of New York City, 1929–1941* (Baltimore: Johns Hopkins University Press, 1978), 15–16. For different average years of school attainment see above note 1.

As early as 1911, almost 50 percent of the students of secondary schools of thirty-seven of the largest cities were of foreign-born parentage; Francesco Cordasco, *Immigrant Children in American Schools: A Classified and Annotated Bibliography with Selected Source Documents* (Fairfield, N.J.: A. M. Kelly, 1976), 27. For the Protestant values of the common-school movement, see Kaestle, *Pillars of the Republic*, pp. 75–103 and passim.

4. In *Rites of Passage: Adolescence in America, 1790 to the Present* (New York: Basic Books, 1977), Joseph F. Kett notes that behind the growing emphasis on adolescence among those who worked with boys lay "the authentic enthusiasm . . . in the early 1900's for a psychological system that subordinated class and religious differences to a principle of biological maturation." (p. 223)

5. Earle Rugg, "Special Types of Activities: Student Participation in School Government," *Twenty-Fifth Yearbook of the National Society for the Study of Education* (Bloomington, Ill.: Public School Publishing Co., 1926), 131, (hereafter, *Twenty-Fifth Yearbook, NSSE*).

6. According to Elbert Fretwell, the premier exponent of extracurricular activities, the 1920s saw a continuous buzzing of interest in their possibilities: "In national and state educational meetings, in sectional and in local conferences, in college and university classes, and in book and magazine publications, there has been, and there is continuing to be, discussion of the theories and plans of extra-curricular activities." Fretwell, *Extra-Curricular Activities*, p. 5. For a brief history of the activities, see Galen Jones, *Extra-Curricular Activities in Relation to the Curriculum*, Contributions to Education (New York: Bureau of Publications, Teachers College, Columbia University, 1935), 13–29.

7. Cubberly in Fretwell, *Extra-Curricular Activities*, p. v; Charles R. Foster,

Extra-Curricular Activities in the High School (Richmond, Va.: Johnson Publishing Co., 1925), 4.

8. Leonard V. Koos, "Analysis of the General Literature on Extra-Curricular Activities," *Twenty-Fifth Yearbook, NSSE* (1926), 10–11.

9. Foster, *Activities in High School*, p. 5; Thomas Briggs quoted in Foster, *Activities in High School*, p. 5; Koos, "Analysis," p. 17.

10. Fretwell, *Extra-Curricular Activities*, pp. ix, 10; Foster, *Activities in High School*, p. 6.

11. Gutowski argues that by the 1920s the aim and reality of extracurricular activities was toward maximum control, see Gutowski, "High School as an Adolescent Raising Institution," pp. 221–238.

12. For surveys, see, for example, Galen Jones, *Activities in Relation to the Curriculum;* William C. Reavis and George E. Van Dyke, *Nonathletic Extracurricular Activities,* Bulletin 1932, no. 17, National Survey of Secondary Education, monograph no. 26(Washington, D.C.: U.S. Government Printing Office, 1933). The quote is from Fretwell, *Extra-Curricular Activities,* p. 116.

13. Native white was defined for the purposes of this study to include all students whose surnames were either British (exclusive of Ireland), Dutch, or French. I have included these non-British groups in this category because historically both the French and Dutch were long settled in New York. It is certainly true that some Germans, Jews, and Irish were also long established in the city, but their numbers were small compared to the large migrations of the late nineteenth and early twentieth centuries. Obviously, those students of French or Dutch ancestry who were part of the newer immigration would have been included among the natives.

Because many individuals of Irish descent have native-sounding names (for example, White), there was probably some undercounting of Irish students. This was probably greatest at Theodore Roosevelt High School where the proportion of Irish was lower than might have been expected from the demographics of the neighborhood.

All German surnamed individuals, except those who were most probably Jews, were included as Germans. The "other" category was composed of a very large variety of individuals, including Hispanics, Scandinavians, Russians and Poles (who were probably not Jews), Chinese, Japanese, and those from the Middle East.

14. Gutowski argues that in Chicago extracurricular participation had become almost a requirement by the 1930s because it was a critical part of the whole way in which high-school education was conceived by educators. Moreover, some extracurricular participation was necessary for election to the honor societies. See Gutowski, "High School as an Adolescent Raising Institution," pp. 211–221. While participation in New York was generally very high, it was not uniform across schools and, therefore, does not appear to have reflected across-the-board policy. At the same time, individual school principals may have made participation almost obligatory.

15. These were: some activity, football, basketball, track, other sports, president of the student body or senior class, yearbook staff, editor in chief of the student newspaper, other news staff, other publication, physics, chemistry, other science, other academic clubs, Arista, glee club, orchestra, drama societies, social clubs, service, religious clubs, other activity, celebrity status.

Two other organizations originally included in the tabulations, debate and program committee, turned out to be unusable because participation was either too low or the activity was not common among enough of the schools.

16. In some cases the categories turned out to be not as comprehensive as in others. At Bay Ridge, the very large number of students engaged in "other activities" suggests that the listed activities only accounted for about one-half of all student participations.

17. The reader should note that small population groups, like blacks and Irish, may be absent from some activities within a school more commonly than larger population groups, like the Jews or Italians.

18. Only at Seward Park were black women overrepresented in "other publications," but this was a statistical fluke since only one black woman was in fact involved (see Appendix 2, Table 2).

19. Jewish men were hardly more conspicuous than native men. This is best seen by looking at the editors. Of twenty-one male editors, thirteen were Jews, although Jews were only 45 percent of the male population. But native men held four of the twenty-one editorial chairs and were more disproportionately represented (13 percent of the male population). Indeed, the apparent success of Jewish men in becoming editors was inflated because of their control of the editorial posts at New Utrecht, where all editors, and many other positions of power were in Jewish hands.

20. For the attitudes of Italians toward women, see Virginia Yans McLaughlin, *Family and Community: Italian Immigrants in Buffalo, 1880–1930* (Ithaca, N.Y.: Cornell University Press, 1977), 147, 149–151, and passim; Kessner, *Golden Door,* pp. 84–85.

21. Bayor, *Neighbors in Conflict,* pp. 17–18.

22. Lieberson observes that conclusions which emphasize the degree to which differences among groups are cultural often suffer from some kind of circular reasoning: "Why are two or more groups different with respect to some characteristic or dependent variable? Presumably, they differ in their values or norms. The argument then frequently involves using the behavioral attribute one is trying to explain as the indicator of the normative or value difference one is trying to use as the explanation. A pure case of circular reasoning!" Lieberson, *A Piece of the Pie,* p. 8.

23. Students were admitted to high schools in New York in the 1930s and '40s technically on the basis of open admission; that is, students could choose what school they wished to attend. In practice, however, except for those who elected to go to the academically exclusive schools like Stuyvesant High School where admission was by test, most students attended high schools in their area.

24. Deborah Dash Moore, *At Home in America: Second Generation New York Jews* (New York: Columbia University Press, 1981), 19–58. Moore emphasizes the role of Jewish builders in the creation of apartment houses and the importance of the apartment house in the culture of the second-generation Jew. But a look at the second-generation neighborhoods in many of the satellite communities suggests that a good many Italians and Jews also chose to own income-producing, two-family, properties. See also Moore, p. 36.

25. In a memoir of depression-era Evander Childs, Shirley Jacoby Paris notes that black students were hardly noticed at Evander Childs because "They were students, integrated with the rest, meeting the same standards as the whites, and given neither adverse nor preferential treatment." Jacoby, "Evander Childs High

School," *The Bronx County Historical Society Journal,* 21(Spring 1984): 5. In fact, whatever their equality in the classroom, blacks were unnoticed in the busy club and activity life because they were so largely absent from these.

26. The Irish appear to have bested the Jews on the news staff according to Appendix 2, Table F, but in fact, the Irish presence was restricted to one individual.

27. Moore, *At Home in America,* pp. 24, 66.

28. Bayor, *Neighbors in Conflict,* p. 162. On the neighborhood conflicts generally and the role of the churches, see pp. 150–163.

29. Some of the natives may in fact have been Irish with English surnames. The high proportion of natives in religious clubs makes this likely as does the low number of Irish at the school.

30. The low level of service participation, especially by men, at Theodore Roosevelt may be a sign of low morale and a by-product of a strained social atmosphere.

31. I used the June 1947 class at Commerce (which was coed) to get a glimpse of patterns when the school contained women, but the numbers were simply too small and the range of activities women joined too limited to be very useful.

32. At Bay Ridge, the secretaries in the principal's office, who had attended the school themselves in the forties, made it clear that parents usually chose to send their daughters to Bay Ridge because it was an exclusively female school and considered safe. This, they told me, was especially true for Italians. Many Catholic families may have used Bay Ridge as a substitute for parochial schools. (See Chapter 6.)

33. There is some reason to believe that religious clubs may have been hidden behind secular names and that therefore the tables underestimate the degree to which certain groups, like the Irish and Italians, joined religious groups. The best evidence comes from a glance at the proportion of Irish and Italian women engaged in "other activities" (Appendix 2, Table N). If a club could not be categorized because its name was obscure, it wound up in the "other activities" category.

34. See Wechsler, *Qualified Student,* pp. 131–185. Wechsler argues, convincingly, that many of the newer criteria for admissions were designed at schools like Columbia to "repel the invasion" of Jewish students. For the infectiousness of collegiate culture in the '20s, see Paula S. Fass, *The Damned and the Beautiful: American Youth in the 1920's* (New York: Oxford University Press, 1977).

35. Covello, *Heart is the Teacher,* p. 47.

36. Milton M. Gordon, *Assimilation in American Life: The Role of Race, Religion and National Origins* (New York: Oxford University Press, 1964), and "Assimilation in America: Theory and Reality," *Daedelus,* 90(1961): 263–285.

4: NEW DAY COMING

1. "Proceedings, The Second National Conference on the Problems of the Negro and Negro Youth," January 12, 13, 14, 1939, National Archives, Judicial, Social, and Fiscal Branch, RG 119, Records of the National Youth Administration (NYA), series 75, Working and Data Files of an NYA History Project, 1935–43 (Administrative History Material), Negro Programs, p. 31 (hereafter, cited as "Proceedings, Second Conference Negro Youth," NYA). All documents from the National

Archives, Records of the National Youth Administration will be cited as NA–NYA.

2. William F. Russell, "Upgrading the Illiterate Registrant for use by the Army," January 12, 1943, p. 4, attached to "Memorandum for the Chief of Staff: Subject: Request for Permission to Proceed with Immediate Operation of Plan for Upgrading Illiterates," 18 January 1943, National Archives, Military Reference Branch, RG 160, Records of Headquarters Army Service Forces, 130.5 (1-1-43 to 11-30-43), entry 153, Director of Military Training, Central Decimal File 1942–46 (Box 277). All documents from the Military Reference Branch of the National Archives will be cited as NA–MRB. Russell's study will hereafter be cited as Russell Report.

3. Harry Zeitlin gives a good, brief introduction to educational activities of the federal government before the New Deal in "Federal Relations in American Education, 1933–43: A Study of New Deal Efforts and Innovations," (Ph.D. dissertation, Teachers College, Columbia University, 1958), Chapter 1. Zeitlin's is also the best and most thorough study of New Deal educational efforts.

4. Recently, Carl F. Kaestle and Marshall S. Smith have denied the significance of national crises for the expansion of federal educational activities. I take issue with this perspective and reject their view of the inevitability of this process. See Kaestle and Smith, "The Federal Role in Elementary and Secondary Education, 1940–1980," *Harvard Educational Review,* 52(November 1982): 384–408. See also the responses to Kaestle and Smith by Marvin Lazerson, Diane Ravitch, and James Q. Wilson in the same issue, pp. 409–418.

5. United States National Advisory Committee on Education, *Federal Relations to Education: Report of the National Advisory Committee on Education,* part I, *Committee Findings and Recommendations* (Washington, D.C.: U.S. Government Printing Office, 1931), 18, 30 (my emphasis). Hereafter, this volume will be cited as USNACE part I. See also part II, *Basic Facts.*

6. USNACE, part I, pp. 12–13.

7. USNACE, part I, pp. 108, 110.

8. USNACE, part I, p. 25.

9. See Paula S. Fass, "Without Design: Education Policy in the New Deal," *American Journal of Education,* 91(November 1982): 51–55; also, David Tyack, Robert Lowe and Elizabeth Hansot, *Public Schools in Hard Times: The Great Depression and Recent Years* (Cambridge: Harvard University Press, 1984).

By 1942, the National Education Association (NEA) had come to acknowledge that the emergency revealed problems that local control had failed to address, but its response was a half-hearted acceptance of relief and a call for a return to normal schooling. By then, of course, it scarcely mattered. See Frank N. Freeman, "Federal Youth Agencies and the Public Schools," *School and Society,* 55(June 20, 1942): 702–704. As late as 1964, the NEA held to its traditional commitment to local control. At its annual convention, the NEA adopted several resolutions. Among them were the following: "That the general federal-support funds be allocated without federal control to state school authorities to be commingled with state public education funds." "That expenditures of the federal funds be only for the purposes for which the states and localities, under their constitutions and statutes, may expend their own public funds." Quoted in Sidney W. Tiedt, *The Role of the Federal Government in Education* (New York: Oxford University Press, 1966), 38.

10. "Address by John Sexon to the Advisory Committee [Advisory Committee

on the Education of Negroes], February 28, 1938," National Archives, Judicial, Social, and Fiscal Branch, RG 12, Records of the Office of Education, inventory item 17, Office File of Ambrose Caliver. Material from this file will hereafter be cited as Caliver Papers.

11. See the heated discussion among black educators concerning the collecting of separate statistics on black high schools for the Office of Education survey of secondary schools, in "National Advisory Committee on the Education of Negroes, Howard University, July 29, 1931," in Caliver Papers.

12. For Roosevelt's attitude toward the Office of Education and the education profession, see Zeitlin, "Federal Relations in American Education," pp. 288–314; for the emergency grant to schools, see Floyd W. Reeves, "Purposes and Functions of the Advisory Committee on Education," in National Education Association, *Proceedings of the Seventy-Fifth Annual Meeting,* 75(Washington, D.C.: National Education Association, 1937), 38 (hereafter, *Proceedings, NEA*); and George F. Zook, "Federal Aid to Education," in *Proceedings, NEA,* 72 (1934), p. 40. In 1950, the Brookings Institution issued a report which condemned the New Deal's relegation of the Office of Education to an insignificant status, yet noted throughout that the Office was a third-rate agency. See the complete report, Hollis P. Allen, *The Federal Government and Education: The Original and Complete Study of Education for the Hoover Commission Task Force on Public Welfare* (New York: McGraw-Hill, 1950), especially 209–212.

13. "Advisory Committee on the Education of Negroes, 1940 meeting, St. Louis, February 27, 1940," in Caliver Papers.

14. The statement of sentiments by blacks is found in "Ninth Meeting of the National Education Committee, Atlantic City, February 17, 1935," in Caliver Papers.

Caliver's liason role is clearly seen in the controversy over the collection of statistics on black secondary schools which emerged at the National Advisory Committee on the Education of Negroes meeting in 1931 at Howard University; see p. 4 of the minutes, "National Advisory Committee, Howard University," in Caliver Papers. When President John P. Davis deplored the proposal for a separate study of black land-grant colleges as "a joke and travesty," he asked pointedly, "just how far are we going to lend our influence, etc. toward a thing like that?" Caliver, as representative of the Office of Education, would not let himself be goaded into taking any more than an expository position.

When the Office of Education assumed a larger share of responsibility for the National Youth Administration after 1940, it was frequently taken to task by interested blacks, like Robert Weaver, in the Roosevelt administration. At one point the Office and Commissioner Studebaker were strongly condemned at a hearing on fair employment practices for their inactivity in providing blacks a fair share of NYA war-related training. The hearing came to a fitting climax when it was noted, "The Committee has found that there is sometimes a correlation between a Federal agency's employment policy in regard to discrimination and its administrative policy in taking effective measures to eliminate discrimination in areas subject to its control." The hiring policy at the Office of Education provided a less than adequate effort, since of its 693 employees, "32 or less than 5% are Negroes, 20 of whom are in custodial positions, with only one in sub-professional and 2 in professional positions." See "Summary of the Hearing on Fair Employment Practice on Discrimination in Defense Training with Findings and Directions Held April

13, 1942," attached to a letter to Commissioner John W. Studebaker from Laurence N. Cramer, July 3, 1942, National Archives, Judicial, Social, and Fiscal Branch, RG 12, Records of the Office of Education, 901 Historical File, Negro Education. This file will hereafter be cited as Office of Education, Historical File–Negro.

15. See C. S. Marsh, "The Educational Program of the Civilian Conservation Corps," *Bulletin of the Department of Secondary School Principals of the National Education Association,* 50(1934): 216–217. For the CCC generally, see George Philip Rawick, "The New Deal and Youth: The Civilian Conservation Corps, the National Youth Administration and the American Youth Congress," (Ph.D. dissertation, University of Wisconsin, 1957); John A. Salmond, *The Civilian Conservation Corps, 1933–1942: A New Deal Case Study* (Durham, N.C.: Duke University Press, 1967). Salmond discusses the educational efforts on pp. 47–54.

16. See Zeitlin, "Federal Relations in American Education," pp. 92, 95; American Council on Education, *The Civilian Conservation Corps: Recommendations of the American Youth Commission of the American Council on Education* (Washington, D.C.: American Council on Education, 1941), 18, 20; United States Advisory Committee on Education, *Report of the Advisory Committee* (Washington, D.C.: U.S. Government Printing Office, 1938), 115–132 (hereafter, cited as USACE, 1938); Allen, *Federal Government and Education,* pp. 93–94.

17. For the number of youth reached by NYA, see Charles H. Judd, "Federal Aid to Education," NA–NYA–Office of Negro Affairs, Office of the Director, series 115, "Files, Address—NYA and General," p. 3. See also Palmer O. Johnson and Oswald L. Harvey, *The National Youth Administration,* Staff Study #13, prepared for the Advisory Committee on Education (Washington, D.C.: U.S. Government Printing Office, 1938), 7; Reeves, "Purposes and Functions of the Advisory Committee," p. 28.

18. The quote is from Johnson and Harvey, *The National Youth Administration,* p. 12. For complaints about the bypassing of traditional institutions, see Lotus D. Coffman, "Federal Support and Social Responsibility for Education," National Education Association, *Proceedings, NEA,* 74(1936):413–421.

The Office of Education was actively hostile to the NYA and by 1938 was campaigning to destroy the agency. In 1940, the Office was given a role in NYA's vocational education program and very soon thereafter it absorbed large portions of NYA's defense industry training program, to which NYA efforts were largely directed by 1941. By this time, NYA was effectively finished as a New Deal agency. See Rawick, "The New Deal and Youth," pp. 252–261.

19. The quote is from Doak S. Campbell, Frederick H. Bair, and Oswald L. Harvey, *Educational Activities of the Works Progress Administration,* Staff Study #14, prepared for the Advisory Committee on Education (Washington, D.C.: U.S. Government Printing Office, 1939), 20–21.

The Brookings Institution made an elaborate critique of New Deal programs and argued that the relief programs did deeply intrude on traditional educational territory; see Allen, *The Federal Government and Education,* pp. 92–99.

20. Campbell, et al., *Works Progress Administration,* p. 157.

21. Hopkins's speech is reprinted in Zeitlin, "Federal Relations in Education," pp. 348–352; the quoted passage appears on p. 349.

22. "Address by Aubrey Williams, text of Inglis Lecture given to Graduate School of Education, Thursday, February 15, 1940," NA–NYA–Office of Negro Affairs,

series 115 "Correspondence, Reports, Information File, 1938–41," pp. 3, 6, 9, 11, 12.

23. Nancy J. Weiss, *Farewell to the Party of Lincoln: Black Politics in the Age of FDR* (Princeton, N.J.: Princeton University Press, 1983), discusses the initial indifference of New Dealers to blacks (pp. 34–59) and Eleanor Roosevelt's important role in changing this situation (pp. 120–135). See also Harvard Sitkoff, *A New Deal for Blacks: The Emergence of Civil Rights as A National Issue—The Depression Decade* (New York: Oxford University Press, 1978), 34–57.

24. "Proceedings, Second Conference Negro Youth," NYA, p. 27.

25. The description of Williams is from Sitkoff, *New Deal for Blacks*, p. 73. The quote is from "Final Report 1943," NA–NYA–Office of Negro Affairs, entry 118, p. 100; the figures on black participation are on p. 212 of this report. For Harold Ickes's role and significance, see John B. Kirby, *Black Americans in the Roosevelt Era: Liberalism and Race* (Knoxville: University of Tennessee Press, 1980), 17–35.

26. The quotation is from Johnson and Harvey, *The National Youth Administration*, pp. 27–28. The figures on graduate-student aid are from "Final Report 1943," NA–NYA–Office of Negro Affairs, entry 118, p. 102.

For Bethune's special commitment to the graduate-student fund, see, for example, "Minutes, Regional Conference College and NYA Officials of the College Work Program, September 6, 1940," NA–NYA–Office of Negro Affairs, series 117, Negro Conferences, p. 6. Bethune kept a sharp eye on this fund. In her instructions to her assistant, she is recorded as saying, "Go over with Dr. Judd very carefully the application of the Special Fund. I don't want anything done so far as allocations are concerned unless I have had a chance to see it. We must be careful in the handling of this fund this year." "Long Distance Call—Mrs. Bethune to Mr. Lanier, August 12, 10:30 A.M. (1940)," NA–NYA–Office of Negro Affairs, series 116, "File of early 'inactive' correspondence 1935–1938."

27. The quote is from Rawick, "The New Deal and Youth," p. 148. See also Salmond, *The Civilian Conservation Corps*, pp. 91–96. For CCC enrollments, see Zeitlin, "Federal Relations in Education," p. 107 and Edgar G. Brown, "Civilian Conservation Corps, Summary—The Program in Action," NA–NYA–Office of Negro Affairs, series 115, "Correspondence, Reports, Information File 1938–1941—Addresses NYA and General." For segregation and personnel in CCC camps, American Council on Education, *The Civilian Conservation Corps*, p. 18. For the experience of blacks in the CCC generally, see Salmond, *The Civilian Conservation Corps*, pp. 88–101.

28. "Proceedings, Second Conference Negro Youth," NYA, p. 62.

29. "Annual Report of the Division of Negro Affairs, July 1, 1936–June 30, 1937," (June 30, 1937), NA–NYA–Office of Negro Affairs, series 115, pp. 2, 8, 9. For Bethune's role, see Kirby, *Black Americans in the Roosevelt Era*, pp. 110–121; Weiss, *Farewell to the Party of Lincoln*, pp. 137–148. For a critical assessment of Bethune's role and views, see B. Joyce Ross, "Mary McLeod Bethune and the National Youth Administration: A Case Study of Power Relationships in the Black Cabinet of Franklin D. Roosevelt," *The Journal of Negro History*, 60(January 1975): 1–28.

30. "Proceedings, Second Conference Negro Youth," NYA, p. 2.

31. "Final Report 1943," NA–NYA–Office of Negro Affairs, entry 118, p. 38.

At a meeting in 1935, Aubrey Williams declared that he hoped "to have a Negro in each state as part of the staff. Cannot guarantee this, but will work on it;" see "Meeting of Negro Leaders," August 8, 1935, entry 118 in the same file with the final report. Bethune worked to hold Williams to his word: "I have been promised a strong stalwart Negro assistant in every State where we had an appreciable number of Negroes. I am holding my executives to that promise;" see Bethune address in "The Conference of the National Youth Administration, September 9–10, 1938, Hay–Adams House," part I, p. 6, NA–NYA, series 75, Working and Data Files of An NYA History Project, 1935–43 (Administrative History Material), Advisory Committees. (Hereafter cited as "Conference, Hay–Adams House," NYA.)

32. "Minutes, Regional Conferences College and NYA, Officials of the College Work Program, September 6, 1940," NA–NYA–Office of Negro Affairs, series 117, pp. 4, 6.

33. "Conference, Hay–Adams House," NYA, pp. 2, 5.

Bethune kept in touch with local officers, through letters, calls, and visits. See, for example, "Mrs. Bethune's Proposed Trip to the West Coast," NA–NYA–Office of Negro Affairs, series 116, "File of early 'inactive' correspondence." For Bethune's style and influence, see Weiss, *Farewell to the Party of Lincoln,* pp. 139–148.

34. "Minutes, Regional Conference College and NYA, Officials of the College Work Program, September 6, 1940," NA–NYA–Office of Negro Affairs, series 117, Negro Conferences, p. 11.

35. "Final Report 1943," NA–NYA–Office of Negro Affairs, entry 118, p. 39. For the Forum Project in Georgia, see Nathaniel P. Tillman, "The Statewide Public Forum Project: An Experiment in Civic Education Among Negroes in Georgia," part I, NA–NYA–Publications, Georgia.

Lasseter was brought to Washington by Williams after an extremely successful period as a Georgia administrator. At one conference he was honored by Bethune and others. "I want to thank Georgia for whatever it did in helping to inspire Mr. Lasseter to look with a straight eye into the problems of Negro youth in this country. I have been with the National Youth Administration since its inception, and we have never had in that office a man who had the keen insight into the problems of the Negro and who has been more willing, as the Deputy Assistant to Mr. Aubrey Williams, to carry on in a fair and just way the program for Negroes, Chinese, Japanese, and everybody that has to be encountered than Mr. Lasseter." "Minutes, Regional Conference College and NYA Officials of the College Work Program, September 6, 1940," NA–NYA–Office of Negro Affairs, series 117, Negro Conferences, p. 20.

36. Inez F. Oliveros, "Final Report of the National Youth Administration for the State of Georgia," NA–NYA–Final State Reports, Florida–Georgia.

37. William H. Shell, "The Negro and the National Youth Administration in Georgia," prepared for the National Conference on the Problems of the Negro and Negro Youth," NA–NYA–Publications, Georgia, pp. 19, 20, 22.

38. Shell, "The Negro and the National Youth Administration in Georgia," pp. 30–31.

39. The quotation about the NYA is from "Final Report 1943," NA–NYA–Office of Negro Affairs, entry 118, p. 214. The students are quoted in *Negro Youth in Kentucky,* NA–NYA–Publications, Kentucky.

The statistic on high-school attendance is from David T. Blose and Ambrose

Caliver, "Statistics of the Education of Negroes," Circular #215, Office of Education, Historical File–Negro.

40. "Final Report 1943," NA–NYA–Office of Negro Affairs, entry 118, pp. 216–217.

41. Johnson is quoted in "Proceedings, Second Conference Negro Youth," NYA, p. 11. Bethune's letter of transmittal appears in "Report of the National Conference on the Problems of the Negro and Negro Youth," held in the government auditorium, Department of Labor, January 6–8, 1937, NA–NYA–Office of Negro Affairs, series 116, "File of 'inactive' correspondence, 1935–38." John Kirby discusses the dissatisfaction that accompanied these successes in *Black Americans in the Roosevelt Era*, pp. 224–225.

42. "Report Concerning First Six Months of the Work of the Office of Negro Affairs," no date, NA–NYA–Office of Negro Affairs, series 116, "File of early 'inactive' correspondence."

43. Quoted in Zeitlin, "Federal Relations in Education," p. 205 note 36.

44. Reeves makes this point in "Purposes and Functions of the Advisory Committee," p. 29.

45. "Proceedings, Second Conference Negro Youth," NYA, p. 32.

46. USACE, 1938, pp. 4–5, 33, 38, 4.

47. USACE, 1938, pp. 19, 39.

48. Doxey A. Wilkerson, *Special Problems of Negro Education*, Staff Study #12, prepared for the Advisory Committee on Education (Washington, D.C.: U.S. Government Printing Office, 1939); USACE, 1938, p. 49.

49. I found *School Money in Black and White*, in NA–NYA–Office of Negro Affairs, series 120—Miscellaneous. The pamphlet was based "Upon statistical material assembled by the Committee on Finance of the National Conference on Fundamental Problems in the Education of Negroes, May 9–10, 1934." The conference was organized by Caliver who also recruited the Julius Rosenwald Fund to finance the publication.

50. See the letter from Secretary of War Henry Stimson to Congressman Francis D. Brown with the attached list of participating colleges and universities, September 30, 1944, NA–MRB, RG 407, Records of the Adjutant General's Office, 353 Training, (9-16-44 to 9-20-44) (Box 2656); "History of Military Training ASF," 5 September 1944, NA–MRB, RG 407, Records of the Adjutant General's Office, AG 353 ASTP (formerly classified) (Box 2204).

The army was deluged with requests by young men and their parents seeking to participate in these benefits. See, for example, letter to the Adjutant General from William A. R. Hawley, 6 May 1943, NA–MRB, RG 407, Records of the Adjutant General's Office, AG 353 (5-6-43 to 5-7-43) (Box 2670) and the letter addressed to President Roosevelt ("Dear Uncle Sam"), 11 May 1943 in the same box. These were but two of hundreds of similar requests.

51. See especially Ulysses Lee, *The Employment of Negro Troops*, in the special studies series by the Office of the Chief of Military History, United States Army, *United States Army in World War II* (Washington, D.C.: U.S. Government Printing Office, 1966); Samuel A. Stouffer, et al., *The American Soldier: Adjustment During Army Life*, vol. I (Princeton, N.J.: Princeton University Press, 1949), Chapter 10.

52. During the early phases of recruitment, the navy was insulated from these problems by the fact that its quotas were filled largely through volunteers rather

than through selective service. Blacks were largely absent from the navy except as messmen. See Lee, *Employment of Negro Troops,* p. 89. After February 1943, the recruitment for various services was integrated, and all men were inducted through selective service; see Lee, *Employment of Negro Troops,* pp. 406–407.

53. Stouffer, et al., *The American Soldier,* p. 494; Lee, *Employment of Negro Troops,* p. 240.

54. "The Training of Negro Troops," Study #36, Historical Section, Army Ground Forces, 1946, p. ii, available at the U.S. Center of Military History, Washington, D.C.

55. "The Training of Negro Troops," pp. iii, 7.

56. The table is drawn from "Command of Negro Troops," War Department Pamphlet No. 20–6, February 29, 1944, as reprinted in Morris J. MacGregor and Bernard C. Nalty (eds.), *Blacks in the United States Armed Forces: Basic Documents,* vol. V, *Black Soldiers in World War II* (Wilmington, Del.: Scholarly Resources, Inc., 1977), 311.

57. "The Training of Negro Troops," p. ii.

58. Samuel Goldberg, *Army Training of Illiterates in World War II,* Contributions to Education (New York: Bureau of Publications, Teachers College, Columbia University, 1951), 12. Hershey is quoted on p. 11.

The Fourth Corp area (Service Command) was composed of Alabama, Florida, Mississippi, North Carolina, South Carolina and Tennessee. The Eighth Corp consisted of Arkansas, Louisiana, New Mexico, Oklahoma, and Texas.

59. Lewis B. Hershey to Secretary of War, 11 September 1941, NA–MRB, RG 407, Records of the Adjutant General's Office, 350.5 "Illiteracy in the Army" (1-1-40 to 12-31-41) (Box 2613).

60. Army Service Forces, Office of the Director of Military Training, "History of Training, Special Training Units, October 1940–December 1944 with Supplement to 31 December 1945," typescript in the U.S. Center of Military History, Washington, D.C. See also "Personnel Utilization: Selection, Classification, and Assignment of Military Personnel in the Army of the United States During World War II," by the staff of the Classification and Replacement Branch, the Adjutant General's Office, September 1947, NA–MRB, RG 319, Records of the Army Staff, ACMH Historical Manuscript File, 4-1.4 BA. Hereafter, this file is cited as "Personnel Utilization."

The voluntary programs are briefly described and evaluated in the Russell Report, pp. 5–9.

61. Russell Report, p. 2; Goldberg, *Training of Illiterates,* p. 14.

62. Bilbo is quoted in Eli Ginzberg and Douglas W. Bray, *The Uneducated* (New York: Columbia University Press, 1953), 65. For the violence in the spring of 1943, see Lee, *Employment of Negro Troops,* pp. 366–379. For complaints about dumping of less desirable personnel in supply units and its consequences, see, for example, the letter to Director of Training, Services of Supply from Brigadier General William R. Dear, 27 October 1942, NA–MRB, RG 407, Records of the Adjutant General's Office, 353 (formerly, Classified Decimal File 1940–42) (Box 501) which notes, "The continued and increased rate of receipt of Grade IV and V individuals in the Medical Replacement Training Centers together with what is an apparent dumping of limited service personnel . . . will seriously impair training and will eventually make the task of adequately maintaining the health of the Army impossible."

63. See "Memorandum for the Adjutant General, Subject: Special Training Units," 10 June 1943, NA–MRB, RG 407, Records of the Adjutant General's Office, 353 (6-10-43 to 6-12-43) (Box 2669).

64. For estimate on blacks in Special Training Units, see "Establishment of Special Training Units," 28 May 1943, NA–MRB, RG 407, Records of the Adjutant General's Office, 353 Training (5-14-43) (Box 2665), p. 2; Lee, *Employment of Negro Troops,* pp. 411–413 discusses black rejections.

65. The quote on federal policy is from Lee, *Employment of Negro Troops,* p. 76; the discussion of Negro units is in "Personnel Utilization," p. 246. Lee discusses the proportional quota issue and the problems with its implementation in *Employment of Negro Troops,* pp. 111–178.

66. In its efforts to sift and place draftees, the army began to use more and more examinations and a variety of interview procedures to determine mental aptitude as well as particular talents. The tests were of both the literate and nonliterate variety and administered individually. These exams were far more exacting than the army Alphas and Betas of World War I. Far more than in the First World War, the army had come to depend on the expertise of psychologists and psychiatrists who had, in the meantime, developed far more sophisticated tests. The army had, in turn, become far more sophisticated in their administration and in the use of follow-ups as the occasion required. Induction procedures now required personal interviews, an assessment of past schooling, an appraisal of literacy where schooling was deficient, and the administration of diverse testing instruments depending on that appraisal. Beyond the initial induction, "The Army Classification system," was in the words of one of its students, a "continuous process during the entire period of an enlisted man's active service." Goldberg, *Training of Illiterates,* p. 41. A description of an individual's qualifications "accompanied each enlisted man throughout his Army career and provided the data for entries on his discharge certificate." See also Lieutenant Colonel Leonard Lerwill, "Personnel Replacement System of the United States Army," Department of the Army Pamphlet #20-211, August 1954, U.S. Center of Military History, Washington, D.C.

67. The figures are from Goldberg, *Training of Illiterates,* Table III, p. 65. There were at least 100,000 men inducted before June 1943 who also received literacy training; see "Memo for Assistant of Staff, Illiterates in the Army," 21 April 1944, NA–MRB, RG 407, Records of the Adjutant General's Office, 327.1 Induction (4-1-44 to 4-30-44). Ginzberg and Bray estimate that approximately 384,000 illiterates were inducted into the army during the entire course of the war. Of these, 164,000 were black. See Ginzberg and Bray, *The Uneducated,* p. 73.

68. Goldberg, *Training of Illiterates,* p. 244. See, also, Stouffer, et al., *The American Soldier,* pp. 492–493.

69. Goldberg, *Training of Illiterates,* Table III, p. 65, Table IIIa, p. 66, Table IIIc, p. 67.

70. The quote is from Ginzberg and Bray, *The Uneducated,* p. 68. For the close supervision of Special Training Units, see, for example, "Method of Training in Special Training Units," NA–MRB, RG 160, Records of the Army Service Forces, entry 153, 350.3, 31 October 1943; also, the directive from Brigadier General W. L. Weible to Commanding General First Service Command, "Methods of Training in Special Training Units," 3 August 1943, NA–MRB, RG 407, Records of the Adjutant General's Office, 357 Responsibility for Training (Box 319).

For segregation of units, see the note attached to a draft memo "Establishment of Special Training Units in Reception Centers," 10 May 1943, NA–MRB, RG 160, Records of the Army Service Forces, entry 153, 350.5 which notes, "It is desired that provision be made to house, mess, train white and colored troops separately, although they may occupy adjacent barracks in the same reception center." The army had problems providing enough Special Training Unit facilities for blacks when they were first reorganized, see "Special Training Units," 27 September 1943, NA–MRB, RG 160, Records of the Army Service Forces, entry 153, 350.5.

71. Army Service Forces, Office of the Director of Military Training, "History of Training, Special Training Units," p. 26; Goldberg, *Training of Illiterates*, pp. 169, 174–175, 232.

72. One report noted that "Spanish is the leading foreign tongue with Italian and Chinese in second and third places. The three account for 70 percent of the language problem. However, it is only in the French (Acadians) and the American Indian groups that we find a high percentage of men who are illiterate in their native tongue as well as English." "Summary of Report of Special Training Units and Literacy Schools of the Army—Period January 16 to February 15, 1943 inclusive," 3 April 1943, NA–MRB, RG 160, Records of the Army Service Forces, entry 153, 350.5.

73. See, for example, "Teaching Devices for Special Training Units," War Department Pamphlet #20.2, 30 December 1943, National Archives Publications; "Publication of W. D. Pamphlet 'Instruction in Special Training Units,' " January 28, 1944, NA–MRB, RG 407, Records of the Adjutant General's Office, 353 (1-26-44 to 1-31-44) (Box 2662), which includes War Department Pamphlet #20.8, 10 April 1944; and RG 407, Records of the Adjutant General's Office, AG 300.7 EM 160 (28 August 1944) which includes the widely used reader, "Meet Private Pete" with specifications for distribution to Special Training Units. The quote is from "Teaching Devices for Special Training Units," p. 3.

74. Eighty percent of all men required sixty days or less; Goldberg, *Training of Illiterates*, p. 274. For honorable discharges of failures, see "Memorandum For the Adjutant General: Subject: Procedure Regarding Illiterates and Grade V Men," 29 September 1943, NA–MRB, RG 407, Records of the Adjutant General's Office, 353 (9-26-43 to 9-27-43) (Box 2665). In at least one instance, the army set up a school for failures of the Special Training Units, see "Report on Special Training School Set Up at Indiantown Gap, Pennsylvania," 2 March 1944, NA–MRB, RG 160, Records of the Army Service Forces, entry 153, 350.5. The memo includes a report on the school by Dunbar S. McLauren dated 9 February 1944. Probably these failures were those judged to have "unusual capacities and civilian skills which the Army needs." See War Department Circular #127, April 1944, U.S. Army Center of Military History, HRC 350.5.

75. For close adherence to regulations, see the letter condemning attempts at subterfuge by Brigadier General R. B. Lovett, Director of Military Training Division, 24 September 1943, NA–MRB, RG 407, Records of the Adjutant General's Office, 353 (5-7-43) (Box 2671). The specific problem was the attempt to artificially raise AGCT scores by repeated administration of the same test.

For appraisals by observers see the report by the American Council on Education by Alonzo G. Grace, *Educational Lessons from Wartime Training: The General*

Report of the Commission on Implications of Armed Services Educational Programs (Washington, D.C.: American Council on Education, 1948).

76. Quoted in Goldberg, *Training of Illiterates*, p. 269.

77. Quoted in Ginzberg and Bray, *The Uneducated*, pp. 129, 130.

78. Quoted in Goldberg, *Training of Illiterates*, pp. 270, 269–270.

79. Ginzberg and Bray, *The Uneducated*, pp. 88–89; the quote is on page 79.

80. Lee, *Employment of Negro Troops*, p. 267. The soldier is quoted in Ginzberg and Bray, *The Uneducated*, p. 130.

81. Ginzberg and Bray, *The Uneducated*, pp. 125, 131.

82. "Command of Negro Troops," War Department Pamphlet No. 20-6, February 29, 1944, in MacGregor and Nalty (eds.), *Blacks in the United States Armed Forces*, p. 311 (my emphasis).

83. "Leadership and the Negro Soldier," Army Service Forces Manual, M5, October 1944, pp. 29–30, 32 (available in U.S. Center of Military History, Washington, D.C.).

84. Ambrose Caliver, "Postwar Education of Negroes, Educational Implications of Army Data and Experiences of Negro Veterans and War Workers," report of a conference sponsored by the United States Office of Education, Office of Education, Historical File—Negro, p. 25 (italics in original).

85. "Instruction in Special Training Units," War Department Pamphlet #20-8, April 10, 1944, NA–MRB, RG 407, Records of the Adjutant General's Office, 353 (1-26-44 to 1-31-44) (Box 2662), p. 6.

86. It should be noted that despite official disclaimers, the army's very heavy dependence on the AGCT scores of soldiers often led to a rationalization of the inadequacy of black units as officers blamed the low "intelligence" of black recruits for a variety of problems. This was always exacerbated by the lack of sufficient black soldiers to fill officer ranks. Since almost all specialist ranks were selected from soldiers scoring fairly high on the AGCT, blacks were frequently excluded from training. Lee discusses the problems and the rationalizations in detail in *Employment of Negro Troops*, pp. 265–274. The dependence on mental testing continued to define the army's endeavors and reflected the infatuation of Americans with what they interpreted as the significance of the capacities tested by these exams. As Lee observes, "In many units the AGCT score became the refrain for a continuous jeremiad used as a fraternal greeting for inspectors. . . . It would be a little silly to assume that all German soldiers are of Class III or better in spite of their claims of superiority. The Russians might lose some of their confidence if they knew the dreadful truth about their mental gradations. . . . Many of our officers are giving the results of these tests more weight than was ever intended." (pp. 273–274)

87. The figures are from Army Service Forces Manual, M5, "Leadership and the Negro Soldier," p. 33; the quote is from "A Study of the Progress Made by 375 Trainees Separated from Special Training Units," 6 August 1943, NA–MRB, RG 160, Records of the Army Service Forces, entry 153, 350.5, p. 10.

88. Army Service Forces Manual, M5, "Leadership and the Negro Soldier," p. 33.

89. H. D. Bond, "To General Brehon B. Somervell, Army Service Forces," 30 June 1945, NA–MRB, RG 160, Records of the Army Service Forces, entry 153, 350.5, Central Decimal File 1942–46 (Box 277). (Italics in original.)

90. Morris J. MacGregor, Jr., *Integration of the Armed Forces,* Defense Studies Series, Center of Military History, United States Army (Washington, D.C.: U.S. Government Printing Office, 1981), 217, 218–219, 216.

Grafenwohr is also briefly discussed in "A Survey of Training in the Occupation Forces, 1 July 1946–30 June 1947," Occupation in Europe Series 1946–47, NA–MRB, RG 319, Records of the Army Staff, 8-3.1/cb3/c1, pp. 21–22.

91. See, Alonzo Grace, *Lessons from Wartime Training,* passim and the article submitted for approval to the Army Service Forces, by Paul A. Witty and A. G. D. Golda Van Buskirk, "The Soldier Learns to Read," 19 November 1943, NA–MRB, RG 160, Records of the Army Service Forces, 350.5 (1-1-43 to 11-30-43). The article was to appear in *National Parent-Teacher Magazine.*

92. See, for example, Caliver's article, "Illiteracy and Manpower Mobilization Needs in the Present Emergency," *School Life,* June 1951, in the Caliver Papers. See also the various efforts by Caliver and the Office of Education for what they called "fundamental education," among them, "Proposed Project on Fundamental Education," with tables and budget proposals and "Some Problems of Fundamental Education in the United States of America," prepared for the Inter-American Cultural Council at its first meeting in Mexico City, September 10–25, 1951, all in Caliver Papers.

Caliver was also involved in preparing the Office of Education's position on desegregation for the Supreme Court; see "Education of Negroes: Segregation Issue Before the Supreme Court," reprinted from *School Life,* February 1954, in Caliver Papers, and the draft implementing "a possible Supreme Court decision abolishing segregation in schools," part II, in Caliver Papers.

93. Ginzberg and Bray, *The Uneducated,* pp. 4, 12.

94. See the discussion of methods of proceeding with educational grants proposed by Johnson's Task Force on Education, as well as the earlier proposal by President Kennedy's Bureau of the Budget in Hugh Davis Graham, *The Uncertain Triumph: Federal Education Policy in the Kennedy and Johnson Years* (Chapel Hill: University of North Carolina Press, 1984), 39–79; also Tiedt, *The Federal Government in Education,* pp. 158–160.

5: THE FEMALE PARADOX

1. Margaret Mead, quoted in Anna L. Rose Hawkins, "Developing Community Leaders," in Leo C. Muller and Ouida G. Muller (eds.), *New Horizons for College Women* (Washington, D.C.: Public Affairs Press, 1960), 61.

2. "Women in A Changing World," *The Education of Women: Information and Research Notes,* issued by the Commission on the Education of Women of the American Council on Education, #3 (October 1958): 2.

3. Millicent Carey McIntosh (President of Barnard College, 1951–63), "The Education of Women in the Modern World," *Education for the Preservation of Democracy,* American Council on Education Studies, series I, no. 35 (April 1949): 80, quoted in George D. Stoddard, *On the Education of Women,* Kappa Delta Phi Lecture Series (New York: Macmillan, 1950), 53.

4. See, for example, National Manpower Council, *Womanpower: A Statement by the National Manpower Council* (New York: Columbia University Press, 1957), 168–188.

5. Mabel Newcomer, *A Century of Higher Education for American Women* (New York: Harper and Bros., 1959), 10–36; Barbara Miller Solomon, *In the Company of Educated Women: A History of Women and Higher Education in America* (New Haven: Yale University Press, 1985), 43–61. For a discussion of the views on women's higher education in the late nineteenth century, see Rosalind Rosenberg, *Beyond Separate Spheres: Intellectual Roots of Modern Feminism* (New Haven: Yale University Press, 1982), 1–27. For statistics on enrollment, see Newcomer, *Higher Education,* Table 2, p. 46.

6. For a discussion of the concern to maximize national resources by developing women's gifts, see National Manpower Council, *Womanpower,* pp. 3–39. Newcomer, *Higher Education,* has the best general introduction to the paradox.

7. Lynn White, Jr., *Educating Our Daughters: A Challenge to the Colleges* (New York: Harper and Bros., 1950), especially 78.

8. Helen Lefkowitz Horowitz, *Alma Mater: Design and Experience in Women's Colleges From Their Nineteenth-Century Beginnings to the 1930s* (New York: Knopf, 1984), especially 9–142. Newcomer, *Higher Education,* pp. 87, 91. It is important to remember that even in the early twentieth century, some women believed that women's education should prepare women better for the realities of their lives, especially for homemaking tasks, as well as for occupational roles. See Marion Talbot, *The Education of Women* (Chicago: University of Chicago Press, 1910). This view was from the start associated with "progressive" educational values and sought to bring to women's household duties the influence of science. See Talbot's approval of James H. Tufts's (the progressive philosopher) views in a footnote on p. 242. See also Solomon, *Company of Educated Women,* pp. 78–93.

9. See Newcomer, *Higher Education,* pp. 89–90; Horowitz, *Alma Mater,* pp. 319–350.

10. Constance Warren, *A New Design for Women's Education* (New York: Frederick A. Stokes, 1940), 2, 6–7. For a discussion of the ferment of new ideas of female roles, as well as the opposition to these views, see Solomon, *Company of Educated Women,* pp. 149–150.

11. Vassar College Catalogue, vol. 40, 1950–51, p. 48. Catholic colleges for women were also adopting the new "progressive" views but assimilated them to older concerns. See Mary Evodine McGrath, "Role of Catholic Colleges in Preparing for Marriage and Family Life," (Ph.D. dissertation, Catholic University, 1952), 17. McGrath found that courses "relating to marriage and family life preparation made their greatest gains in Catholic Colleges in the Midwest" in the decade 1930–40. See also my discussion below in Chapter 6, pp. 212–217.

12. Warren, *New Design,* p. 6. Ernest Havemann and Patricia Salter West, *They Went to College: The College Graduate in America Today,* based on a survey of U.S. college graduates made by *Time* magazine and analyzed by the Columbia University Bureau of Applied Social Research (New York: Harcourt, Brace & Co., 1952), 61.

13. Robert G. Foster and Pauline Park Wilson, *Women After College: A Study of the Effectiveness of Their Education* (New York: The Merrill-Palmer School, Columbia University, 1942), 93, 113, 220.

14. Foster and Wilson, *Women After College,* pp. 252, 260, 186.

15. National Manpower Council, *Education and Manpower,* edited by Henry David (New York: Columbia University Press, 1960), Table 8, p. 267.

16. "Higher Education: The *Fortune* Survey," conducted by Elmo Roper, *Fortune* (September 1949), 6, 7, 12, 15.

17. Havemann and West, *They Went to College*, pp. 64–65.
In a very large poll conducted by the American Association of University Women, most of the 30,000 respondents complained about the lack of practical education in their college experience. Most women asked for more actual practice. In the words of Margaret Mead, who presented an initial analysis of the poll, "Most of the expressed wishes are for . . . practice in homemaking, in child care, in public speaking, in teaching live children rather than merely learning the history of education out of a book, more chance to speak the foreign languages which are studied. . . . There is a strong expression, in these replies, that all women should be prepared to marry and bring up children, to participate in the community, to use leisure well." See Margaret Mead, "The Higher Education Survey, A First Report of Findings from the Questionnaire," *Journal of the American Association of University Women,* 43(Fall 1949): 8–12, the quote is on p. 10.

18. See Lynn White, Jr., *Medieval Technology and Social Change* (London: Oxford University Press; 1962).

19. See also George D. Stoddard, *On the Education of Women*. Stoddard, president of the University of Illinois, noted, "At this critical time, women should start to restore the balance. The things within their sphere of interest and competence are the starting point of a new culture. The general run of men, having done the work they like best to do, seem incapable of progress in new directions. The masculine component multiplies and runs to the ground everything connected with technology. Uncontrolled technology goes the way of the Mussolinis, Hitlers, and Stalins. . . . Instead of fretting about the disinterest of women in the mechanical, we should be thankful for this saving part of our culture." (p. 51)

20. David (ed.), *Education and Manpower*, Table 8, p. 267.

21. National Manpower Council, *Womanpower*, pp. 46, 65–70. For the pattern, see, for example, Marguerite W. Zapoleon, "The Myth of the Marriage-Career Conflict," in Muller and Muller (eds.), *New Horizons*, pp. 79–87.

22. National Manpower Council, *Womanpower*, pp. 74, 75; the quotes are on pp. 125, 132.

23. Both men are quoted in Commission on the Education of Women, *The Education of Women,* #1(March 1958): 5, 7; Althea K. Hottel, *How Fare American Women? A Report of the Commission on the Education of Women of the American Council on Education* (Washington, D.C.: American Council on Education, 1955), 20. Related publications sponsored by the Commission on the Education of Women of the American Council on Education and published by that organization include: *The Span of a Woman's Life and Learning: A Statement of the Commission on the Education of Women of the American Council on Education* (Washington, D.C., 1960); Lawrence E. Dennis (ed.), *Education and a Woman's Life* (Washington, D.C., 1963); Opal D. David (ed.), *The Education of Women: Signs for the Future* (Washington, D.C., 1959); *The Education of Women, Information and Research Notes of the Commission on the Education of Women,* #1-13 (March 1958–September 1961).

24. O. Meredith Wilson, "A Woman is a Woman is a Woman," in Dennis (ed.), *Education and a Woman's Life*, p. 7. For the various programs see, pp. 93–124 in the same volume.

25. Hottel, *How Fare American Women?*, pp. 30–36; "The Multiple Roles of Women," *The Education of Women,* #1(March 1958): 6.

26. "Young Wives," *Newsweek* (March 7, 1960), 59.

27. Hottel, *How Fare American Women?*, p. 36.

28. Agnes Scott College Catalogue, 1957, p. 32; Bulletin of Connecticut College, New London, Connecticut, March 1951, pp. 47, 111.

29. For women's choices of courses, see National Manpower Council, *Womanpower*, pp. 174–182. The quote is from Eunice C. Roberts, "Comment," in David (ed.), *The Education of Women*, p. 118.

30. Vassar College Catalogue, 1955–56, p. 46; Vassar College Catalogue, 1958–59, p. 145.

31. The conclusions of the Commission on the Education of Women can be found in Esther Lloyd-Jones, "Women Today and Their Education," *Teachers College Record*, 57(April 1956): 434; Dorothy Woodward is quoted in "Goals of Women in Higher Education," *Journal of the American Association of University Women*, 48(March 1955): 161, 162, 163.

32. Karem J. Monsour, "Education and a Woman's Life," in Dennis (ed.), *Education and A Woman's Life*, p. 19; Felice Schwartz, "Discussion," in the same volume, p. 23; C. Easton Rothwell, "The Milieu of the Educated Women," in the same volume, p. 37.

33. Esther Peterson, "Needs and Opportunities in Our Society for the Educated Woman," in Dennis (ed.), *Education and A Woman's Life*, p. 55; Mary H. Donlon, "The Scope of the Problem," in David (ed.), *The Education of Women*, p. 13. See also Mary H. Donlon, "Women's Education Today," *Educational Record*, 39(July 1958): 246–252. For the arguments supporting women's education in the early republic, see Linda K. Kerber, *Women of the Republic: Intellect and Ideology in Revolutionary America* (Chapel Hill: University of North Carolina Press, 1980), 189–231, and Linda K. Kerber, "Daughters of Columbia: Educating Women for the Republic, 1787–1805," in Stanley Elkins and Eric McKitrick (eds.), *The Hofstadter Aegis: A Memorial* (New York: Knopf, 1974), 36–59.

34. The President's Commission on the Status of Women, *Report of the Committee on Education*, October 1963 (Washington, D.C.: U.S. Government Printing Office, 1963), 2, 41. See also the discussion of the Kennedy Commission in Cynthia E. Harrison, "A 'New Frontier' for Women: The Public Policy of the Kennedy Administration," in Jean E. Friedman and William G. Shade (eds.), *Our American Sisters: Women in American Life and Thought* (Lexington, Mass.: D. C. Heath, 1982), 541–562.

35. Alan Simpson is quoted in *The Vassarian*, 1965, no page.

36. Paul Heist, "The Uncommitted Majority," Spring 1961, p. 1 (paper on file at the Center for Studies in Higher Education, University of California, Berkeley; formerly, Center for the Study of Higher Education).

37. M. Eunice Hilton, "What Changes Should There Be in Our Educational Programs to Meet the Special Needs and Interests of Women More Adequately?" in G. Kerry Smith (ed.), *Current Issues in Higher Education, 1955*, Proceedings of the Tenth Annual National Conference on Higher Education, Chicago, Illinois, February 28–March 2, 1955 (Washington, D.C.: Association for Higher Education, 1955), 118, 121–122.

38. Kate Hevner Mueller, "What's It All About?" in Muller and Muller (eds.), *New Horizons for College Girls*, p. 9.

39. Mueller, "What's It All About?" pp. 9, 14.

40. Bancroft Beatley, "Another Look At Women's Education," in *What Education for Women?*, Bevier Lecture Series 1950 (University of Illinois, 1950), 25,

26; Mervin B. Freedman, "The Passage Through College," *Journal of Social Issues,* 12(1956): 15; Carl Binger, "The Pressure on College Girls Today," *Atlantic,* 207(February 1961): 40.

Simmons was established in 1902 with the specific goal of providing vocational education for women. According to the 1922–23 Simmons College Bulletin, instruction at Simmons was to be in "art, science, and industry best calculated to enable the scholars to acquire an independent livelihood." Simmons gave an education to women in areas of work suitable to women, as secretaries, librarians, dieticians, business accountants, laboratory assistants, teachers of home economics, social workers, and public health nurses.

41. Jane Berry, "Life Plans of College Women," *Journal of the National Association of Women Deans and Counselors,* 18(January 1955): 77.

42. Paul Heist, "The Motivation of College Women Today," pp. 1, 3, 4 (paper on file at the Center for Studies in Higher Education, University of California, Berkeley); Heist, "A Commentary on the Motivation and Education of College Women," *Journal of the National Association of Women Deans and Counselors,* 25(January 1962): 56.

43. Heist, "The Motivation of College Students," p. 7; Heist, "A Commentary on the Motivation of College Women," p. 59. For the interest of the Berkeley center to locate the motivational forces that would impel talented women to continue their education, see Paul A. Heist, "Research on Talented Women: Problems of Appropriate and Adequate Sources of Data," 1964 (paper on file at the Center for Studies in Higher Education, Berkeley).

44. John H. Bushnell, "Student Culture at Vassar," in Nevitt Sanford (ed.), *The American College: A Psychological and Social Interpretation* (New York: John Wiley and Sons, 1962), 509, 510.

45. The Educational Testing Service (ETS) conducted a study of high-school seniors in 1954–55 which found that 35 percent of girls pursued a commercial program and another 6 percent were in a vocational curriculum, a far larger proportion than boys. The study also suggested that women stayed away from science and mathematics courses. The ETS study found that men were far more likely to have college-going plans than high-school women of equal ability. Women seemed to prepare themselves much more frequently for direct entry into the workplace than men, not, however, because their goals were to work but because "The anticipation of becoming wives, mothers and homemakers appears to be strong enough to turn them away from investing in more education." See, National Manpower Council, *Womanpower,* pp. 174–182. The quote is on p. 180.

46. Elizabeth Douvan and Carol Kaye, "Motivational Factors in College Entrance," in Sanford (ed.), *The American College,* pp. 202, 204, 205, 206. See also Elizabeth M. Douvan, "Adolescent Girls: Their Attitudes Toward Education," in David (ed.), *The Education of Women,* pp. 23–29.

47. Margaret Clapp, "Comment," in David (ed.), *The Education of Women,* p. 121.

48. Mirra Komarovsky, *Women in the Modern World: Their Education and Dilemmas* (New York: Little Brown, 1953), 214, 216.

49. Komarovsky, *Women in the Modern World,* pp. 66–67, 288.

50. Komarovsky, *Women in the Modern World,* p. 105.

51. Heist, "A Commentary on the Motivation and Education of College Women," p. 56; Heist, "The Uncommitted Majority," p. 2; Berry, "Life Plan of College Women," p. 79.

52. See, for example, Erik H. Erikson, "Eight Ages of Man," in his *Childhood and Society* (New York: Norton, 1950; revised edition 1963), 247–265.

In 1966, sociologist Walter Wallace presented an interesting variation on this difference. Based on the Parsonian differentiation between men's orientation to manipulating things and women's attention to persons, Wallace argued that for women, choices made in college (of future mates) were more singular and status-determining than those made by men. "A woman's selection of a husband can thus be a more significant, because more singular, rite of passage to adult status than is a man's selection of occupation." In this sense, college for women was a turning point rather than a preparation. This view also supports the more jagged sense of women's life course which is punctuated by significant events rather than integrated through linear progressions. See Walter L. Wallace, *Student Culture: Social Structure and Continuity in a Liberal Arts College*, National Opinion Research Center Monographs in Social Research (Chicago: Aldine Publishing Co., 1966), 132.

53. Freedman, "The Passage Through College," pp. 26, 27; Komarovsky, *Women in the Modern World*, p. 56.

54. Komarovsky, *Women in the Modern World*, pp. 78, 92, 82; Paul Heist, "Motivation of College Women Today: A Closer Look," 1962 (paper on file at the Center for Studies in Higher Education, Berkeley), p. 6; David Riesman, "Two Generations," in Robert Jay Lifton (ed.), *The Woman in America* (Boston: Beacon Press, 1964), 91; Bushnell, "Student Culture at Vassar," p. 508.

In an essay comparing women college students in the fifties and sixties to those of the previous generation, David Riesman observed, "In a woman's college today, the students are conscious simultaneously of the future and of the boys they hope to marry, even if the boys are not physically present, and of *what sorts of intellectuality these boys may welcome or resent.* As a teacher in coeducational settings, I have watched some of the ways in which women students presently cope with the dilemma of being at once women and students. For some, to have men around is a liberation and provides an impetus to a more dialectical and less docile approach than might be the case if they attended college only with their own sex. But perhaps the greater number, *trained to be socially conscious since childhood, are intensely aware of how their class performance is viewed both by boys and by their own sex;* facing a double audience, they are inhibited from expressing themselves as rambunctiously or spiritedly as they might do in a single-sex setting." "Two Generations," p. 90 (my emphases).

55. Harold Webster, Mervin Freedman, and Paul Heist, "Personality Changes in College Students," in Sanford (ed.), *The American College*, p. 839.

56. Vera Schletzer, "The Minnesota Plan," in Dennis (ed.), *Education and A Woman's Life*, pp. 121, 122; Mervin B. Freedman, "Studies of College Alumni," in Sanford (ed.), *The American College*, p. 873; Riesman, "Two Generations," pp. 91–92.

57. Wallace, *Student Culture*, p. 133; Florence Kluckholn, "Women in America," in *What Education For Women?*, Bevier Lecture Series 1950 (University of Illinois, 1950), 12.

58. Of women college graduates in June 1955, 60 percent received a teaching certificate and another 13 percent had taken some education courses. A very large majority of those who were employed as teachers taught in elementary schools. Jean A. Wells, "Employment of June 1955 Women College Graduates," *Monthly Labor Review*, 79, no. 9(September 1956): 1060.

59. The Radcliffe Committee on Graduate Education for Women, *Graduate Education for Women: The Radcliffe Ph.D., A Report by a Faculty-Trustee Committee* (Cambridge: Harvard University Press, 1956), 58, 59.

60. Newcomer, *Higher Education,* p. 3.

61. Quoted in Komarovsky, *Women in the Modern World,* p. 14.

62. Wallace, *Student Culture,* p. 135.

In 1956, for example, 80 percent of all June 1955 women college graduates were employed. Of these, 61 percent were employed as teachers, while 15 percent were clerical workers. Thus, fully three-quarters were in one of these two occupations. It is also worth noting that 35 percent of the graduates had been education majors in school, while 20 percent had majored in humanities and art; and another 11 percent in the social sciences. As noted by Jean A. Wells of the Women's Bureau who issued the report in which these conclusions were presented, "Relatively few of the women graduates other than teachers trained in fields of manpower shortages in 1955." Wells, "Employment of June 1955 Women College Graduates," pp. 1057, 1058.

6: IMITATION AND AUTONOMY

1. Supreme Court decision on Oregon law, 268 U.S. 510, p. 535 (1925).

2. John J. Fallon (Monsignor), "Multi-Grade Classrooms," *Bulletin,* National Catholic Education Association, 33(November 1936): 398. This journal will hereafter be cited as *Bulletin,* NCEA.

3. The best general discussions of these issues are Carl F. Kaestle, *Pillars of the Republic: Common Schools and American Society, 1780–1860* (New York: Hill and Wang, 1983), 62–135; and Lawrence A. Cremin, *American Education: The National Experience, 1783–1876* (New York: Harper and Row, 1980), 103–147.

4. For the early battles, see Cremin, *American Education,* pp. 166–170; Diane Ravitch, *The Great School Wars: New York City, 1805–1973* (New York: Basic Books, 1974), 20–76. An excellent discussion of Catholic-school development focusing on Chicago, which had the largest parochial school system, is James W. Sanders, *The Education of an Urban Minority: Catholics in Chicago, 1833–1965* (New York: Oxford University Press, 1977), especially 18–39.

My figures for Catholic-school enrollment are derived from Andrew M. Greeley and Peter H. Rossi, *The Education of Catholic Americans* (Chicago: Aldine Publishing Co., 1966), 1; and The Notre Dame Study of Catholic Elementary and Secondary Schools in the United States, *Catholic Schools in Action,* edited by Reginald A. Neuwien (Notre Dame, Ind.: University of Notre Dame Press; 1966), 33. Catholic schools were most important in those cities that had drawn the strongest current of European immigration—New York, Chicago, Detroit, Pittsburgh, Buffalo, Boston, Milwaukee, Cincinnati—where they enrolled from one quarter to nearly one-half of all school children. See Neil G. McCluskey (S.J.), *Catholic Viewpoint on Education* (Garden City, N.J.: Hanover House, 1959), 46.

5. See, for example, McCluskey, *Catholic Viewpoint,* pp. 59–79; James Michael Lee, "Catholic Education in the United States," in Lee (ed.), *Catholic Education in the Western World* (Notre Dame, Ind.: University of Notre Dame Press, 1967), 257–311. The quote is from William McGucken (S.J.), "The Philosophy of Catholic Education," in National Society for the Study of Education, *Philosophies of*

Education, the 41st Yearbook, part I (Chicago: University of Chicago Press, 1942), 268.

6. Notre Dame Study, *Catholic Schools in Action,* p. 17.

7. McCluskey, *Catholic Viewpoint,* p. 82; Edward F. Spiers (Reverend), *The Central Catholic High School: A Survey of their History and Status in the United States* (Washington, D.C.: Catholic University of America Press, 1951), 18.

8. Spiers, *Central Catholic High School,* pp. 20, 21; Mary Janet (Sister), *Catholic Secondary Education: A National Survey* (Washington, D.C.: National Catholic Welfare Conference, 1949), 7.

9. Mary Janet, *Catholic Secondary Education,* pp. 20, 13. The quote is from Mary Janet Miller (Sister), "General Education in the American Catholic Secondary School" (Ph.D. dissertation, Catholic University, 1952), 11.

10. Sanders, *Education of an Urban Minority,* p. 91. See also George Johnson (Reverend), "The Catholic Church and Secondary Education," in Roy J. Deferrari (ed.), *Vital Problems of Catholic Education* (Washington, D.C.: Catholic University of America Press, 1939), 76.

11. Mary Janet, *Secondary Education,* pp. 13, 20; Notre Dame Study, *Catholic Schools in Action,* pp. 33, 46; The Reverend John P. Breheny is quoted in Spiers, *Central Catholic High School,* p. 142.

12. Mary Janet, *Secondary Education,* p. 19; Notre Dame Study, *Catholic Schools in Action,* pp. 46, 47.

13. Notre Dame Study, *Catholic Schools in Action,* p. 33.

14. Greeley and Rossi, *Education of Catholic Americans,* pp. 27, 36–37; the quote is on p. 34. Canon law 1374 of the revised code, 1918, makes the requirement; see McCluskey, *Catholic Viewpoint,* p. 100.

Tuition costs in Catholic schools varied widely, but until recently they were quite modest. In 1949, Sister Mary Janet found that parental contributions toward secondary-school tuition (public and private) ranged from nothing to $400, with the prevailing range from $20 to $40 yearly. "Students are not rejected, however, because of inability to meet the request," *Secondary Education,* p. 21.

At one school in the 1950s, probably less expensive than most, the "cost per family (not per child) is ten dollars a year. Added to this is the book rental of five dollars," and various lesser and very small fees. The total tuition for the first child was $17.25 per year, with $7.25 for each additional child; Joseph Henry Fichter (S.J.), *Parochial School: A Sociological Study* (Notre Dame, Ind.: University of Notre Dame Press, 1958), 351.

A large survey of secondary schools in the 1950s found that in "46.3 percent of the large diocesan high schools, there is no tuition charge to the pupils." For the rest, tuition ranged from $10 to $175 per year, with the median at $50. In small, diocesan high schools, 37.9 percent had no tuition, while the median for the rest was $40 per year; Mary Pauline Degan (Sister), "Student Admittance and Placement in Regional Catholic High Schools" (Ph.D. dissertation, Catholic University, 1950), 17.

In the mid-1960s, Greeley and Rossi reported that 62 percent of respondents spent less than $100 a year in tuition and only 20 percent spent more than $200; *Education of Catholic Americans,* p. 206. By the mid-1970s, however, the cost of tuition had become far more of a grievance to parents and more frequently cited as a reason for nonattendance at Catholic schools. By 1974, the average cost was $373 per year. See Andrew M. Greeley, William C. McCready, and Kathleen

McCourt, *Catholic Schools in a Declining Church* (Kansas City, Kans.: Sheed and Ward, 1976), 37, 247.

15. During the 1970s, the sharp decline in Catholic-school enrollments appears to have been the result of two factors: the migration of Catholics to suburbs where no church schools existed and the decision by the Catholic hierarchy to disinvest their resources from parochial schools. Andrew Greeley argues that lay Catholics still strongly supported the schools and would have liked to have their children attend. See Andrew M. Greeley, *The American Catholic: A Social Portrait* (New York: Basic Books, 1977), 167–170, 173, 185. In fact, two other issues may also have influenced enrollments: the disarray in the church following Vatican II and the declining importance of ethnicity.

16. See McCluskey, *Catholic Viewpoint,* p. 83; Degan, "Student Admittance and Placement," p. 43. The surveys cited are: William Mang (Brother), "The Curriculum of the Catholic High School for Boys" (Ph.D. dissertation, University of Chicago, 1940), 51; Spiers, *Central Catholic High School,* p. 130, 127; James S. Coleman, Thomas Hoffer, and Sally Kilgore, *High School Achievement: Public, Catholic, and Private Schools Compared* (New York: Basic Books, 1982), 90–94. See also Thomas Joseph Frain (Reverend), "Administrative and Instructional Provisions for Rapid and Slow Learners in Catholic Secondary Schools" (Ph.D. dissertation, Catholic University, 1956), passim.

For the lack of vocational courses, see, among others, Thomas F. Jordan, "The Problem of Vocational Education and the Catholic Secondary School" (Ph.D. dissertation, Catholic University, 1942); Degan, "Student Admittance and Placement," pp. 77, 104, 148, and passim; Mary Janet Miller, "General Education," p. 44; Spiers, *Central Catholic High School,* pp. 169–170; Mary Janet, *Secondary Education,* p. 89.

17. George Johnson, "The Curriculum of the Catholic Elementary School: A Discussion of Its Psychological and Social Foundations" (Ph.D. dissertation, Catholic University, 1919), viii; John F. O'Dwyer (Reverend), "An Evaluation of the Catholic High-School Curriculum in the Light of Practical Life Situations," *Bulletin,* NCEA 33(November 1936): 283.

18. See, for example, John T. O'Dowd (Reverend), *Standardization and Its Influence on Catholic Secondary Education in the United States,* Catholic University of America, Educational Research Monographs, vol. 9 #1(January 2, 1936).

19. John J. Fallon (Monsignor), "The Instructional Problem of Individual Differences of Students in Our Schools," *Bulletin,* NCEA, 33(August 1937): 290, 297.

20. On the college preparatory curriculum in private schools, see McCluskey, *Catholic Viewpoint,* p. 83. For selectivity, see Mary Janet, *Secondary Education,* p. 54; quote is on p. 53; Degan, "Student Admittance and Placement," pp. 9–10; quote is on p. 102; Frain, "Rapid and Slow Learners" p. 16.

21. Notre Dame Study, *Catholic Schools in Action,* pp. 40–41, 38, 78.

22. Degan, "Student Admittance and Placement," pp. 14, 13, 60.

23. Mang, "Catholic High School for Boys," pp. 89–90.

The dominance of the Irish at higher levels of Catholic education continued at the college level. At the time of the investigations of the United States Immigration Commission in 1908, 59.8 percent of all students in Catholic colleges surveyed were Irish and another 19.1 percent were German in background. In 1928, at Duquesne University in Pittsburgh, a large center for new immigrant Catholics, the

Irish were still dominant with Germans in second place. See Bernard J. Weiss, "Duquesne University: A Case Study of the Catholic University and the Urban Ethnic, 1878–1928" in Weiss (ed.), *American Education and the European Immigrant 1840–1940* (Urbana: University of Illinois Press, 1982), 179–180.

24. Weiss, "Duquesne University," p. 180; Sanders, *Education of an Urban Minority,* p. 155.

25. Catholic secondary education did indeed permit the Irish to realize their social and economic ambitions, since the Irish have in the twentieth century equalled and even bested other groups in realizing economic and educational mobility and in achieving financial success. See Andrew M. Greeley, *Ethnicity, Denomination, and Inequality* (Beverly Hills: Sage Publications, 1976), 17, 45–56, 70, and Greeley, *American Catholic,* pp. 50–67.

According to data assembled by the National Opinion Research Center in 1963–64, there continued to be considerable differences in the attendance at Catholic high schools by Catholic youth of different ethnic backgrounds, although far more now came from newer immigrant groups. The following table is adapted from Table 25, p. 91 in Harold J. Abramson, *Ethnic Diversity in Catholic America* (New York: John Wiley & Sons, 1973). See table.

Percent Attending Catholic High School by Age and Ethnicity

| Age in 1963–64 | Ethnic Background | | | | | | | |
	Total	Irish	French-Canadian	German	Polish	East European	Italian	Spanish
20s to 30s (High school after WWII)	35	57	53	42	33	25	14	6
40s to 50s (High school between WWI & WWII)	20	34	21	30	21	16	6	0

26. The sense of failure may have been especially keen for the Irish who set great store by Catholic high-school attendance and probably not great for the Italians who chose not to attend even lower-level Catholic schools. See Abramson, *Ethnic Diversity,* pp. 89–90. According to Abramson, the possibility of graduating and/or going to college was very significantly higher for the Irish if they attended Catholic high schools rather than public high schools. Although other groups, like the Germans, French-Canadians, and Poles also showed some measurable difference, the difference was much greater for the Irish. For them, going to Catholic high schools or not literally meant the difference between succeeding and not succeeding educationally. Abramson observes, "Clearly . . . the merger of educational success and parochial school involvement is a distinctly Irish phenomenon." (p. 89)

27. Greeley and Rossi, *Education of Catholic Americans,* p. 192, quote is on p. 48; also Greeley, McCready, and McCourt, *Catholic Schools in a Declining Church,* pp. 196–198. See, however, the attack on Catholic education as anti-intellectual

and inhibiting to genuine achievement and creativity in James W. Trent, with Jenette Golds, *Catholics in College: Religious Commitment and the Intellectual Life* (Chicago: University of Chicago Press, 1967).

28. George Johnson, "Catholic Church and Secondary Education," p. 82. For success of Catholic high-school graduates, see Greeley, *Ethnicity*, p. 40. In 1968, Greeley found that of all Catholic college graduates, those with both a Catholic primary- and secondary-school education had the highest proportion (45 percent), earning more than $11,000 a year.

29. Mary Janet, *Secondary Education*, p. 9; Johnson, "Curriculum of the Catholic Elementary School," pp. 107, 115.

30. Policies Committee, Secondary School Department, National Catholic Education Association, "The Objectives of Catholic Secondary Education in the United States," *Catholic High School Quarterly Bulletin*, 2(April 1944): 21.

31. M. Juliana (Mother), "Provision for the Poor in the Catholic High School," *Bulletin*, NCEA, 28(November 1931): 344–345; Mary Janet, *Secondary Education*, pp. 131, 133.

32. Louis J. Faerber (Brother), "Provisions for Low-Ability Pupils in Catholic High Schools" (Ph.D. dissertation, Catholic University, 1948), xxiii, xxiv (italics in original); the workshop is quoted in Miller, *General Education*, p. 60; Michael J. McKeough, "The Curriculum—Its Nature and Philosophy," in McKeough (ed.), *The Curriculum of the Catholic Secondary School*, proceedings of the workshop on the curriculum of the Catholic secondary school conducted at the Catholic University of America from June 11 to June 22, 1948 (Washington, D.C.: Catholic University, 1949), 3.

33. Faerber, "Low-Ability Pupils," p. 3; Degan, "Student Admittance and Placement," p. xii.

34. Mary Janet, *Secondary Education*, pp. 54, 56; Faerber, "Low Ability Pupils," pp. 11–12; Degan, "Student Admittance and Placement," pp. 36, 37.

35. See, for example, Spiers, *Central Catholic High School*, p. 126; Mary Janet, *Secondary Education*, pp. 60–64.

36. Paul E. Campbell (Reverend), "What of the High School," *Bulletin* NCEA, 32(November 1935): 209; Kilian J. Heinrich (Reverend), "The Place of Vocational Guidance in the Whole Guidance Program: Ways and Means to Promote It," *Bulletin* NCEA, 29(November 1932): 226–227; Faerber, "Low-Ability Pupils," p. 79.

37. George Johnson (Reverend), "Equalizing Educational Opportunity for Whom?" *Bulletin* NCEA, 35(August 1938): 69, 70.

38. Fichter, *Parochial School*, p. 387. See also Edward Riley, "Extracurricular Activities Programs in the Catholic High Schools" (Ph.D. dissertation, Catholic University, 1954).

39. Mary Margarita Geartts (Sister), "A Critical Analysis of the Objectives of Extracurricular Activities Programs in Catholic High Schools," abstract of a dissertation (Catholic University, 1960), pp. 2, 14, 21.

40. Geartts, "Extracurricular Activities Programs," p. 1; the second quote is from Fichter, *Parochial School*, p. 387.

41. Sanders, *Education of an Urban Minority*, p. 145; the Third Plenary Council is quoted in McCluskey, *Catholic Viewpoint*, note 7, pp. 103–104; Joseph E. Hamill, "The Junior High School: Its Feasibility in the Catholic Educational System" (Ph.D. dissertation, Catholic University, 1922), p. 96; McCluskey, *Catholic Viewpoint*, p. 109.

42. Notre Dame Study, *Catholic Schools in Action,* pp. 16–17.

43. Laurence J. O'Connell, *Are Catholic Schools Progressive?* (St. Louis: B. Herder Book Co., 1946), vii–viii, 134. For an even more vociferous attack on Dewey, see McCluskey, *Catholic Viewpoint,* pp. 47–51. Most studies merely stopped momentarily to disassociate themselves from Dewey's philosophy. See, for example, Faerber, "Low-Ability Students," p. 75. By the late 1960s, Catholic education became more comfortable with the ideas of progressive education, and even Dewey became marginally acceptable. See Notre Dame Study, *Catholic Schools in Action,* pp. 146–153.

44. Don Sharkey, *These Young Lives: A Review of Catholic Education in the United States,* sponsored by the Department of Superintendents of the National Catholic Education Association (Chicago: W. H. Sadlier, 1950), 41.

45. McCluskey, *Catholic Viewpoint,* p. 96; quote is on p. 97. Mary Janet, *Secondary Education,* p. 23; Notre Dame Study, *Catholic Schools in Action,* p. 62.

46. Cited in Trent, *Catholics in College,* p. 53; quote on p. 53.

47. M. Madeleva (Sister), *Conversations with Cassandra: Who Believes in Education?* (New York: Macmillan, 1961), 25, 22; Grace Dammann (Mother), "The American Catholic College for Women," in Deferrari (ed.), *Essays in Catholic Education,* p. 187.

48. Robert Hassenger, "Portrait of a Catholic Women's College," in Hassenger (ed.), *The Shape of Catholic Higher Education* (Chicago: University of Chicago Press, 1967), 83; Mary Rosalia Flaherty (Sister), "Patterns of Administration in Catholic Colleges for Women in the United States" (Ph.D. dissertation, Catholic University, 1960), 4; Dammann, "Catholic College for Women," p. 181; quote in Dammann, p. 47.

49. Dammann, "Catholic College for Women," p. 189; the brochures and bulletins are quoted in Helen B. McMurray, *Personnel Services in Catholic Four Year Colleges for Women,* Catholic University of America, Educational Research Monographs, 24 #4(June 1, 1957): 41, see also 105, 108.

50. Mary Evodine McGrath (Sister), "The Role of the Catholic College in Preparing for Marriage and Family Life" (Ph.D. dissertation, Catholic University, 1952), 17; M. Redempta Prose, "The Liberal Arts Ideal in Catholic Colleges for Women in the United States" (Ph.D. dissertation, Catholic University, 1943), 159.

51. Mary Audrey Bourgeois (Sister), "A Study of the Preparation for the Role of Parent-As-Educator in Selected Catholic Women's Colleges" (Ph.D. dissertation, Catholic University, 1961), 24; McGrath, "Preparing for Marriage and Family Life," pp. 17, 22–24, 42, 111–118. For religion courses and family preparation, see McGrath, p. 17 and Bourgeois, p. 34.

52. Bourgeois, "Parent-As-Educator," pp. 26, 31; for conferences on family life, see, pp. 91–93.

53. Dammann, "Catholic College for Women," p. 193. The objectives were present throughout the twentieth century. In 1932, Sister Eveline observed, "Every college-bred woman has at her command rich intellectual resources in science, literature, and art. These she can reorganize, and thus adjust herself in life to any social or economic status in which she may be placed. This status will be the measuring instrument of society which cannot hope to reach a higher level than its mothers." "Objectives of Catholic Colleges for Women," *Bulletin* NCEA, 29(November 1932): 139.

54. McGrath, "Preparing for Marriage and Family," pp. 124, 127.

55. McGrath, "Preparing for Marriage and Family," p. 131.

56. Mary Janet, *Secondary Education*, p. 32. For the number of black Catholics, Cecilia Marie Leonard (Sister), "The Negro Question," *Catholic High School Quarterly Bulletin*, 4(January 1947): 4–5.

57. The papal encyclical is quoted in Leonard, "Negro Question," p. 5; for non-Catholics at Catholic schools, Louis Cavell (Brother), "What Shall We Do About the Religious Instruction for Non-Catholics Attending Our Schools?" *Catholic High School Quarterly Bulletin*, 5(October 1947): 21. The quotes on racial exclusion are from Degan, "Student Admittance and Placement," pp. 4–5, 6; for admission of non-Catholics, p. 6. For enrollment figures in 1971, see Francis Flanigan, "Integration of Catholic Schools: What is Possible? What is Working?" in Mary von Euler and Gail Lambers (eds.), *The Catholic Community and the Integration of Public and Catholic Schools*, papers from a conference sponsored by the National Conference for Interracial Justice in cooperation with the National Catholic Education Association and the Department of Education of the United States Catholic Conference, May 1978, (Washington, D.C.: U.S. Department of Health, Education and Welfare, National Institute of Education, U.S. Government Printing Office, 1979), p. 20.

58. Father Moore is quoted in Flanigan, "Integration in Catholic Schools," p. 19; Mary Innocent Montay (Sister), "The History of Catholic Secondary Education in the Archdiocese of Chicago" (Ph.D. dissertation, Catholic University, 1953), 263, 264, 280. For the industrial school, see p. 20. See also, Sanders, *Education of an Urban Minority*, p. 207.

59. Mary de Sales Harris (Sister), "A History of Catholic Elementary Schools for the Negro in North Carolina" (Masters essay, Catholic University, 1965), ii, 41, 45.

60. Loretta M. Butler, "A History of Catholic Elementary Education for Negroes in the Diocese of Lafayette, Louisiana" (Ph.D. dissertation, Catholic University, 1963), 52, 73, 89–90; quote on p. 52.

61. Butler, "Negroes in the Diocese of Lafayette," pp. 133, 168.

62. Sanders, *Education of an Urban Minority*, pp. 219, 220, 221; quote on p. 221.

63. See Sanders, *Education of an Urban Minority*, pp. 40–71; Montay, "Education in the Archdiocese of Chicago," p. 328; Marie Patrice Gallagher, "The History of Catholic Elementary Education in the Diocese of Buffalo, 1847–1944" (Ph.D. dissertation, Catholic University, 1945), 59, 64, 65, 70, 85, 92, 153–154; Mary Xaveria Sullivan (Sister), "The History of Catholic Secondary Education in the Archdiocese of Boston" (Ph.D. dissertation, Catholic University, 1946), 39.

64. Peter H. Rossi and Alice S. Rossi, "Some Effects of Parochial School Education in America," *Daedalus*, 90(Spring 1961): 300–328. Quote is in Degan, "Student Admittance and Placement," p. 148.

65. For the church's experience in the late nineteenth century in the context of mass immigration, see James Hennessey (S.J.), *American Catholics: A History of the Roman Catholic Community in the United States* (New York: Oxford University Press, 1981), 184–220. Sanders, *Education of an Urban Minority*, pp. 47, 48.

66. Hennessey, *American Catholics*, p. 209.

67. For the erosion of ethnic parish lines, Sanders, *Education of an Urban Minority*, pp. 105–120. Abramson gives statistics on parochial-school attendance by

various ethnic groups in the 1960s, *Ethnic Diversity*, pp. 119–120. For the tendency to increased attendance with longer residence, see Greeley and Rossi, *Education of Catholic Americans*, p. 38. In fact, the data on ethnic support in the recent past are extremely confusing. Compare Greeley, McCready, and McCourt, *Catholic Schools in a Declining Church*, p. 223 with Notre Dame Study, *Catholic Schools in Action*, p. 259. Often ethnic comparisons are presented in a way that does not give a sense of meaning of the statistics in the context of the larger demographic patterns of the society. For the nativity of parents of parochial-school students, see Notre Dame Study, *Catholic Schools in Action*, Table 105, p. 259.

68. Rossi and Rossi, "Parochial School Education in America," pp. 183, 182; quote is on p. 182; Greeley and Rossi, *Education of Catholic Americans*, p. 50.

69. For the reasons for Catholic-school attendance, see Fichter, *Parochial School*, p. 298; Notre Dame Study, *Catholic Schools in Action*, pp. 262–265. The recent study by Greeley, McCready, and McCourt suggests that the significance of religion declines with parental age and that the younger, better-educated Catholic parents value Catholic education for their children above all because they believe the schools to be better; see *Catholic Schools in a Declining Church*, pp. 227–229.

70. The class factors should not, however, be exaggerated. Greeley and Rossi note that "It was the poor and the poorly educated who disproportionately did not send their children to Catholic schools. An eighth grade education and a job which put one in the lower middle class . . . were enough" to make the likelihood of a Catholic-school education for one's children approximate the average; *Education of Catholic Americans*, p. 42.

71. Will Herberg, *Protestant-Catholic-Jew* (Garden City, N.Y.: Doubleday, 1955). For a recent evaluation of the importance of ethnicity among Catholics, see Abramson, *Ethnic Diversity*, passim. For an example of how attendance at Catholic institutions reflects mobility aspirations and how these have changed for different groups, see Hassenger, "Portrait of a Catholic Women's College."

72. St. Clair Drake and Horace R. Cayton, *Black Metropolis: A Study of Negro Life in a Northern City*, 2 vols. (New York: Harper, 1962), 413–414. Some of the results of the massive Coleman investigation of public and private schools as they relate to blacks are presented by James S. Coleman in "Quality and Equality in American Education: Public and Catholic Schools," *Phi Delta Kappan*, 63(November 1981): 160. See also Andrew M. Greeley, *Catholic High Schools and Minority Students* (New Brunswick, N.J.: Transaction Books, 1982), 21.

The question of the Catholic church's obligation to inner-city students, most of them black or Hispanic, and especially to those (largely black) who are not Catholic, has been the subject of debates in the church and has inevitably been intermixed with the issue of public funding for parochial schools. In the 1970s, as Catholic schools were increasingly financed out of tuition charges rather than parish funds and as whites abandoned inner-city schools, Catholic schools often faced the dilemma of whether to close schools—overwhelmingly populated by poor minorities—or to keep them open at the church's expense. See the discussion in Thomas Vitullo-Martin, *Catholic Inner-City Schools: The Future* (Washington, D.C.: United States Catholic Conference, 1979). See also the church's self-evaluation of its success with minority students in National Catholic Education Association, *Catholic High Schools: Their Impact on Low-Income Students* (Washington, D.C.: National Catholic Education Association, 1986).

73. See Notre Dame Study, *Catholic Schools in Action*, pp. 263, 231, 262–263, 271, 280, 267. See also Greeley, McCready, and McCourt, *Catholic Schools in a Declining Church*, pp. 227–228.

74. See Michael Katz's classic study, *The Irony of Early School Reform: Educational Innovation in Mid-Nineteenth Century Massachusetts* (Boston: Beacon, 1968).

75. As Greeley notes, "Catholic school students are at least twice as likely to rate the quality of instruction, teacher interest, and the effectiveness and fairness of discipline as 'excellent.' They are much more satisfied with their experience in the schools they attend." See, Greeley, *Minority Students*, pp. 42–43. Coleman found similar results, with Catholic high-school students registering much higher levels of satisfaction than public-school students. See, Coleman, et al., *High School Achievement*, pp. 99–103.

76. See, for example, parents' evaluation of the success of Catholic schools in various academic and religious matters in Notre Dame Study, *Catholic Schools in Action*, pp. 262–283.

77. Coleman, "Quality and Equality," passim; Coleman, et al., *High School Achievement*, p. 144. See also Greeley, *Minority Students*, pp. 79–80.

78. Coleman attempted to eliminate various factors and to define the specific contribution of Catholic schools to achievement, see *High School Achievement*, pp. 137–178.

79. At the time of this writing the secretary of education, William Bennett, has gone considerably beyond even this to suggest that the Catholic schools be given the responsibility to educate the worst 5 to 10 percent of public-school students. He made this suggestion at an address to the National Catholic Education Association. See *San Francisco Chronicle*, April 8, 1988, p. A14.

It it significant that Deacon Norman Phillips, a spokesman for the San Francisco archdiocese was reported by the *Chronicle* as noting that parochial schools "might not want the public schools' problem children in their schools." "Parents," Phillips explained "are footing the bills because *they want a school a certain way*" (emphasis mine).

80. Coleman, "Quality and Equality," p. 164.

Index